T0190379

AN INTRODUCTION TO PARTIAL DIFFERENTIAL EQUATIONS

A complete introduction to partial differential equations, this textbook provides a rigorous yet accessible guide to students in mathematics, physics and engineering. The presentation is lively and up to date, with particular emphasis on developing an appreciation of underlying mathematical theory.

Beginning with basic definitions, properties and derivations of some fundamental equations of mathematical physics from basic principles, the book studies first-order equations, the classification of second-order equations, and the one-dimensional wave equation. Two chapters are devoted to the separation of variables, whilst others concentrate on a wide range of topics including elliptic theory, Green's functions, variational and numerical methods.

A rich collection of worked examples and exercises accompany the text, along with a large number of illustrations and graphs to provide insight into the numerical examples.

Solutions and hints to selected exercises are included for students whilst extended solution sets are available to lecturers from solutions@cambridge.org.

AN INTRODUCTION TO PARTIAL DIFFERENTIAL EQUATIONS

YEHUDA PINCHOVER AND JACOB RUBINSTEIN

CAMBRIDGE
UNIVERSITY PRESS

CAMBRIDGE
UNIVERSITY PRESS

University Printing House, Cambridge CB2 8BS, United Kingdom

Cambridge University Press is part of the University of Cambridge.

It furthers the University's mission by disseminating knowledge in the pursuit of education, learning and research at the highest international levels of excellence.

www.cambridge.org
Information on this title: www.cambridge.org/9780521613231

First published 2005
8th printing 2013

A catalogue record for this publication is available from the British Library

ISBN 978-0-521-84886-2 Hardback
ISBN 978-0-521-61323-1 Paperback

To our parents

The equation of heaven and earth remains unsolved.
(Yehuda Amichai)

הַמִּשְׁוָאָה שֶׁל שָׁמַיִם וָאָרֶץ
נִשְׁאָרֶת בְּלִי פִּתְרוֹן.
(יהודה עמיחי)

Contents

Preface		*page* xi
1	Introduction	1
	1.1 Preliminaries	1
	1.2 Classification	3
	1.3 Differential operators and the superposition principle	3
	1.4 Differential equations as mathematical models	4
	1.5 Associated conditions	17
	1.6 Simple examples	20
	1.7 Exercises	21
2	First-order equations	23
	2.1 Introduction	23
	2.2 Quasilinear equations	24
	2.3 The method of characteristics	25
	2.4 Examples of the characteristics method	30
	2.5 The existence and uniqueness theorem	36
	2.6 The Lagrange method	39
	2.7 Conservation laws and shock waves	41
	2.8 The eikonal equation	50
	2.9 General nonlinear equations	52
	2.10 Exercises	58
3	Second-order linear equations in two indenpendent variables	64
	3.1 Introduction	64
	3.2 Classification	64
	3.3 Canonical form of hyperbolic equations	67
	3.4 Canonical form of parabolic equations	69
	3.5 Canonical form of elliptic equations	70
	3.6 Exercises	73

4   The one-dimensional wave equation   76
     4.1   Introduction   76
     4.2   Canonical form and general solution   76
     4.3   The Cauchy problem and d'Alembert's formula   78
     4.4   Domain of dependence and region of influence   82
     4.5   The Cauchy problem for the nonhomogeneous wave equation   87
     4.6   Exercises   93
5   The method of separation of variables   98
     5.1   Introduction   98
     5.2   Heat equation: homogeneous boundary condition   99
     5.3   Separation of variables for the wave equation   109
     5.4   Separation of variables for nonhomogeneous equations   114
     5.5   The energy method and uniqueness   116
     5.6   Further applications of the heat equation   119
     5.7   Exercises   124
6   Sturm–Liouville problems and eigenfunction expansions   130
     6.1   Introduction   130
     6.2   The Sturm–Liouville problem   133
     6.3   Inner product spaces and orthonormal systems   136
     6.4   The basic properties of Sturm–Liouville eigenfunctions
           and eigenvalues   141
     6.5   Nonhomogeneous equations   159
     6.6   Nonhomogeneous boundary conditions   164
     6.7   Exercises   168
7   Elliptic equations   173
     7.1   Introduction   173
     7.2   Basic properties of elliptic problems   173
     7.3   The maximum principle   178
     7.4   Applications of the maximum principle   181
     7.5   Green's identities   182
     7.6   The maximum principle for the heat equation   184
     7.7   Separation of variables for elliptic problems   187
     7.8   Poisson's formula   201
     7.9   Exercises   204
8   Green's functions and integral representations   208
     8.1   Introduction   208
     8.2   Green's function for Dirichlet problem in the plane   209
     8.3   Neumann's function in the plane   219
     8.4   The heat kernel   221
     8.5   Exercises   223

9	Equations in high dimensions	226
	9.1 Introduction	226
	9.2 First-order equations	226
	9.3 Classification of second-order equations	228
	9.4 The wave equation in $\mathbb{R}^2$ and $\mathbb{R}^3$	234
	9.5 The eigenvalue problem for the Laplace equation	242
	9.6 Separation of variables for the heat equation	258
	9.7 Separation of variables for the wave equation	259
	9.8 Separation of variables for the Laplace equation	261
	9.9 Schrödinger equation for the hydrogen atom	263
	9.10 Musical instruments	266
	9.11 Green's functions in higher dimensions	269
	9.12 Heat kernel in higher dimensions	275
	9.13 Exercises	279
10	Variational methods	282
	10.1 Calculus of variations	282
	10.2 Function spaces and weak formulation	296
	10.3 Exercises	306
11	Numerical methods	309
	11.1 Introduction	309
	11.2 Finite differences	311
	11.3 The heat equation: explicit and implicit schemes, stability, consistency and convergence	312
	11.4 Laplace equation	318
	11.5 The wave equation	322
	11.6 Numerical solutions of large linear algebraic systems	324
	11.7 The finite elements method	329
	11.8 Exercises	334
12	Solutions of odd-numbered problems	337
	A.1 Trigonometric formulas	361
	A.2 Integration formulas	362
	A.3 Elementary ODEs	362
	A.4 Differential operators in polar coordinates	363
	A.5 Differential operators in spherical coordinates	363
	References	364
	Index	366

Preface

This book presents an introduction to the theory and applications of partial differential equations (PDEs). The book is suitable for all types of basic courses on PDEs, including courses for undergraduate engineering, sciences and mathematics students, and for first-year graduate courses as well.

Having taught courses on PDEs for many years to varied groups of students from engineering, science and mathematics departments, we felt the need for a textbook that is concise, clear, motivated by real examples and mathematically rigorous. We therefore wrote a book that covers the foundations of the theory of PDEs. This theory has been developed over the last 250 years to solve the most fundamental problems in engineering, physics and other sciences. Therefore we think that one should not treat PDEs as an abstract mathematical discipline; rather it is a field that is closely related to real-world problems. For this reason we strongly emphasize throughout the book the relevance of every bit of theory and every practical tool to some specific application. At the same time, we think that the modern engineer or scientist should understand the basics of PDE theory when attempting to solve specific problems that arise in applications. Therefore we took great care to create a balanced exposition of the theoretical and applied facets of PDEs.

The book is flexible enough to serve as a textbook or a self-study book for a large class of readers. The first seven chapters include the core of a typical one-semester course. In fact, they also include advanced material that can be used in a graduate course. Chapters 9 and 11 include additional material that together with the first seven chapters fits into a typical curriculum of a two-semester course. In addition, Chapters 8 and 10 contain advanced material on Green's functions and the calculus of variations. The book covers all the classical subjects, such as the separation of variables technique and Fourier's method (Chapters 5, 6, 7, and 9), the method of characteristics (Chapters 2 and 9), and Green's function methods (Chapter 8). At the same time we introduce the basic theorems that guarantee that the problem at hand is well defined (Chapters 2–10), and we took care to include modern ideas such as variational methods (Chapter 10) and numerical methods (Chapter 11).

The first eight chapters mainly discuss PDEs in two independent variables. Chapter 9 shows how the methods of the first eight chapters are extended and enhanced to handle PDEs in higher dimensions. Generalized and weak solutions are presented in many parts of the book.

Throughout the book we illustrate the mathematical ideas and techniques by applying them to a large variety of practical problems, including heat conduction, wave propagation, acoustics, optics, solid and fluid mechanics, quantum mechanics, communication, image processing, musical instruments, and traffic flow.

We believe that the best way to grasp a new theory is by considering examples and solving problems. Therefore the book contains hundreds of examples and problems, most of them at least partially solved. Extended solutions to the problems are available for course instructors using the book from solutions@cambridge.org. We also include dozens of drawing and graphs to explain the text better and to demonstrate visually some of the special features of certain solutions.

It is assumed that the reader is familiar with the calculus of functions in several variables, with linear algebra and with the basics of ordinary differential equations. The book is almost entirely self-contained, and in the very few places where we cannot go into details, a reference is provided.

The book is the culmination of a slow evolutionary process. We wrote it during several years, and kept changing and adding material in light of our experience in the classroom. The current text is an expanded version of a book in Hebrew that the authors published in 2001, which has been used successfully at Israeli universities and colleges since then.

Our cumulative expertise of over 30 years of teaching PDEs at several universities, including Stanford University, UCLA, Indiana University and the Technion – Israel Institute of Technology guided to us to create a text that enhances not just technical competence but also deep understanding of PDEs. We are grateful to our many students at these universities with whom we had the pleasure of studying this fascinating subject. We hope that the readers will also learn to enjoy it.

We gratefully acknowledge the help we received from a number of individuals. Kristian Jenssen from North Carolina State University, Lydia Peres and Tiferet Saadon from the Technion – Israel Institute of Technology, and Peter Sternberg from Indiana University read portions of the draft and made numerous comments and suggestions for improvement. Raya Rubinstein prepared the drawings, while Yishai Pinchover and Aviad Rubinstein assisted with the graphs. Despite our best efforts, we surely did not discover all the mistakes in the draft. Therefore we encourage observant readers to send us their comments at pincho@techunix.technion.ac.il. We will maintain a webpage with a list of errata at http://www.math.technion.ac .il/~pincho/PDE.pdf.

1

Introduction

1.1 Preliminaries

A partial differential equation (PDE) describes a relation between an unknown function and its partial derivatives. PDEs appear frequently in all areas of physics and engineering. Moreover, in recent years we have seen a dramatic increase in the use of PDEs in areas such as biology, chemistry, computer sciences (particularly in relation to image processing and graphics) and in economics (finance). In fact, in each area where there is an interaction between a number of independent variables, we attempt to define functions in these variables and to model a variety of processes by constructing equations for these functions. When the value of the unknown function(s) at a certain point depends only on what happens in the vicinity of this point, we shall, in general, obtain a PDE. The general form of a PDE for a function $u(x_1, x_2, \ldots, x_n)$ is

$$F(x_1, x_2, \ldots, x_n, u, u_{x_1}, u_{x_2}, \ldots, u_{x_{11}}, \ldots) = 0, \tag{1.1}$$

where $x_1, x_2, \ldots, x_n$ are the independent variables, u is the unknown function, and u_{x_i} denotes the partial derivative $\partial u / \partial x_i$. The equation is, in general, supplemented by additional conditions such as initial conditions (as we have often seen in the theory of ordinary differential equations (ODEs)) or boundary conditions.

The analysis of PDEs has many facets. The classical approach that dominated the nineteenth century was to develop methods for finding explicit solutions. Because of the immense importance of PDEs in the different branches of physics, every mathematical development that enabled a solution of a new class of PDEs was accompanied by significant progress in physics. Thus, the method of characteristics invented by Hamilton led to major advances in optics and in analytical mechanics. The Fourier method enabled the solution of heat transfer and wave propagation, and Green's method was instrumental in the development of the theory of electromagnetism. The most dramatic progress in PDEs has been achieved in the last 50 years with the introduction of numerical methods that allow the use of computers to solve PDEs of virtually every kind, in general geometries and under arbitrary external conditions (at least in theory; in practice there are still a large number of hurdles to be overcome).

The technical advances were followed by theoretical progress aimed at understanding the solution's structure. The goal is to discover some of the solution's properties before actually computing it, and sometimes even without a complete solution. The theoretical analysis of PDEs is not merely of academic interest, but rather has many applications. It should be stressed that there exist very complex equations that cannot be solved even with the aid of supercomputers. All we can do in these cases is to attempt to obtain qualitative information on the solution. In addition, a deep important question relates to the formulation of the equation and its associated side conditions. In general, the equation originates from a model of a physical or engineering problem. It is not automatically obvious that the model is indeed consistent in the sense that it leads to a solvable PDE. Furthermore, it is desired in most cases that the solution will be unique, and that it will be stable under small perturbations of the data. A theoretical understanding of the equation enables us to check whether these conditions are satisfied. As we shall see in what follows, there are many ways to solve PDEs, each way applicable to a certain class of equations. Therefore it is important to have a thorough analysis of the equation before (or during) solving it.

The fundamental theoretical question is whether the problem consisting of the equation and its associated side conditions is well posed. The French mathematician Jacques Hadamard (1865–1963) coined the notion of *well-posedness*. According to his definition, a problem is called well-posed if it satisfies all of the following criteria

1. **Existence** The problem has a solution.
2. **Uniqueness** There is no more than one solution.
3. **Stability** A small change in the equation or in the side conditions gives rise to a small change in the solution.

If one or more of the conditions above does not hold, we say that the problem is *ill-posed*. One can fairly say that the fundamental problems of mathematical physics are all well-posed. However, in certain engineering applications we might tackle problems that are ill-posed. In practice, such problems are unsolvable. Therefore, when we face an ill-posed problem, the first step should be to modify it appropriately in order to render it well-posed.

1.2 Classification

We pointed out in the previous section that PDEs are often classified into different types. In fact, there exist several such classifications. Some of them will be described here. Other important classifications will be described in Chapter 3 and in Chapter 9.

- **The order of an equation**
  The first classification is according to the *order* of the equation. The order is defined to be the order of the highest derivative in the equation. If the highest derivative is of order k, then the equation is said to be of order k. Thus, for example, the equation $u_{tt} - u_{xx} = f(x, t)$ is called a second-order equation, while $u_t + u_{xxxx} = 0$ is called a fourth-order equation.
- **Linear equations**
  Another classification is into two groups: linear versus nonlinear equations. An equation is called *linear* if in (1.1), F is a linear function of the unknown function u and its derivatives. Thus, for example, the equation $x^7 u_x + e^{xy} u_y + \sin(x^2 + y^2)u = x^3$ is a linear equation, while $u_x^2 + u_y^2 = 1$ is a nonlinear equation. The nonlinear equations are often further classified into subclasses according to the type of the nonlinearity. Generally speaking, the nonlinearity is more pronounced when it appears in a higher derivative. For example, the following two equations are both nonlinear:

$$u_{xx} + u_{yy} = u^3, \tag{1.2}$$

$$u_{xx} + u_{yy} = |\nabla u|^2 u. \tag{1.3}$$

Here $|\nabla u|$ denotes the norm of the gradient of u. While (1.3) is nonlinear, it is still linear as a function of the highest-order derivative. Such a nonlinearity is called *quasilinear*. On the other hand in (1.2) the nonlinearity is only in the unknown function. Such equations are often called *semilinear*.
- **Scalar equations versus systems of equations**
  A single PDE with just one unknown function is called a *scalar equation*. In contrast, a set of m equations with l unknown functions is called a *system* of m equations.

1.3 Differential operators and the superposition principle

A function has to be k times differentiable in order to be a solution of an equation of order k. For this purpose we define the set $C^k(D)$ to be the set of all functions that are k times continuously differentiable in D. In particular, we denote the set of continuous functions in D by $C^0(D)$, or $C(D)$. A function in the set C^k that satisfies a PDE of order k, will be called a *classical* (or *strong*) solution of the PDE. It should be stressed that we sometimes also have to deal with solutions that are not classical. Such solutions are called *weak* solutions. The possibility of weak solutions and their physical meaning will be discussed on several occasions later, see for example Sections 2.7 and 10.2. Note also that, in general, we are required to solve a problem that consists of a PDE and associated conditions. In order for a strong solution of the PDE to also be a strong solution of the full problem, it is required to satisfy the additional conditions in a smooth way.

Mappings between different function sets are called *operators*. The operation of an operator L on a function u will be denoted by $L[u]$. In particular, we shall deal in this book with operators defined by partial derivatives of functions. Such operators, which are in fact mappings between different C^k classes, are called *differential operators*.

An operator that satisfies a relation of the form

$$L[a_1u_1 + a_2u_2] = a_1L[u_1] + a_2L[u_2],$$

where a_1 and a_2 are arbitrary constants, and u_1 and u_2 are arbitrary functions is called a *linear operator*. A linear differential equation naturally defines a linear operator: the equation can be expressed as $L[u] = f$, where L is a linear operator and f is a given function.

A linear differential equation of the form $L[u] = 0$, where L is a linear operator, is called a *homogeneous equation*. For example, define the operator $L = \partial^2/\partial x^2 - \partial^2/\partial y^2$. The equation

$$L[u] = u_{xx} - u_{yy} = 0$$

is a homogeneous equation, while the equation

$$L[u] = u_{xx} - u_{yy} = x^2$$

is an example of a *nonhomogeneous equation*.

Linear operators play a central role in mathematics in general, and in PDE theory in particular. This results from the important property (which follows at once from the definition) that if for $1 \le i \le n$, the function u_i satisfies the linear differential equation $L[u_i] = f_i$, then the linear combination $v := \sum_{i=1}^{n} \alpha_i u_i$ satisfies the equation $L[v] = \sum_{i=1}^{n} \alpha_i f_i$. In particular, if each of the functions $u_1, u_2, \ldots, u_n$ satisfies the homogeneous equation $L[u] = 0$, then every linear combination of them satisfies that equation too. This property is called the *superposition principle*. It allows the construction of complex solutions through combinations of simple solutions. In addition, we shall use the superposition principle to prove uniqueness of solutions to linear PDEs.

1.4 Differential equations as mathematical models

PDEs are woven throughout science and technology. We shall briefly review a number of canonical equations in different areas of application. The fundamental laws of physics provide a mathematical description of nature's phenomena on a variety of scales of time and space. Thus, for example, very large scale phenomena (astronomical scales) are controlled by the laws of gravity. The theory of electromagnetism controls the scales involved in many daily activities, while quantum mechanics is used to describe phenomena on the atomic scale. It turns out, however, that many important problems involve interaction between a large number of objects, and thus it is difficult to use the basic laws of physics to describe them. For example, we do not fall to the floor when we sit on a chair. Why? The fundamental reason lies in the electric forces between the atoms constituting the chair. These forces endow the chair with high rigidity. It is clear, though, that it is not feasible to solve the equations of electromagnetism (Maxwell's equations) to describe the interaction between such a vast number of objects. As another example, consider the flow of a gas. Each molecule obeys Newton's laws, but we cannot in practice solve for the evolution of an Avogadro number of individual molecules. Therefore, it is necessary in many applications to develop simpler models.

The basic approach towards the derivation of these models is to define new quantities (temperature, pressure, tension,. . .) that describe average macroscopic values of the fundamental microscopic quantities, to assume several fundamental principles, such as conservation of mass, conservation of momentum, conservation of energy, etc., and to apply the new principles to the macroscopic quantities. We shall often need some additional ad-hoc assumptions to connect different macroscopic entities. In the optimal case we would like to start from the fundamental laws and then average them to achieve simpler models. However, it is often very hard to do so, and, instead, we shall sometimes use experimental observations to supplement the basic principles. We shall use x, y, z to denote spatial variables, and t to denote the time variable.

1.4.1 The heat equation

A common way to encourage scientific progress is to confer prizes and awards. Thus, the French Academy used to set up competitions for its prestigious prizes by presenting specific problems in mathematics and physics. In 1811 the Academy chose the problem of heat transfer for its annual prize. The prize was awarded to the French mathematician Jean Baptiste Joseph Fourier (1768–1830) for two important contributions. (It is interesting to mention that he was not an active scientist at that time, but rather the governor of a region in the French Alps – actually a politician!). He developed, as we shall soon see, an appropriate differential equation, and, in addition developed, as we shall see in Chapter 5, a novel method for solving this equation.

The basic idea that guided Fourier was conservation of energy. For simplicity we assume that the material density and the heat capacity are constant in space and time, and we scale them to be 1. We can therefore identify heat energy with temperature. Let D be a fixed spatial domain, and denote its boundary by ∂D. Under these conditions we shall write down the change in the energy stored in D between time t and time $t + \Delta t$:

$$\int_D [u(x, y, z, t + \Delta t) - u(x, y, z, t)] \, dV$$

$$= \int_t^{t+\Delta t} \int_D q(x, y, z, t, u) dV \, dt - \int_t^{t+\Delta t} \int_{\partial D} \vec{B}(x, y, z, t) \cdot \hat{n} dS dt, \quad (1.4)$$

where u is the temperature, q is the rate of heat production in D, $\vec{B}$ is the heat flux through the boundary, dV and dS are space and surface integration elements, respectively, and $\hat{n}$ is a unit vector pointing in the direction of the outward normal to ∂D. Notice that the heat production can be negative (a refrigerator, an air conditioner), as can the heat flux.

In general the heat production is determined by external sources that are independent of the temperature. In some cases (such as an air conditioner controlled by a thermostat) it depends on the temperature itself but not on its derivatives. Hence we assume $q = q(x, y, z, t, u)$. To determine the functional form of the heat flux, Fourier used the experimental observation that 'heat flows from hotter places to colder places'. Recall from calculus that the direction of maximal growth of a function is given by its gradient. Therefore, Fourier postulated

$$\vec{B} = -k(x, y, z) \vec{\nabla} u. \quad (1.5)$$

The formula (1.5) is called *Fourier's law of heat conduction*. The (positive!) function k is called the *heat conduction (or Fourier) coefficient*. The value(s) of k depend on the medium in which the heat diffuses. In a homogeneous domain k is expected to be constant. The assumptions on the functional dependence of q and $\vec{B}$ on u are called *constitutive laws*.

We substitute our formula for q and $\vec{B}$ into (1.4), approximate the t integrals using the mean value theorem, divide both sides of the equation by Δt, and take the limit $\Delta t \to 0$. We obtain

$$\int_D u_t dV = \int_D q(x, y, z, t, u) dV + \int_{\partial D} k(x, y, z) \vec{\nabla} u \cdot \hat{n} dS. \quad (1.6)$$

Observe that the integration in the second term on the right hand side is over a different set than in the other terms. Thus we shall use Gauss' theorem to convert the surface integral into a volume integral:

$$\int_D [u_t - q - \vec{\nabla} \cdot (k\vec{\nabla}u)]dV = 0, \tag{1.7}$$

where $\vec{\nabla}\cdot$ denotes the divergence operator. The following simple result will be used several times in the book.

Lemma 1.1 *Let $h(x, y, z)$ be a continuous function satisfying $\int_\Omega h(x, y, z)dV = 0$ for every domain Ω. Then $h \equiv 0$.*

Proof Let us assume to the contrary that there exists a point $P = (x_0, y_0, z_0)$ where $h(P) \neq 0$. Assume without loss of generality that $h(P) > 0$. Since h is continuous, there exists a domain (maybe very small) D_0, containing P and $\epsilon > 0$, such that $h > \epsilon > 0$ at each point in D_0. Therefore $\int_{D_0} hdV > \epsilon \text{Vol}(D_0) > 0$ which contradicts the lemma's assumption. □

Returning to the energy integral balance (1.7), we notice that it holds for any domain D. Assuming further that all the functions in the integrand are continuous, we obtain the PDE

$$u_t = q + \vec{\nabla} \cdot (k\vec{\nabla}u). \tag{1.8}$$

In the special (but common) case where the diffusion coefficient is constant, and there are no heat sources in D itself, we obtain the classical heat equation

$$u_t = k\Delta u, \tag{1.9}$$

where we use Δu to denote the important operator $u_{xx} + u_{yy} + u_{zz}$. Observe that we have assumed that the solution of the heat equation, and even some of its derivatives are continuous functions, although we have not solved the equation yet. Therefore, in principle we have to reexamine our assumptions a posteriori. We shall see examples later in the book in which solutions of a PDE (or their derivatives) are *not* continuous. We shall then consider ways to provide a meaning for the seemingly absurd process of substituting a discontinuous function into a differential equation. One of the fundamental ways of doing so is to observe that the integral balance equation (1.6) provides a more fundamental model than the PDE (1.8).

1.4.2 Hydrodynamics and acoustics

Hydrodynamics is the physical theory of fluid motion. Since almost any conceivable volume of fluid (whether it is a cup of coffee or the Pacific Ocean) contains a huge number of molecules, it is not feasible to describe the fluid using the law of electromagnetism or quantum mechanics. Hence, since the eighteenth century scientists have developed models and equations that are appropriate to macroscopic entities such as temperature, pressure, effective velocity, etc. As explained above, these equations are based on conservation laws.

The simplest description of a fluid consists of three functions describing its state at any point in space-time:

- the density (mass per unit of volume) $\rho(x, y, z, t)$;
- the velocity $\vec{u}(x, y, z, t)$;
- the pressure $p(x, y, z, t)$.

To be precise, we must also include the temperature field in the fluid. But to simplify matters, it will be assumed here that the temperature is a known constant. We start with conservation of mass. Consider a fluid element occupying an arbitrary spatial domain D. We assume that matter neither is created nor disappears in D. Thus the total mass in D does not change:

$$\frac{\partial}{\partial t} \int_D \rho \mathrm{d}V = 0. \tag{1.10}$$

The motion of the fluid boundary is given by the component of the velocity $\vec{u}$ in the direction orthogonal to the boundary ∂D. Thus we can write

$$\int_D \frac{\partial}{\partial t} \rho \mathrm{d}V + \int_{\partial D} \rho \vec{u} \cdot \hat{n} \mathrm{d}S = 0, \tag{1.11}$$

where we denoted the unit external normal to ∂D by $\hat{n}$. Using Gauss' theorem we obtain

$$\int_D [\rho_t + \vec{\nabla} \cdot (\rho \vec{u})] \mathrm{d}V = 0. \tag{1.12}$$

Since D is an arbitrary domain we can use again Lemma 1.1 to obtain the mass *transport equation*

$$\rho_t + \vec{\nabla} \cdot (\rho \vec{u}) = 0. \tag{1.13}$$

Next we require the fluid to satisfy the momentum conservation law. The forces acting on the fluid in D are gravity, acting on each point in the fluid, and the pressure applied at the boundary of D by the rest of the fluid outside D. We denote the density per unit mass of the gravitational force by $\vec{g}$. For simplicity we neglect the friction forces between adjacent fluid molecules. Newton's law of motion implies an equality between the change in the fluid momentum and the total forces acting on the fluid. Thus

$$\frac{\partial}{\partial t} \int_D \rho \vec{u} \mathrm{d}V = -\int_{\partial D} p \hat{n} \mathrm{d}s + \int_D \rho \vec{g} \mathrm{d}V. \tag{1.14}$$

Let us interchange again the t differentiation with the spatial integration, and use (1.13) to obtain the integral balance

$$\int_D [\rho \vec{u}_t + \rho(\vec{u} \cdot \vec{\nabla})\vec{u}] dV = \int_D (-\vec{\nabla}p + \rho \vec{g}) dV. \tag{1.15}$$

From this balance we deduce the PDE

$$\vec{u}_t + (\vec{u} \cdot \vec{\nabla})\vec{u} = -\frac{1}{\rho}\vec{\nabla}p + \vec{g}. \tag{1.16}$$

So far we have developed two PDEs for three unknown functions $(\rho, \vec{u}, p)$. We therefore need a third equation to complete the system. Notice that conservation of energy has already been accounted for by assuming that the temperature is fixed. In fact, the additional equation does not follow from a conservation law, rather one imposes a constitutive relation (like Fourier's law from the previous subsection). Specifically, we postulate a relation of the form

$$p = f(\rho), \tag{1.17}$$

where the function f is determined by the specific fluid (or gas). The full system comprising (1.13), (1.16) and (1.17) is called the *Euler fluid flow equations*. These equations were derived in 1755 by the Swiss mathematician Leonhard Euler (1707–1783).

If one takes into account the friction between the fluid molecules, the equations acquire an additional term. This friction is called *viscosity*. The special case of viscous fluids where the density is essentially constant is of particular importance. It characterizes, for example, most phenomena involving the flow of water. This case was analyzed first in 1822 by the French engineer Claude Navier (1785–1836), and then studied further by the British mathematician George Gabriel Stokes (1819–1903). They derived the following set of equations:

$$\rho(\vec{u}_t + (\vec{u} \cdot \vec{\nabla})\vec{u}) = \mu \Delta \vec{u} - \vec{\nabla}p, \tag{1.18}$$

$$\vec{\nabla} \cdot \vec{u} = 0. \tag{1.19}$$

The parameter μ is called the fluid's viscosity. Notice that (1.18)–(1.19) form a quasilinear system of equations. The Navier–Stokes system lies at the foundation of hydrodynamics. Enormous computational efforts are invested in solving them under a variety of conditions and in a plurality of applications, including, for example, the design of airplanes and ships, the design of vehicles, the flow of blood in arteries, the flow of ink in a printer, the locomotion of birds and fish, and so forth. Therefore it is astonishing that the well-posedness of the Navier–Stokes equations has not yet been established. Proving or disproving their well-posedness is one of the most important open problems in mathematics. A prize of one million dollars awaits the person who solves it.

An important phenomenon described by the Euler equations is the propagation of sound waves. In order to construct a simple model for sound waves, let us look at the Euler equations for a gas at rest. For simplicity we neglect gravity. It is easy to check that the equations have a solution of the form

$$\vec{u} = 0,$$
$$\rho = \rho_0, \qquad (1.20)$$
$$p = p_0 = f(\rho_0),$$

where ρ_0 and p_0 are constants describing uniform pressure and density. Let us perturb the gas by creating a localized pressure (for example by producing a sound out of our throats, or by playing a musical instrument). Assume that the perturbation is small compared with the original pressure p_0. One can therefore write

$$\vec{u} = \epsilon \vec{u}^1,$$
$$\rho = \rho^0 + \epsilon \rho^1, \qquad (1.21)$$
$$p = p_0 + \epsilon p^1 = f(\rho^0) + \epsilon f'(\rho^0)\rho^1,$$

where we denoted the perturbation to the density, velocity and pressure by $\vec{u}^1$, ρ^1, and p^1, respectively, ϵ denotes a small positive parameter, and we used (1.17). Substituting the expansion (1.21) into the Euler equations, and retaining only the terms that are linear in ϵ, we find

$$\rho_t^1 + \rho_o \vec{\nabla} \cdot \vec{u}^1 = 0,$$
$$\vec{u}_t^1 + \frac{1}{\rho^0} \vec{\nabla} p^1 = 0. \qquad (1.22)$$

Applying the operator $\vec{\nabla} \cdot$ to the second equation in (1.22), and substituting the result into the time derivative of the first equation leads to

$$\rho_{tt}^1 - f'(\rho^0)\Delta \rho^1 = 0. \qquad (1.23)$$

Alternatively we can use the linear relation between p^1 and ρ^1 to write a similar equation for the pressure

$$p_{tt}^1 - f'(\rho^0)\Delta p^1 = 0. \qquad (1.24)$$

The equation we have obtained is called a *wave equation*. We shall see later that this equation indeed describes waves propagating with speed $c = \sqrt{f'(\rho^0)}$. In particular, in the case of waves in a long narrow tube, or in a long and narrow tunnel, the pressure only depends on time and on a single spatial coordinate x along the tube. We then obtain the one-dimensional wave equation

$$p_{tt}^1 - c^2 p_{xx}^1 = 0. \tag{1.25}$$

Remark 1.2 Many problems in chemistry, biology and ecology involve the spread of some substrate being convected by a given velocity field. Denoting the concentration of the substrate by $C(x, y, z, t)$, and assuming that the fluid's velocity does not depend on the concentration itself, we find that (1.13) in the formulation

$$C_t + \vec{\nabla} \cdot (C\vec{u}) = 0 \tag{1.26}$$

describes the spread of the substrate. This equation is naturally called the *convection equation*. In Chapter 2 we shall develop solution methods for it.

1.4.3 Vibrations of a string

Many different phenomena are associated with the vibrations of elastic bodies. For example, recall the wave equation derived in the previous subsection for the propagation of sound waves. The *generation* of sound waves also involves a wave equation – for example the vibration of the sound chords, or the vibration of a string or a membrane in a musical instrument.

Consider a uniform string undergoing transversal motion whose amplitude is denoted by $u(x, t)$, where x is the spatial coordinate, and t denotes time. We also use ρ to denote the mass density per unit length of the string. We shall assume that ρ is constant. Consider further a small interval $(-\delta, \delta)$. Just as in the previous subsection, we shall consider two forces acting on the string: an external given force (e.g. gravity) acting only in the transversal (y) direction, whose density is denoted by $f(x, t)$, and an internal force acting between adjacent string elements. This internal force is called *tension*. It will be denoted by $\vec{T}$. The tension acts on the string element under consideration at its two ends. A tension $\vec{T}_+$ acts at the right hand end, and a tension $\vec{T}_-$ acts at the left hand end. We assume that the tension is in the direction tangent to the string, and that it is proportional to the string's elongation. Namely, we assume the constitutive law

$$\vec{T} = d\sqrt{1 + u_x^2}\,\hat{e}_\tau, \tag{1.27}$$

where d is a constant depending on the material of which the string is made, and $\hat{e}_\tau$ is a unit vector in the direction of the string's tangent. It is an empirical law, i.e. it stems from experimental observations. Projecting the momentum conservation equation (Newton's second law) along the y direction we find:

$$\int_{-\delta}^{\delta} \rho u_{tt} dl = \int_{-\delta}^{\delta} f(x,t) dl + \hat{e}_2 \cdot (\vec{T}_+ - \vec{T}_-) = \int_{-\delta}^{\delta} f(x,t) dl + \int_{-\delta}^{\delta} (\hat{e}_2 \cdot \vec{T})_x dx,$$

where dl denotes a length element, and $\hat{e}_2 = (0,1)$. Using the constitutive law for the tension and the following formula for the tangent vector $\hat{e}_\tau = (1, u_x)/\sqrt{1 + u_x^2}$, we can write

$$\hat{e}_2 \cdot \vec{T} = d\sqrt{1 + u_x^2} \hat{e}_2 \cdot \hat{e}_\tau = d u_x.$$

Substituting this equation into the momentum equation we obtain the integral balance

$$\int_{-\delta}^{\delta} \rho u_{tt} \sqrt{1 + u_x^2} dx = \int_{-\delta}^{\delta} \left[f\sqrt{1 + u_x^2} + d u_{xx} \right] dx.$$

Since this equation holds for arbitrary intervals, we can use Lemma 1.1 once again to obtain

$$u_{tt} - \frac{c^2}{\sqrt{1 + u_x^2}} u_{xx} = \frac{f(x,t)}{\rho}, \tag{1.28}$$

where the wave speed is given by $c = \sqrt{d/\rho}$. A different string model will be derived in Chapter 10. The two models are compared in Remark 10.5.

In the case of weak vibrations the slopes of the amplitude are small, and we can make the simplifying assumption $|u_x| \ll 1$. We can then write an approximate equation:

$$u_{tt} - c^2 u_{xx} = \frac{1}{\rho} f(x,t). \tag{1.29}$$

Thus, the wave equation developed earlier for sound waves is also applicable to describe certain elastic waves. Equation (1.29) was proposed as early as 1752 by the French mathematician Jean d'Alembert (1717–1783). We shall see in Chapter 4 how d'Alembert solved it.

Remark 1.3 We have derived an equation for the transversal vibrations of a string. What about its longitudinal vibrations? To answer this question, project the momentum equation along the tangential direction, and again use the constitutive law. We find that the density of the tension force in the longitudinal direction is given by

$$\frac{\partial}{\partial x} \left(d \frac{\sqrt{1 + u_x^2}}{\sqrt{1 + u_x^2}} \right) = 0.$$

This implies that the constitutive law we used is equivalent to assuming the string does not undergo longitudinal vibrations!

1.4.4 Random motion

Random motion of minute particles was first described in 1827 by the British biologist Robert Brown (1773–1858). Hence this motion is called *Brownian motion*. The first mathematical model to describe this motion was developed by Einstein in 1905. He proposed a model in which a particle at a point (x, y) in the plane jumps during a small time interval δt to a nearby point from the set $(x \pm \delta x, y \pm \delta x)$. Einstein showed that under a suitable assumption on δx and δt, the probability that the particle will be found at a point (x, y) at time t satisfies the heat equation. His model has found many applications in physics, biology, chemistry, economics etc. We shall demonstrate now how to obtain a PDE from a typical problem in the theory of Brownian motion.

Consider a particle in a two-dimensional domain D. For simplicity we shall limit ourselves to the case where D is the unit square. Divide the square into N^2 identical little squares, and denote their vertices by $\{(x_i, y_j)\}$. The size of each edge of a small square will be denoted by δx. A particle located at an internal vertex (x_i, y_j) jumps during a time interval δt to one of its nearest neighbors with equal probability. When the particle reaches a boundary point it dies.

Question What is the life expectancy $u(x, y)$ of a particle that starts its life at a point (x, y) in the limit

$$\delta x \to 0, \quad \delta t \to 0, \quad \frac{(\delta x)^2}{2\delta t} = k? \tag{1.30}$$

We shall answer the question using an intuitive notion of the expectancy. Obviously a particle starting its life at a boundary point dies at once. Thus

$$u(x, y) = 0, \quad (x, y) \in \partial D. \tag{1.31}$$

Consider now an internal point (x, y). A particle must have reached this point from one of its four nearest neighbors with equal probability for each neighbor. In addition, the trip from the neighboring point lasted a time interval δt. Therefore u satisfies the difference equation

$$u(x, y) = \delta t + \frac{1}{4}[u(x - \delta x, y) + u(x + \delta x, y) + u(x, y - \delta x) + u(x, y + \delta x)]. \tag{1.32}$$

We expand all functions on the right hand side into a Taylor series, assuming $u \in C^4$. Dividing by δt and taking the limit (1.30) we obtain (see also Chapter 11)

$$\Delta u = -\frac{1}{k}, \quad (x, y) \in D. \tag{1.33}$$

An equation of the type (1.33) is called a *Poisson equation*. We shall elaborate on such equations in Chapter 7.

The model we just investigated has many applications. One of them relates to the analysis of variations in stock prices. Many models in the stock market are based on assuming that stocks prices vary randomly. Assume for example that a broker buys a stock at a certain price m. She decides in advance to sell it if its price reaches an upper bound m_2 (in order to cash in her profit) or a lower bound m_1 (to minimize losses in case the stock dives). How much time on average will the broker hold the stock, assuming that the stock price performs a Brownian motion? This is a one-dimensional version of the model we derived. The equation and the associated boundary conditions are

$$ku''(m) = -1, \quad u(m_1) = u(m_2) = 0. \tag{1.34}$$

The reader will be asked to solve the equation in Exercise 1.6.

1.4.5 Geometrical optics

We have seen two derivations of the wave equation – one for sound waves, and another one for elastic waves. Yet there are many other physical phenomena controlled by wave propagation. Two notable examples are electromagnetic waves and water waves. Although there exist many analytic methods for solving wave equations (we shall learn some of them later), it is not easy to apply them in complex geometries. One might be tempted to proceed in such cases to numerical methods (see Chapter 11). The problem is that in many applications the waves are of very high frequency (or, equivalently, of very small wavelength). To describe such waves we need a resolution that is considerably smaller than a single wavelength. Consider for example optical phenomena. They are described by a wave equation; a typical wavelength for the visible light part of the spectrum is about half a micron. Assuming that we use five points per wavelength to describe the wave, and that we deal with a three-dimensional domain with linear dimension of 10^{-1} meters, we conclude that we need altogether about 10^{17} points! Even storing the data is a difficult task, not to mention the formidable complexity of solving equations with so many unknowns (Chapter 11).

Fortunately it is possible to turn the problem around and actually *use* the short wavelength to derive approximate equations that are much simpler to solve, and, yet, provide a fair description of optics. Consider for this purpose the wave equation in $\mathbb{R}^3$:

$$v_{tt} - c^2(\vec{x})\Delta v = 0. \tag{1.35}$$

Notice that the wave's speed need not be constant. We expect solutions that are oscillatory in time (see Chapter 5). Therefore we seek solutions of the form

$$v(x, y, z, t) = e^{i\omega t}\psi(x, y, z).$$

It is convenient to introduce at this stage the notation $k = \omega/c_0$ and $n = c_0/c(x)$, where c_0 is an average wave velocity in the medium. Substituting v into (1.35) yields

$$\Delta\psi + k^2 n^2(\vec{x})\psi = 0. \tag{1.36}$$

The function $n(x)$ is called the *refraction index*. The parameter k is called the *wave number*. It is easy to see that k^{-1} has the dimension of length. In fact, the wavelength is given by $2\pi k^{-1}$. As was explained above, the wavelength is often much smaller than any other length scale in the problem. For example, spectacle lenses involve scales such as 5 mm (thickness), 60 mm (radius of curvature) or 40 mm (frame size), all of them far greater than half a micron which is a typical wavelength. We therefore assume that the problem is scaled with respect to one of the large scales, and hence k is a very large number. To use this fact we seek a solution to (1.36) of the form:

$$\psi(x, y, z) = A(x, y, z; k)e^{ikS(x,y,z)}. \tag{1.37}$$

Substituting (1.37) into (1.36), and assuming that A is bounded with respect to k, we get

$$A[|\vec{\nabla}S|^2 - n^2(\vec{x})] = O\left(\frac{1}{k}\right).$$

Thus the function S satisfies the *eikonal equation*

$$|\vec{\nabla}S| = n(\vec{x}). \tag{1.38}$$

This equation, postulated in 1827 by the Irish mathematician William Rowan Hamilton (1805–1865), provides the foundation for *geometrical optics*. It is extremely useful in many applications in optics, such as radar, contact lenses, projectors, mirrors, etc. In Chapter 2 we shall develop a method for solving eikonal equations. Later, in Chapter 9, we shall encounter the eikonal equation from a different perspective.

1.4.6 Further real world equations

• **The Laplace equation**
  Many of the models we have examined so far have something in common – they involve the operator

$$\Delta u = \frac{\partial^2 u}{\partial x^2} + \frac{\partial^2 u}{\partial y^2} + \frac{\partial^2 u}{\partial z^2}.$$

This operator is called the *Laplacian*. Probably the 'most important' PDE is the *Laplace equation*

$$\Delta u = 0. \tag{1.39}$$

The equation, which is a special case of the Poisson equation we introduced earlier, was proposed in 1780 by the French mathematician Pierre-Simon Laplace (1749–1827) in his work on gravity. Solutions of the Laplace equation are called *harmonic functions*. Laplace's equation can be found everywhere. For example, in the heat conduction problems that were introduced earlier, the temperature field is harmonic when temporal equilibrium is achieved. The equation is also fundamental in mechanics, electromagnetism, probability, quantum mechanics, gravity, biology, etc.

- **The minimal surface equation**
  When we dip a narrow wire in a soap bath, and then lift the wire gently out of the bath, we can observe a thin membrane spanning the wire. The French mathematician Joseph-Louis Lagrange (1736–1813) showed in 1760 that the surface area of the membrane is smaller than the surface area of any other surface that is a small perturbation of it. Such special surfaces are called *minimal surfaces*. Lagrange further demonstrated that the graph of a minimal surface satisfies the following second-order nonlinear PDE:

$$(1 + u_y^2)u_{xx} - 2u_x u_y u_{xy} + (1 + u_x^2)u_{yy} = 0. \tag{1.40}$$

When the slopes of the minimal surface are small, i.e. $u_x, u_y \ll 1$, we see at once that (1.40) can be approximated by the Laplace equation. We shall return to the minimal surface equation in Chapter 10.

- **The biharmonic equation**
  The equilibrium state of a thin elastic plate is provided by its amplitude function $u(x, y)$, which describes the deviation of the plate from its horizontal position. It can be shown that the unknown function u satisfies the equation

$$\Delta^2 u = \Delta(\Delta u) = u_{xxxx} + 2u_{xxyy} + u_{yyyy} = 0. \tag{1.41}$$

For an obvious reason this equation is called the *biharmonic equation*. Notice that in contrast to all the examples we have seen so far, it is a fourth-order equation. We further point out that almost all the equations we have seen here, and also other important equations such as Maxwell's equations, the Schrödinger equation and Newton's equation for the gravitational field are of second order. We shall return to the plate equation in Chapter 10.

- **The Schrödinger equation**
  One of the fundamental equations of quantum mechanics, derived in 1926 by the Austrian physicist Erwin Schrödinger (1887–1961), governs the evolution of the wave function u of a particle in a potential field V:

$$i\hbar \frac{\partial u}{\partial t} = -\frac{\hbar}{2m} \Delta u + V u. \tag{1.42}$$

Here V is a known function (potential), m is the particle's mass, and $\hbar$ is Planck's constant divided by 2π. We shall consider the Schrödinger equation for the special case of an electron in the hydrogen atom in Chapter 9.

- **Other equations**

There are many other PDEs that are central to the study of different problems in science and technology. For example we mention: the Maxwell equations of electromagnetism; reaction–diffusion equations that model chemical reactions; the equations of elasticity; the Korteweg–de Vries equation for solitary waves; the nonlinear Schrödinger equation in nonlinear optics and in superfluids; the Ginzburg–Landau equations of superconductivity; Einstein's equations of general relativity, and many more.

1.5 Associated conditions

PDEs have in general infinitely many solutions. In order to obtain a unique solution one must supplement the equation with additional conditions. What kind of conditions should be supplied? It turns out that the answer depends on the type of PDE under consideration. In this section we briefly review the common conditions, and explain through examples their physical significance.

1.5.1 Initial conditions

Let us consider the transport equation (1.26) in one spatial dimension as a prototype for equations of first order. The unknown function $C(x, t)$ is a surface defined over the (x, t) plane. It is natural to formulate a problem in which one supplies the concentration at a given time t_0, and then to deduce from the equation the concentration at later times. Namely, we solve the problem consisting of the convection equation

$$C_t + \vec{\nabla} \cdot (C\vec{u}) = 0,$$

and the condition

$$C(x, t_0) = C_0(x). \tag{1.43}$$

This problem is called an *initial value problem*. Geometrically speaking, condition (1.43) determines a curve through which the solution surface must pass. We can generalize (1.43) by imposing a curve Γ that must lie on the solution surface, so that the projection of Γ on the (x, t) plane is not necessarily the x axis. In Chapter 2 we shall show that under suitable assumptions on the equation and Γ, there indeed exists a unique solution.

Another case where it is natural to impose initial conditions is the heat equation (1.9). Here we provide the temperature distribution at some initial time (say $t = 0$), and solve for its distribution at later times, namely, the initial condition for (1.9) is of the form $u(x, y, z, 0) = u_0(x, y, z)$.

The last two examples involve PDEs with just a first derivative with respect to t. In analogy with the theory of initial value problems for ODEs, we expect that equations that involve second derivatives with respect to t will require two initial conditions. Indeed, let us look at the wave equation (1.29). As explained in the previous section, this equation is nothing but Newton's second law, equating the mass times the acceleration and the forces acting on the string. Therefore it is natural to supply two initial conditions, one for the initial location of the string, and one for its initial velocity:

$$u(x, 0) = u_0(x), \quad u_t(x, 0) = u_1(x). \tag{1.44}$$

We shall indeed prove in Chapter 4 that these conditions, together with the wave equation lead to a well-posed problem.

1.5.2 Boundary conditions

Another type of constraint for PDEs that appears in many applications is called *boundary conditions*. As the name indicates, these are conditions on the behavior of the solution (or its derivative) at the boundary of the domain under consideration. As a first example, consider again the heat equation; this time, however, we limit ourselves to a given spatial domain Ω:

$$u_t = k\Delta u \quad (x, y, z) \in \Omega, \ t > 0. \tag{1.45}$$

We shall assume in general that Ω is *bounded*. It turns out that in order to obtain a unique solution, one should provide (in addition to initial conditions) information on the behavior of u on the boundary $\partial\Omega$. Excluding rare exceptions, we encounter in applications three kinds of boundary conditions. The first kind, where the values of the temperature on the boundary are supplied, i.e.

$$u(x, y, z, t) = f(x, y, z, t) \quad (x, y, z) \in \partial\Omega, \ t > 0, \tag{1.46}$$

is called a *Dirichlet condition* in honor of the German mathematician Johann Lejeune Dirichlet (1805–1859). For example, this condition is used when the boundary temperature is given through measurements, or when the temperature distribution is examined under a variety of external heat conditions.

Alternatively one can supply the normal derivative of the temperature on the boundary; namely, we impose (as usual we use here the notation ∂_n to denote the outward normal derivative at $\partial\Omega$)

$$\partial_n u(x, y, z, t) = f(x, y, z, t) \quad (x, y, z) \in \partial\Omega, \ t > 0. \tag{1.47}$$

This condition is called a *Neumann condition* after the German mathematician Carl Neumann (1832–1925). We have seen that the normal derivative $\partial_n u$ describes the flux through the boundary. For example, an insulating boundary is modeled by condition (1.47) with $f = 0$.

A third kind of boundary condition involves a relation between the boundary values of u and its normal derivative:

$$\alpha(x, y, z)\partial_n u(x, y, z, t) + u(x, y, z, t) = f(x, y, z, t) \quad (x, y, z) \in \partial D, \quad t > 0.$$
$$(1.48)$$

Such a condition is called *a condition of the third kind*. Sometimes it is also called the Robin condition.

Although the three types of boundary conditions defined above are by far the most common conditions seen in applications, there are exceptions. For example, we can supply the values of u at some parts of the boundary, and the values of its normal derivative at the rest of the boundary. This is called a *mixed boundary condition*. Another possibility is to generalize the condition of the third kind and replace the normal derivative by a (smoothly dependent) directional derivative of u in any direction that is not tangent to the boundary. This is called an *oblique boundary condition*. Also, one can provide a *nonlocal boundary condition*. For example, one can provide a boundary condition relating the heat flux at each point on the boundary to the integral of the temperature over the whole boundary.

To illustrate further the physical meaning of boundary conditions, let us consider again the wave equation for a string:

$$u_{tt} - c^2 u_{xx} = f(x, t) \quad a < x < b, \, t > 0. \tag{1.49}$$

When the locations of the end points of the string are known, we supply Dirichlet boundary conditions (Figure 1.1(a)):

$$u(a, t) = \beta_1(t), \quad u(b, t) = \beta_2(t), \quad t > 0. \tag{1.50}$$

Another possibility is that the tension at the end points is given. From our derivation of the string equation in Subsection 1.4.3 it follows that this case involves a

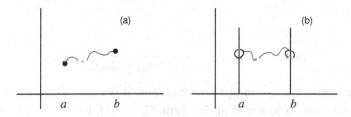

Figure 1.1 Illustrating boundary conditions for a string.

Neumann condition:

$$u_x(a, t) = \beta_1(t), \quad u_x(b, t) = \beta_2(t), \quad t > 0. \tag{1.51}$$

Thus, for example, when the end points are free to move in the transversal direction (Figure 1.1(b)), we shall use a homogeneous Neumann condition, i.e. $\beta_1 = \beta_2 = 0$.

1.6 Simple examples

Before proceeding to develop general solution methods, let us warm up with a few very simple examples.

Example 1.4 Solve the equation $u_{xx} = 0$ for an unknown function $u(x, y)$. We can consider the equation as an ODE in x, with y being a parameter. Thus the general solution is $u(x, y) = A(y)x + B(y)$. Notice that the solution space is huge, since $A(y)$ and $B(y)$ are arbitrary functions.

Example 1.5 Solve the equation $u_{xy} + u_x = 0$. We can transform the problem into an ODE by setting $v = u_x$. The new function $v(x, y)$ satisfies the equation $v_y + v = 0$. Treating x as a parameter, we obtain $v(x, y) = C(x)e^{-y}$. Integrating v we construct the solution to the original problem: $u(x, y) = D(x)e^{-y} + E(y)$.

Example 1.6 Find a solution of the wave equation $u_{tt} - 4u_{xx} = \sin t + x^{2000}$. Notice that we are asked to find *a* solution, and not the most general solution. We shall exploit the linearity of the wave equation. According to the superposition principle, we can split $u = v + w$, such that v and w are solutions of

$$v_{tt} - 4v_{xx} = \sin t, \tag{1.52}$$

$$w_{tt} - 4w_{xx} = x^{2000}. \tag{1.53}$$

The advantage gained by this step is that solutions for each of these equations can be easily obtained:

$$v(x, t) = -\sin t, \qquad w(x, t) = -\frac{1}{4 \times 2001 \times 2002} x^{2002}.$$

Thus

$$u(x, t) = -\sin t - \frac{1}{4 \times 2001 \times 2002} x^{2002}.$$

There are many other solutions. For example, it is easy to check that if we add to the solution above a function of the form $f(x - 2t)$, where $f(s)$ is an arbitrary twice differentiable function, a new solution is obtained.

Unfortunately one rarely encounters real problems described by such simple equations. Nevertheless, we can draw a few useful conclusions from these examples. For instance, a commonly used method is to seek a transformation from the original variables to new variables in which the equation takes a simpler form. Also, the superposition principle, which enables us to decompose a problem into a set of far simpler problems, is quite general.

1.7 Exercises

1.1 Show that each of the following equations has a solution of the form $u(x, y) = f(ax + by)$ for a proper choice of constants a, b. Find the constants for each example.

  (a) $u_x + 3u_y = 0$.
  (b) $3u_x - 7u_y = 0$.
  (c) $2u_x + \pi u_y = 0$.

1.2 Show that each of the following equations has a solution of the form $u(x, y) = e^{\alpha x + \beta y}$. Find the constants α, β for each example.

  (a) $u_x + 3u_y + u = 0$.
  (b) $u_{xx} + u_{yy} = 5e^{x-2y}$.
  (c) $u_{xxxx} + u_{yyyy} + 2u_{xxyy} = 0$.

1.3 (a) Show that there exists a unique solution for the system

$$u_x = 3x^2 y + y,$$
$$u_y = x^3 + x, \tag{1.54}$$

together with the initial condition $u(0, 0) = 0$.
(b) Prove that the system

$$u_x = 2.999999x^2 y + y,$$
$$u_y = x^3 + x \tag{1.55}$$

has no solution at all.

1.4 Let $u(x, y) = h(\sqrt{x^2 + y^2})$ be a solution of the minimal surface equation.
  (a) Show that $h(r)$ satisfies the ODE

$$rh'' + h'(1 + (h')^2) = 0.$$

  (b) What is the general solution to the equation of part (a)?

1.5 Let $p : \mathbb{R} \to \mathbb{R}$ be a differentiable function. Prove that the equation

$$u_t = p(u)u_x \qquad t > 0$$

has a solution satisfying the functional relation $u = f(x + p(u)t)$, where f is a differentiable function. In particular find such solutions for the following equations:

(a) $u_t = ku_x$.

(b) $u_t = uu_x$.

(c) $u_t = u\sin(u)u_x$.

1.6 Solve (1.34), and compute the average time for which the broker holds the stock. Analyze the result in light of the financial interpretation of the parameters (m_1, m_2, k).

1.7 (a) Consider the equation $u_{xx} + 2u_{xy} + u_{yy} = 0$. Write the equation in the coordinates $s = x, t = x - y$.

(b) Find the general solution of the equation.

(c) Consider the equation $u_{xx} - 2u_{xy} + 5u_{yy} = 0$. Write it in the coordinates $s = x + y$, $t = 2x$.

2

First-order equations

2.1 Introduction

A first-order PDE for an unknown function $u(x_1, x_2, \ldots, x_n)$ has the following general form:

$$F(x_1, x_2, \ldots, x_n, u, u_{x_1}, u_{x_2}, \ldots, u_{x_n}) = 0, \tag{2.1}$$

where F is a given function of $2n + 1$ variables. First-order equations appear in a variety of physical and engineering processes, such as the transport of material in a fluid flow and propagation of wavefronts in optics. Nevertheless they appear less frequently than second-order equations. For simplicity we shall limit the presentation in this chapter to functions in two variables. The reason for this is not just to simplify the algebra. As we shall soon observe, the solution method is based on the geometrical interpretation of u as a surface in an $(n + 1)$-dimensional space. The results will be generalized to equations in any number of variables in Chapter 9.

We thus consider a surface in $\mathbb{R}^3$ whose graph is given by $u(x, y)$. The surface satisfies an equation of the form

$$F(x, y, u, u_x, u_y) = 0. \tag{2.2}$$

Equation (2.2) is still quite general. In many practical situations we deal with equations with a special structure that simplifies the solution process. Therefore we shall progress from very simple equations to more complex ones. There is a common thread to all types of equations – the geometrical approach. The basic idea is that since $u(x, y)$ is a surface in $\mathbb{R}^3$, and since the normal to the surface is given by the vector $(u_x, u_y, -1)$, the PDE (2.2) can be considered as an equation relating the surface to its normal (or alternatively its tangent plane). Indeed the main solution method will be a direct construction of the solution surface.

23

2.2 Quasilinear equations

We consider first a special class of nonlinear equations where the nonlinearity is confined to the unknown function u. The derivatives of u appear in the equation linearly. Such equations are called *quasilinear*. The general form of a quasilinear equation is

$$a(x, y, u)u_x + b(x, y, u)u_y = c(x, y, u). \qquad (2.3)$$

An important special case of quasilinear equations is that of *linear* equations:

$$a(x, y)u_x + b(x, y)u_y = c_0(x, y)u + c_1(x, y), \qquad (2.4)$$

where a, b, c_0, c_1 are given functions of (x, y).

Before developing the general theory for quasilinear equations, let us warm up with a simple example.

Example 2.1

$$u_x = c_0 u + c_1. \qquad (2.5)$$

In this example we set $a = 1, b = 0, c_0$ is a constant, and $c_1 = c_1(x, y)$. Since (2.5) contains no derivative with respect to the y variable, we can regard this variable as a parameter. Recall from the theory of ODEs that in order to obtain a unique solution we must supply an additional condition. We saw in Chapter 1 that there are many ways to supply additional conditions to a PDE. The natural condition for a first-order PDE is a curve lying on the solution surface. We shall refer to such a condition as an *initial condition*, and the problem will be called an *initial value problem* or a *Cauchy problem* in honor of the French mathematician Augustin Louis Cauchy (1789–1857). For example, we can supplement (2.5) with the initial condition

$$u(0, y) = y. \qquad (2.6)$$

Since we are actually dealing with an ODE, the solution is immediate:

$$u(x, y) = e^{c_0 x} \left[\int_0^x e^{-c_0 \xi} c_1(\xi, y) d\xi + y \right]. \qquad (2.7)$$

A basic approach for solving the general case is to seek special variables in which the equation is simplified (actually, similar to (2.5)). Before doing so, let us draw a few conclusions from this simple example.

(1) Notice that we integrated along the x direction (see Figure 2.1) from each point on the y axis where the initial condition was given, i.e. we actually solved an infinite set of ODEs.

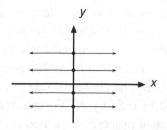

Figure 2.1 Integration of (2.5).

(2) Is there always a solution to (2.5) and an initial condition? At a first sight the answer seems positive; we can write a general solution for (2.5) in the form

$$u(x, y) = e^{c_0 x} \left[\int_0^x e^{-c_0 \xi} c_1(\xi, y) d\xi + T(y) \right], \tag{2.8}$$

where the function $T(y)$ is determined by the initial condition. There are examples, however, where such a function does not exist at all! For instance, consider the special case of (2.5) in which $c_1 \equiv 0$. The solution (2.8) now becomes $u(x, y) = e^{c_0 x} T(y)$. Replace the initial condition (2.6) with the condition

$$u(x, 0) = 2x. \tag{2.9}$$

Now $T(y)$ must satisfy $T(0) = 2x e^{-c_0 x}$, which is of course impossible.

(3) We have seen so far an example in which a problem had a unique solution, and an example where there was no solution at all. It turns out that an equation might have infinitely many solutions. To demonstrate this possibility, let us return to the last example, and replace the initial condition (2.6) by

$$u(x, 0) = 2e^{c_0 x}. \tag{2.10}$$

Now $T(y)$ should satisfy $T(0) = 2$. Thus every function $T(y)$ satisfying $T(0) = 2$ will provide a solution for the equation together with the initial condition. Therefore, (2.5) with $c_1 = 0$ has infinitely many solutions under the initial condition (2.10).

We conclude from Example 2.1 that the solution process must include the step of checking for existence and uniqueness. This is an example of the well-posedness issue that was introduced in Chapter 1.

2.3 The method of characteristics

We solve first-order PDEs by the method of *characteristics*. This method was developed in the middle of the nineteenth century by Hamilton. Hamilton investigated the propagation of light. He sought to derive the rules governing this propagation from a purely geometric theory, akin to Euclidean geometry. Hamilton was well aware of the wave theory of light, which was proposed by the Dutch physicist Christian Huygens (1629–1695) and advanced early in the nineteenth century by the English scientist Thomas Young (1773–1829) and the French physicist Augustin Fresnel (1788–1827). Yet, he chose to base his theory on the principle of least time that was proposed in 1657 by the French scientist (and lawyer!) Pierre de Fermat (1601–1665). Fermat proposed a unified principle, according to which light rays travel from a point A to a point B in an orbit that takes the least amount of time. Hamilton showed that this principle can serve as a foundation of a dynamical theory of rays. He thus derived an axiomatic theory that provided equations of motion for light rays. The main building block in the theory is a function that completely characterizes any given optical medium. Hamilton called it the *characteristic function*. He showed that Fermat's principle implies that his characteristic function must satisfy a certain first-order nonlinear PDE. Hamilton's characteristic function and characteristic equation are now called the *eikonal* function and eikonal equation after the Greek word $\epsilon\iota\kappa\omega\nu$ (or $\epsilon\iota\kappa o\nu$) which means "an image".

Hamilton discovered that the eikonal equation can be solved by integrating it along special curves that he called *characteristics*. Furthermore, he showed that in a uniform medium, these curves are exactly the straight light rays whose existence has been assumed since ancient times. In 1911 it was shown by the German physicists Arnold Sommerfeld (1868–1951) and Carl Runge (1856–1927) that the eikonal equation, proposed by Hamilton from his geometric theory, can be derived as a small wavelength limit of the wave equation, as was shown in Chapter 1. Notice that although the eikonal equation is of first order, it is in fact fully nonlinear and not quasilinear. We shall treat it separately later.

We shall first develop the method of characteristics heuristically. Later we shall present a precise theorem that guarantees that, under suitable assumptions, the equation together with its associated condition has a unique solution. The characteristics method is based on 'knitting' the solution surface with a one-parameter family of curves that intersect a given curve in space. Consider the general linear equation (2.4), and write the initial condition parameterically:

$$\Gamma = \Gamma(s) = (x_0(s), y_0(s), u_0(s)), \quad s \in I = (\alpha, \beta). \tag{2.11}$$

The curve Γ will be called the *initial curve*.

The linear equation (2.4) can be rewritten as

$$(a, b, c_0 u + c_1) \cdot (u_x, u_y, -1) = 0. \tag{2.12}$$

Since $(u_x, u_y, -1)$ is normal to the surface u, the vector $(a, b, c_0 u + c_1)$ is in the tangent plane. Hence, the system of equations

$$\frac{dx}{dt}(t) = a(x(t), y(t)),$$

$$\frac{dy}{dt}(t) = b(x(t), y(t)),$$ (2.13)

$$\frac{du}{dt}(t) = c(x(t), y(t)))u(t) + c_1(x(t), y(t))$$

defines spatial curves lying on the solution surface (conditioned so that the curves start on the surface). This is a system of first-order ODEs. They are called the *system of characteristic equations* or, for short, the *characteristic equations*. The solutions are called *characteristic curves* of the equation. Notice that equations (2.13) are autonomous, i.e. there is no explicit dependence upon the parameter t. In order to determine a characteristic curve we need an initial condition. We shall require the initial point to lie on the initial curve Γ. Since each curve $(x(t), y(t), u(t))$ emanates from a different point $\Gamma(s)$, we shall explicitly write the curves in the form $(x(t, s), y(t, s), u(t, s))$. The initial conditions are written as:

$$x(0, s) = x_0(s), \quad y(0, s) = y_0(s), \quad u(0, s) = u_0(s). \quad (2.14)$$

Notice that we selected the parameter t such that the characteristic curve is located on Γ when $t = 0$. One may, of course, select any other parameterization. We also notice that, in general, the parameterization $(x(t, s), y(t, s), u(t, s))$ represents a surface in $\mathbb{R}^3$.

One can readily verify that the method of characteristics applies to the quasilinear equation (2.3) as well. Namely, each point on the initial curve Γ is a starting point for a characteristic curve. The characteristic equations are now

$$x_t(t) = a(x, y, u),$$
$$y_t(t) = b(x, y, u),$$ (2.15)
$$u_t(t) = c(x, y, u),$$

supplemented by the initial condition

$$x(0, s) = x_0(s), \quad y(0, s) = y_0(s), \quad u(0, s) = u_0(s). \quad (2.16)$$

The problem consisting of (2.3) and initial conditions (2.16) is called the *Cauchy problem* for quasilinear equations.

The main difference between the characteristic equations (2.13) derived for the linear equation, and the set (2.15) is that in the former case the first two equations of (2.13) are independent of the third equation and of the initial conditions. We shall observe later the special role played by the projection of the characteristic curves on the (x, y) plane. Therefore, we write (for the linear case) the equation for this

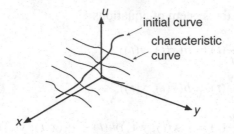

Figure 2.2 Sketch of the method of characteristics.

projection separately:

$$x_t = a(x, y), \quad y_t = b(x, y). \tag{2.17}$$

In the quasilinear case, this uncoupling of the characteristic equations is no longer possible, since the coefficients a and b depend upon u. We also point out that in the linear case, the equation for u is always linear, and thus it is guaranteed to have a global solution (provided that the solutions $x(t)$ and $y(t)$ exist globally).

To summarize the preliminary presentation of the method of characteristics, let us consult Figure 2.2. In the first step we identify the initial curve Γ. In the second step we select a point s on Γ and solve the characteristic equations (2.13) (or (2.15)), using the point we selected on Γ as an initial point. After performing these steps for all points on Γ we obtain a portion of the solution surface (also called the *integral surface*) that consists of the union of the characteristic curves. Philosophically speaking, one might say that the characteristic curves take with them an initial piece of information from Γ, and propagate it with them. Furthermore, each characteristic curve propagates independently of the other characteristic curves.

Let us demonstrate the method for a very simple case.

Example 2.2 Solve the equation

$$u_x + u_y = 2$$

subject to the initial condition $u(x, 0) = x^2$.

The characteristic equations and the parametric initial conditions are

$$x_t(t, s) = 1, \quad y_t(t, s) = 1, \quad u_t(t, s) = 2,$$
$$x(0, s) = s, \quad y(0, s) = 0, \quad u(0, s) = s^2.$$

It is a simple matter to solve for the characteristic curves:

$$x(t, s) = t + f_1(s), \quad y(t, s) = t + f_2(s), \quad u(t, s) = 2t + f_3(s).$$

Upon substituting into the initial conditions, we find

$$x(t, s) = t + s, \quad y(t, s) = t, \quad u(t, s) = 2t + s^2.$$

We have thus obtained a parametric representation of the integral surface. To find an explicit representation of the surface u as a function of x and y we need to invert the transformation $(x(t, s), y(t, s))$, and to express it in the form $(t(x, y), s(x, y))$, namely, we have to solve for (t, s) as functions of (x, y). In the current example the inversion is easy to perform:

$$t = y, \quad s = x - y.$$

Thus the explicit representation of the integral surface is given by

$$u(x, y) = 2y + (x - y)^2.$$

This simple example might lead us to think that each initial value problem for a first-order PDE possesses a unique solution. But we have already seen that this is not the case. What, therefore, are the obstacles we might face? Is (2.3) equipped with initial conditions (2.14) well-posed? For simplicity we shall discuss in this chapter two aspects of well-posedness: existence and uniqueness. Thus the question is whether there exists a unique integral surface for (2.3) that contains the initial curve.

(1) Notice that even if the PDE is linear, the characteristic equations are nonlinear! We know from the theory of ODEs that in general one can only establish local existence of a unique solution (assuming that the coefficients of the equation are smooth functions). In other words, the solutions of nonlinear ODEs might develop singularities within a short distance from the initial point even if the equation is very smooth. It follows that one can expect at most a local existence theorem for a first-order PDE, even if the PDE is linear.

(2) The parametric representation of the integral surface might hide further difficulties. We shall demonstrate this in the sequel by obtaining naive-looking parametric representations of singular surfaces. The difficulty lies in the inversion of the transformation from the plane (t, s) to the plane (x, y). Recall that the implicit function theorem implies that such a transformation is invertible if the Jacobian $J = \partial(x, y)/\partial(t, s) \neq 0$. But we observe that while the dependence of the characteristic curves on the variable t is derived from the PDE itself, the dependence on the variable s is derived from the initial condition. Since the equation and the initial condition do not depend upon each other, it follows that for any given equation there exist initial curves for which the Jacobian vanishes, and the implicit function theorem does not hold.

The functional problem we just described has an important geometrical interpretation. An explicit computation of the Jacobian at points located on the initial curve Γ, using the characteristic equations, gives

$$J = \frac{\partial x}{\partial t} \frac{\partial y}{\partial s} - \frac{\partial x}{\partial s} \frac{\partial y}{\partial t} = \begin{vmatrix} a & b \\ (x_0)_s & (y_0)_s \end{vmatrix} = (y_0)_s a - (x_0)_s b, \qquad (2.18)$$

where $(x_0)_s = dx_0/ds$. Thus the Jacobian vanishes at some point if and only if the vectors (a, b) and $((x_0)_s, (y_0)_s)$ are linearly dependent. Hence the geometrical meaning of a vanishing Jacobian is that the projection of Γ on the (x, y) plane is tangent at this point to the projection of the characteristic curve on that plane. As a rule, in order for a first-order quasilinear PDE to have a unique solution near the initial curve, we must have $J \neq 0$. This condition is called the *transversality condition*.

(3) So far we have discussed local problems. One can also encounter global problems. For example, a characteristic curve might intersect the initial curve more than once. Since the characteristic equation is well-posed for a single initial condition, then in such a situation the solution will, in general, develop a singularity. We can think about this situation in the following way. Recall that a characteristic curve 'carries' with it along its orbit a charge of information from its intersection point with Γ. If a characteristic curve intersects Γ more than once, these two 'information charges' might be in conflict.

A similar global problem is the intersection of the projection on the (x, y) plane of different characteristic curves with each other. Such an intersection is problematic for the same reason as the intersection of a characteristic curve with the initial curve. Each characteristic curve carries with it a different information charge, and a conflict might arise at such an intersection.

(4) Another potential problem relates to a lack of uniqueness of the solution to the characteristic equation. We should not worry about this possibility if the coefficients of the equations are smooth (Lipschitz continuous, to be precise). But when considering a nonsmooth problem, we should pay attention to this issue. We shall demonstrate such a case below.

In Section 2.5 we shall formulate and prove a precise theorem (Theorem 2.10) that will include all the problems discussed above. Before doing so, let us examine a few examples.

2.4 Examples of the characteristics method

Example 2.3 Solve the equation $u_x = 1$ subject to the initial condition $u(0, y) = g(y)$.

The characteristic equations and the associated initial conditions are given by

$$x_t = 1, \quad y_t = 0, \quad u_t = 1, \qquad (2.19)$$

$$x(0, s) = 0, \quad y(0, s) = s, \quad u(0, s) = g(s), \qquad (2.20)$$

respectively. The parametric integral surface is $(x(t, s), y(t, s), u(t, s)) = (t, s, t + g(s))$. It is easy to deduce from here the explicit solution $u(x, y) = x + g(y)$.

On the other hand, if we keep the equation unchanged, but modify the initial conditions into $u(x, 0) = h(x)$, the picture changes dramatically. In this case the parametric solution is

$$(x(t, s), y(t, s), u(t, s)) = (t + s, 0, t + h(s)).$$

Now, however, the transformation $(x(t, s), y(t, s))$ cannot be inverted. Geometrically speaking, the reason is simple: the projection of the initial curve is precisely the x axis, but this is also the projection of a characteristic curve. In the special case where $h(x) = x + c$ for some constant c, we obtain $u(t, s) = s + t + c$. Then it is not necessary to invert the mapping $(x(t, s), y(t, s))$, since we find at once $u = x + c + f(y)$ for every differentiable function $f(y)$ that vanishes at the origin. But for any other choice of h the problem has no solution at all.

We note that for the initial conditions $u(x, 0) = h(x)$ we could have foreseen the problem through a direct computation of the Jacobian:

$$J = \begin{vmatrix} a & b \\ (x_0)_s & (y_0)_s \end{vmatrix} = \begin{vmatrix} 1 & 0 \\ 1 & 0 \end{vmatrix} = 0. \tag{2.21}$$

Whenever the Jacobian vanishes along an interval (like in the example we are considering), the problem will, in general, have no solution at all. If a solution does exist, we shall see that this implies the existence of infinitely many solutions.

Because of the special role played by the projection of the characteristic curves on the (x, y) plane we shall use the term *characteristics* to denote them for short. There are several ways to compute the characteristics. One of them is to solve the full characteristic equations, and then to project the solution on the (x, y) plane. We note that the projection of a characteristic curve is given by the condition $s = $ constant. Substituting this condition into the equation $s = s(x, y)$ determines an explicit equation for the characteristics. An alternative method is valid whenever the PDE is linear. The linearity implies that the first two characteristic equations are independent of u. Thus they can be solved directly for the characteristics themselves without solving first for the parametric integral surface. Furthermore, since the characteristic equations are autonomous (i.e. they do not explicitly include the variable t), it follows that the equations for the characteristics can be written simply as the first-order ODE

$$\frac{dy}{dx} = \frac{b(x, y)}{a(x, y)}.$$

Example 2.4 The current example will be useful for us in Chapter 3, where we shall need to solve linear equations of the form

$$a(x, y)u_x + b(x, y)u_y = 0. \tag{2.22}$$

The equations for the characteristic curves

$$\frac{dx}{dt} = a, \quad \frac{dy}{dt} = b, \quad \frac{du}{dt} = 0$$

imply at once that the solution u is constant on the characteristics that are determined by

$$\frac{dy}{dx} = \frac{b(x, y)}{a(x, y)}. \tag{2.23}$$

For instance, when $a = 1, b = \sqrt{-x}$ (see Example 3.7) we obtain that u is constant along the lines $\frac{3}{2}y + (-x)^{3/2} = $ constant.

Example 2.5 Solve the equation $u_x + u_y + u = 1$, subject to the initial condition

$$u = \sin x, \quad \text{on} \quad y = x + x^2, \quad x > 0.$$

The characteristic equations and the associated initial conditions are given by

$$x_t = 1, \quad y_t = 1, \quad u_t + u = 1, \tag{2.24}$$

$$x(0, s) = s, \quad y(0, s) = s + s^2, \quad u(0, s) = \sin s, \tag{2.25}$$

respectively. Let us compute first the Jacobian along the initial curve:

$$J = \begin{vmatrix} 1 & 1 \\ 1 & 1 + 2s \end{vmatrix} = 2s. \tag{2.26}$$

Thus we anticipate a unique solution at each point where $s \neq 0$. Since we are limited to the regime $x > 0$ we indeed expect a unique solution.

The parametric integral surface is given by

$$(x(t, s), y(t, s), u(t, s)) = (s + t, s + s^2 + t, 1 - (1 - \sin s)e^{-t}).$$

In order to invert the mapping $(x(t, s), y(t, s))$, we substitute the equation for x into the equation for y to obtain $s = (y - x)^{1/2}$. The sign of the square root was selected according to the condition $x > 0$. Now it is easy to find $t = x - (y - x)^{\frac{1}{2}}$, whence the explicit representation of the integral surface

$$u(x, y) = 1 - [1 - \sin(y - x)^{\frac{1}{2}}]e^{-x+(y-x)^{\frac{1}{2}}}.$$

Notice that the solution exists only in the domain

$$D = \{(x, y) \mid 0 < x < y\} \cup \{(x, y) \mid x \leq 0 \text{ and } x + x^2 < y\},$$

and in particular it is not differentiable at the origin of the (x, y) plane. To see the geometrical reason for this, consult Figure 2.3. We see that the slope of characteristic passing through the origin equals 1, which is exactly the slope of the projection of

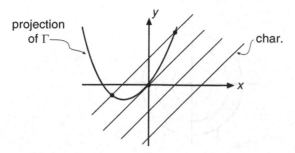

Figure 2.3 The characteristics and projection of Γ for Example 2.5.

the initial curve there. Namely, the transversality condition does not hold there (a fact we already expected from our computation of the Jacobian above). Indeed the violation of the transversality condition led to nonuniqueness of the solution near the curve

$$\{(x, y) \mid x < 0 \quad \text{and} \quad y = x + x^2\},$$

which is manifested in the ambiguity of the sign of the square root.

Example 2.6 Solve the equation $-yu_x + xu_y = u$ subject to the initial condition $u(x, 0) = \psi(x)$.

The characteristic equations and the associated initial conditions are given by

$$x_t = -y, \quad y_t = x, \quad u_t = u, \tag{2.27}$$

$$x(0, s) = s, \quad y(0, s) = 0, \quad u(0, s) = \psi(s). \tag{2.28}$$

Let us examine the transversality condition:

$$J = \begin{vmatrix} 0 & s \\ 1 & 0 \end{vmatrix} = -s. \tag{2.29}$$

Thus we expect a unique solution (at least locally) near each point on the initial curve, except, perhaps, the point $x = 0$.

The solution of the characteristic equations is given by

$$(x(t, s), y(t, s), u(t, s))$$
$$= (f_1(s) \cos t + f_2(s) \sin t, \; f_1(s) \sin t - f_2(s) \cos t, \; e^t f_3(s)).$$

Substituting the initial condition into the solution above leads to the parametric integral surface

$$(x(t, s), y(t, s), u(t, s)) = (s \cos t, \; s \sin t, \; e^t \psi(s)).$$

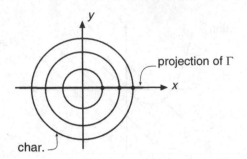

Figure 2.4 The characteristics and projection of Γ for Example 2.6.

Isolating s and t we obtain the explicit representation

$$u(x, y) = \psi(\sqrt{x^2 + y^2}) \exp\left[\arctan\left(\frac{y}{x}\right)\right].$$

It can be readily verified that the characteristics form a one-parameter family of circles around the origin (see Figure 2.4). Therefore, each one of them intersects the projection of the initial curve (the x axis) twice. We also saw that the Jacobian vanishes at the origin. So how is it that we seem to have obtained a unique solution? The mystery is easily resolved by observing that in choosing the positive sign for the square root in the argument of ψ, we effectively reduced the solution to the ray $\{x > 0\}$. Indeed, in this region a characteristic intersects the projection of the initial curve only once.

Example 2.7 Solve the equation $u_x + 3y^{2/3}u_y = 2$ subject to the initial condition $u(x, 1) = 1 + x$.

The characteristic equations and the associated initial conditions are given by

$$x_t = 1, \quad y_t = 3y^{2/3}, \quad u_t = 2, \tag{2.30}$$

$$x(0, s) = s, \quad y(0, s) = 1, \quad u(0, s) = 1 + s. \tag{2.31}$$

In this example we expect a unique solution in a neighborhood of the initial curve since the transversality condition holds:

$$J = \begin{vmatrix} 1 & 3 \\ 1 & 0 \end{vmatrix} = -3 \neq 0. \tag{2.32}$$

The parametric integral surface is given by

$$x(t, s) = s + t, \quad y(t, s) = (t + 1)^3, \quad u(t, s) = 2t + 1 + s.$$

Before proceeding to compute an explicit solution, let us find the characteristics. For this purpose recall that each characteristic curve passes through a specific s value. Therefore, we isolate t from the equation for x, and substitute it into the expression

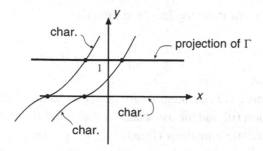

Figure 2.5 Self-intersection of characteristics.

for y. We obtain $y = (x + 1 - s)^3$, and, thus, for each fixed s this is an equation for a characteristic. A number of characteristics and their intersection with the projection of the initial curve $y = 1$ are sketched in Figure 2.5. While the picture indicates no problems, we were not careful enough in solving the characteristic equations, since the function $y^{2/3}$ is not Lipschitz continuous at the origin. Thus the characteristic equations might not have a unique solution there! In fact, it can be easily verified that $y = 0$ is also a solution of $y_t = 3y^{2/3}$. But, as can be seen from Figure 2.5, the well behaved characteristics near the projection of the initial curve $y = 1$ intersect at some point the extra characteristic $y = 0$. Thus we can anticipate irregular behavior near $y = 0$. Inverting the mapping $(x(t, s), y(t, s))$ we obtain

$$t = y^{1/3} - 1, s = x + 1 - y^{1/3}.$$

Hence the explicit solution to the PDE is $u(x, y) = x + y^{1/3}$, which is indeed singular on the x axis.

Example 2.8 Solve the equation $(y + u)u_x + yu_y = x - y$ subject to the initial conditions $u(x, 1) = 1 + x$.

This is an example of a quasilinear equation. The characteristic equations and the initial data are:

$$\text{(i) } x_t = y + u, \quad \text{(ii) } y_t = y, \quad \text{(iii) } u_t = x - y,$$
$$x(0, s) = s, \quad y(0, s) = 1, \quad u(0, s) = 1 + s.$$

Let us examine the transversality condition. Notice that while u is yet to be found, the transversality condition only involves the values of u on the initial curve Γ. It is easy to verify that on Γ we have $a = 2 + s$, $b = 1$. It follows that the tangent to the characteristic has a nonzero component in the direction of the y axis. Thus it is nowhere tangent to the projection of the initial curve (the x axis, in this case).

Alternatively, we can compute the Jacobian directly:

$$J = \begin{vmatrix} 2+s & 1 \\ 1 & 0 \end{vmatrix} = -1 \neq 0. \tag{2.33}$$

We conclude that there exists an integral surface at least at the vicinity of Γ. From the characteristic equation (ii) and the associated initial condition we find $y(t,s) = e^t$. Adding the characteristic equations (i) and (iii) we get $(x+u)_t = x+u$. Therefore, $u+x = (1+2s)e^t$. Returning to (i) we obtain $x(t,s) = (1+s)e^t - e^{-t}$ and $u(t,s) = se^t + e^{-t}$. Observing that $x - y = se^t - e^{-t}$, we finally get $u = 2/y + (x-y)$. The solution is *not* global (it becomes singular on the x axis), but it is well defined near the initial curve.

2.5 The existence and uniqueness theorem

We shall summarize the discussion on linear and quasilinear equations into a general theorem. For this purpose we need the following definition.

Definition 2.9 Consider a quasilinear equation (2.3) with initial conditions (2.16) defining an initial curve for the integral surface. We say that the equation and the initial curve satisfy the *transversality condition* at a point s on Γ, if the characteristic emanating from the projection of $\Gamma(s)$ intersects the projection of Γ nontangentially, i.e.

$$J \mid_{t=0} = x_t(0,s)y_s(0,s) - y_t(0,s)x_s(0,s) = \begin{vmatrix} a & b \\ (x_0)_s & (y_0)_s \end{vmatrix} \neq 0.$$

Theorem 2.10 *Assume that the coefficients of the quasilinear equation (2.3) are smooth functions of their variables in a neighborhood of the initial curve (2.16). Assume further that the transversality condition holds at each point s in the interval $(s_0 - 2\delta, s_0 + 2\delta)$ on the initial curve. Then the Cauchy problem (2.3), (2.16) has a unique solution in the neighborhood $(t,s) \in (-\epsilon, \epsilon) \times (s_0 - \delta, s_0 + \delta)$ of the initial curve. If the transversality condition does not hold for an interval of s values, then the Cauchy problem (2.3), (2.16) has either no solution at all, or it has infinitely many solutions.*

Proof The existence and uniqueness theorem for ODEs, applied to (2.15) together with the initial data (2.16), guarantees the existence of a unique characteristic curve for each point on the initial curve. The family of characteristic curves forms a parametric representation of a surface. The transversality condition implies that the parametric representation provides a smooth surface. Let us verify now that the surface thus constructed indeed satisfies the PDE (2.3). We write

$$\tilde{u} = \tilde{u}(x, y) = u(t(x, y), s(x, y)),$$

and compute

$$a\tilde{u}_x + b\tilde{u}_y = a(u_t t_x + u_s s_x) + b(u_t t_y + u_s s_y) = u_t(at_x + bt_y) + u_s(as_x + bs_y).$$

But the characteristic equations and the chain rule imply

$$1 = t_t = at_x + bt_y, \quad 0 = s_t = as_x + bs_y.$$

Hence $a\tilde{u}_x + b\tilde{u}_y = u_t = c$, i.e. $\tilde{u}$ satisfies (2.3).

To show that there are no further integral surfaces, we prove that the characteristic curves we constructed must lie on an integral surface. Since the characteristic curve starts on the integral surface, we only have to show that it remains there. This is intuitively clear, since the characteristic curve is, by definition, orthogonal at every point to the surface normal. On the other hand, clearly for a curve starting on some surface to leave the surface, its tangent must at some point have a nonzero projection on the normal to the surface. This simple geometrical reasoning can be supported through an explicit computation; for this purpose we write a given integral surface in the form $u = f(x, y)$. Let $(x(t), y(t), u(t))$ be a characteristic curve. We assume $u(0) = f(x(0), y(0))$. Define the function

$$\Psi(t) = u(t) - f(x(t), y(t)).$$

Differentiating by t we write

$$\Psi_t = u_t - f_x(x, y)x_t - f_y(x, y)y_t.$$

Substituting the system (2.15) into the above equations for Ψ_t we obtain

$$\Psi_t = c(x, y, \Psi + f) - f_x(x, y)a(x, y, \Psi + f) - f_y(x, y)b(x, y, \Psi + f).$$

$$(2.34)$$

But the initial condition implies $\Psi(0) = 0$. It is easy to check (using (2.3)) that $\Psi(t) \equiv 0$ solves the ODE (2.34). Since that equation has smooth coefficients, it has a unique solution. Thus $\Psi \equiv 0$ is the only solution, and the curve $(x(t), y(t), u(t))$ indeed lies on the integral surface. Therefore the integral surface we constructed earlier through the parametric representation induced by the characteristic equations is unique.

When the transversality condition does not hold along an interval of s values, the characteristic there is the same as the projection of Γ. If the solution of the characteristic equation is a curve that is not identical to the initial curve, then the tangent (vector) to the initial curve at some point cannot be at that point tangential to any integral surface. In other words, the initial condition contradicts the equation and thus there can be no solution to the Cauchy problem. If, on the other hand, the characteristic curve agrees with the initial curve at that point, there are infinitely many ways to extend it into a compatible integral surface that contains it. Therefore, in this case we have infinitely many solutions to the Cauchy problem. We now present a method for constructing this family of solutions. Select an arbitrary point $P_0 = (x_0, y_0, u_0)$ on Γ. Construct a new initial curve Γ', passing through P_0, which is not tangent to Γ at P_0. Solve the new Cauchy problem consisting of (2.3) with Γ' as initial curve. Since, by construction, the transversality condition holds now, the first part of the theorem guarantees a unique solution. Since there are infinitely many ways of selecting such an initial curve Γ', we obtain infinitely many solutions. $\qquad\qquad\qquad\qquad\qquad\qquad\qquad\qquad\qquad\qquad\qquad\qquad\qquad\square$

The following simple example demonstrates the case where the transversality condition fails along some interval.

Example 2.11 Consider the Cauchy problem

$$u_x + u_y = 1, \quad u(x, x) = x.$$

Show that it has infinitely many solutions.

The transversality condition is violated identically. However the characteristic direction is $(1, 1, 1)$, and so is the direction of the initial curve. Hence, the initial curve is itself a characteristic curve. Thus there exist infinitely many solutions. To find these solutions, set the problem

$$u_x + u_y = 1, \quad u(x, 0) = f(x),$$

for an arbitrary f satisfying $f(0) = 0$. The solution is easily found to be $u(x, y) = y + f(x - y)$.

Notice that the Cauchy problem

$$u_x + u_y = 1, \quad u(x, x) = 1,$$

on the other hand, is not solvable. To see this observe that the transversality condition fails again, but now the initial curve is *not* a characteristic curve. Thus there is no solution.

Remark 2.12 There is one additional possibility not covered by Theorem 2.10. This is the case where the transversality condition does not hold on isolated points (as was indeed the case in some of the preceding examples). It is difficult to formulate universal statements here. Instead, each such case has to be analyzed separately.

2.6 The Lagrange method

First-order quasilinear equations were in fact studied by Lagrange even before Hamilton. Lagrange developed a solution method that is also geometric in nature, albeit less general than Hamilton's method. The main advantage of Lagrange's method is that it provides general solutions for the equation, regardless of the initial data.

Let us reconsider (2.15). The set of all solutions to this system forms a two-parameter set of curves. To justify this assertion, notice that since the system (2.15) is autonomous, it is equivalent to the system

$$y_x = b(x, y, u)/a(x, y, u), \quad u_x = c(x, y, u)/a(x, y, u). \tag{2.35}$$

Since (2.35) is a system of two first-order ODEs in the (y, u) plane, where x is a parameter, it follows that the set of solutions is determined by two initial conditions.

Lagrange assumed that the two-parameter set of solution curves for (2.15) can be represented by the intersection of two families of integral surfaces

$$\psi(x, y, u) = \alpha, \quad \phi(x, y, u) = \beta. \tag{2.36}$$

When we vary the parameters α and β we obtain (through intersecting the surfaces ψ and ϕ) the two-parameter set of curves that are generated by the intersection. Recall that a solution surface of (2.3) passing through an initial curve is obtained from a one-parameter family of curves solving the characteristic equation (2.15). Each such one-parameter subfamily describes a curve in the parameter space (α, β). Since such a curve can be expressed in the form $F(\alpha, \beta) = 0$, it follows that every solution of (2.3) and (2.16) is given by

$$F(\psi(x, y, u), \phi(x, y, u)) = 0. \tag{2.37}$$

Since the surfaces ψ and ϕ were determined by the equation itself, (2.37) defines a general solution to the PDE. When we solve for a particular initial curve, we just have to use this curve to determine the specific functional form of F associated with that initial curve.

We are still left with one "little" problem: how to find the surfaces ψ and ϕ. In the theory of ODEs one solves first-order equations by the method of integration factors. While this method is always feasible in theory, it involves great technical difficulties. In fact, it is possible to find integration factors only in special cases. In a sense, the Lagrange method is a generalization of the integration factor method for ODEs, as we have to find solution surfaces for the two first-order ODEs (2.35). Hence it is not surprising that the method is applicable only in special cases.

We proceed by introducing a method for computing the surfaces ψ and ϕ, and then apply the method to a specific example.

Example 2.13 Recall that by definition, the surfaces $\psi = \alpha$, $\phi = \beta$ contain the characteristic curves. Assume there exist two independent vector fields $\vec{P}_1 = (a_1, b_1, c_1)$ and $\vec{P}_2 = (a_2, b_2, c_2)$ (i.e. they are nowhere tangent to each other) that are both orthogonal to the vector $\vec{P} = (a, b, c)$ (the vector defining the characteristic equations). This means $aa_1 + bb_1 + cc_1 = 0 = aa_2 + bb_2 + cc_2$. Let us assume further that the vector fields $\vec{P}_1$ and $\vec{P}_2$ are exact, i.e. $\nabla \times \vec{P}_1 = 0 = \nabla \times \vec{P}_2$. This implies that there exist two potentials ψ and ϕ satisfying $\nabla \psi = \vec{P}_1$ and $\nabla \phi = \vec{P}_2$. By construction it follows that $d\psi = \nabla \psi \cdot \vec{P} = 0 = \nabla \phi \cdot \vec{P} = d\phi$, namely, ψ and ϕ are constant on every characteristic curve, and form the requested two-parameter integral surfaces.

Let us apply this method to find the general solution to the equation

$$-yu_x + xu_y = 0. \tag{2.38}$$

The characteristic equations are

$$x_t = -y, \quad y_t = x, \quad u_t = 0.$$

In this example $\vec{P} = (-y, x, 0)$. It is easy to guess orthogonal vector fields $\vec{P}_1 = (x, y, 0)$ and $\vec{P}_2 = (0, 0, 1)$. The reader can verify that they are indeed exact vector fields. The associated potentials are

$$\psi(x, y, u) = \frac{1}{2}(x^2 + y^2), \quad \phi(x, y, u) = (0, 0, u).$$

Therefore, the general solution of (2.38) is given by

$$F(x^2 + y^2, u) = 0, \tag{2.39}$$

or

$$u = g(x^2 + y^2). \tag{2.40}$$

To find the specific solution of (2.38) that satisfies a given initial condition, we shall use that condition to eliminate g. For example, let us compute the solution of (2.38) satisfying $u(x, 0) = \sin x$ for $x > 0$. Substituting the initial condition into (2.40) yields $g(\xi) = \sin \sqrt{\xi}$; hence $u(x, y) = \sin \sqrt{x^2 + y^2}$ is the required solution.

While the Lagrange method has an advantage over the characteristics method, since it provides a *general* solution to the equation, valid for all initial conditions, it also has a number of disadvantages.

(1) We have already explained that ψ and ϕ can only be found under special circumstances. Many tricks for this purpose have been developed since Lagrange's days, yet, only a limited number of equations can be solved in this way.

(2) It is difficult to deduce from the Lagrange method any potential problems arising from the interaction between the equation and the initial data.

(3) The Lagrange method is limited to quasilinear equations. Its generalization to arbitrary nonlinearities is very difficult. On the other hand, as we shall soon see, the characteristics method can be naturally extended to a method that is applicable to general nonlinear PDEs.

It would be fair to say that the main value of the Lagrange method is historical, and in supplying general solutions to certain canonical equations (such as in the example above).

2.7 Conservation laws and shock waves

The existence theorem for quasilinear equations only guarantees (under suitable conditions) the existence of a local solution. Nevertheless, there are cases of interest where we need to compute the solution of a physical problem beyond the point where the solution breaks down. In this section we shall discuss such a situation. For simplicity we shall perform the analysis in some detail for a canonical prototype of quasilinear equations given by

$$u_y + u u_x = 0. \tag{2.41}$$

This equation plays an important role in hydrodynamics. It models the flow of mass with concentration u, where the speed of the flow depends on the concentration. The variable y has the physical interpretation of time. We shall show that the solutions to this equation often develop a special singularity that is called a *shock wave*. In hydrodynamics the equation is called the *Euler equation* (cf. Chapter 1; the reader may be baffled by now by the multitude of differential equations that are called after Euler. We have to bear in mind that Euler was a highly prolific mathematician who published over 800 papers and books). Towards the end of the section we shall generalize the analysis that is performed for (2.41) to a larger family of equations, and in particular, we shall apply the theory to study traffic flow.

As a warm-up we start with the simple linear equation

$$u_y + c u_x = 0. \tag{2.42}$$

The difference between this equation and (2.41), is that in (2.42) the flow speed is given by the positive constant c. The initial condition

$$u(x, 0) = h(x) \tag{2.43}$$

will be used for both equations. Solving the characteristic equations for the linear equation (2.42) we get

$$(x, y, u) = (s + ct, t, h(s)).$$

Eliminating s and t yields the explicit solution $u = h(x - cy)$. The solution implies that the initial profile does not change; it merely moves with speed c along the positive x axis, namely, we have a fixed wave, moving with a speed c while preserving the initial shape.

Euler's equation (2.41) is solved similarly. The characteristic equations are

$$x_t = u, \quad y_t = 1, \quad u_t = 0,$$

and their solution is

$$(x, y, u) = (s + h(s)t, t, h(s)),$$

where we used the parameterization $x(0, s) = s$, $y(0, s) = 0$, $u(0, s) = h(s)$ for the initial data. Therefore, the solution of the PDE is

$$u = h(x - uy), \tag{2.44}$$

except that this time this solution is actually implicit. In order to analyze this solution further we eliminate the y variable from the equations for the characteristics (the projection of the characteristic curves on the (x, y) plane):

$$x = s + h(s)y. \tag{2.45}$$

The third characteristic equation implies that for each fixed s, i.e. along each characteristic, u preserves its initial value $u = h(s)$. The other characteristic equations imply, then, that the characteristics are straight lines.

Since different characteristics have different slopes that are determined by the initial values of u, they might intersect. Such an intersection has an obvious physical interpretation that can be seen from (2.45): The initial data $h(s)$ determine the speed of the characteristic emanating from a given s. Therefore, if a characteristic leaving the point s_1 has a higher speed than a characteristic leaving the point s_2, and if $s_1 < s_2$, then after some (positive) time the faster characteristic will overtake the slower one.

As we explained above, the solution is not well defined at points where characteristic curves intersect. To see the resulting difficulty from an algebraic perspective, we differentiate the implicit solution with respect to x to get $u_x = h'(1 - yu_x)$, implying

$$u_x = \frac{h'}{1 + yh'}. \tag{2.46}$$

Recalling that physically the variable y stands for time, we consider the ray $y > 0$. We conclude that the solution's derivative blows up at the critical time

$$y_c = -\frac{1}{h'(s)}. \tag{2.47}$$

Hence the classical solution is not defined for $y > y_c$. This conclusion is consistent with the heuristic physical interpretation presented above. Indeed a necessary condition for a singularity formation is that $h'(s) < 0$ at least at one point, such that a faster characteristic will start from a point behind a slower characteristic. If $h(s)$ is never decreasing, there will be no singularity; however, such data are exceptional. Observe that the solution becomes singular at the first time y that satisfies (2.47); such a value is achieved for the value s, where $h'(s)$ is minimal.

Equation (2.41) arises in the investigation of a fundamental physical problem. Thus we cannot end our analysis when a singularity forms. In other words, while the solution becomes singular at the critical time (2.47), the fluid described by the equations keeps flowing unaware of our mathematical troubles! Therefore we must find a means of extending the solution beyond y_c.

Extending singular solutions is not a simple matter. There are several ways to construct such extensions, and we must select a method that conforms with fundamental physical principles. The basic idea is to define a new problem. This new problem is formulated so as to be satisfied by each classical solution of the Euler equation, and such that each continuously differentiable solution of the new problem will satisfy the Euler equation. Yet, the new problem will also have nonsmooth solutions. A solution of the new extended problem is called a *weak solution*, and the new problem itself is called the weak formulation of the original PDE. We shall see that sometimes there exist more than one weak solution, and this will require upgrading the weak formulation to include a selection principle.

We choose to formulate the weak problem by replacing the differential equation with an integral balance. In fact, we have already discussed in Chapter 1 the connection between an integral balance and the associated differential relation emerging from it. We explained that the integral balance is more fundamental, and can only be transformed into a differential relation when the functions involved are sufficiently smooth. To apply the integral balance method we rewrite (2.41) in the form

$$\partial_y u + \frac{1}{2}\frac{\partial}{\partial x}(u^2) = 0, \tag{2.48}$$

and integrate (with respect to x, and for a fixed y) over an arbitrary interval $[a, b]$ to obtain

$$\partial_y \int_a^b u(\xi, y)\mathrm{d}\xi + \frac{1}{2}[u^2(b, y) - u^2(a, y)] = 0. \tag{2.49}$$

It is clear that every solution of the PDE satisfies the integral relation (2.49) as well. Also, since a and b are arbitrary, any function $u \in C^1$ that satisfies (2.49) would also satisfy the PDE. Nevertheless, the integral balance is also well defined for functions not in C^1; actually, (2.49) is even defined for functions with finitely many discontinuities.

We now demonstrate the construction of a weak solution that is a smooth function (continuously differentiable) except for discontinuities along a curve $x = \gamma(y)$. Since the solution is smooth on both sides of γ, it satisfies the equation there. It remains to compute γ. For this purpose we write the weak formulation in the form

$$\partial_y \left[\int_a^{\gamma(y)} u(\xi, y)d\xi + \int_{\gamma(y)}^b u(\xi, y)d\xi \right] + \frac{1}{2}[(u^2(b, y) - u^2(a, y)] = 0.$$

Differentiating the integrals with respect to y and using the PDE itself leads to

$$\gamma_y(y)u^- - \gamma_y(y)u^+ - \frac{1}{2}\left[\int_a^{\gamma(y)} (u^2(\xi, y))_\xi d\xi + \int_{\gamma(y)}^b (u^2(\xi, y))_\xi d\xi \right]$$
$$+ \frac{1}{2}[u^2(b, y) - u^2(a, y)] = 0.$$

Here we used u^- and u^+ to denote the values of u when we approach the curve γ from the left and from the right, respectively. Performing the integration we obtain

$$\gamma_y(y) = \frac{1}{2}(u^- + u^+), \tag{2.50}$$

namely, the curve γ moves at a speed that is the average of the speeds on the left and right ends of it.

Example 2.14 Consider the Euler equation (2.41) with the initial conditions

$$u(x, 0) = h(x) = \begin{cases} 1 & x \le 0, \\ 1 - x/\alpha & 0 < x < \alpha, \\ 0 & x \ge \alpha. \end{cases} \tag{2.51}$$

Since $h(x)$ is not monotone increasing, the solution will develop a singularity at some finite (positive) time. Formula (2.47) implies $y_c = \alpha$. For all $y < \alpha$ the (smooth) solution is given by

$$u(x, y) = \begin{cases} 1 & x \le y, \\ \dfrac{x - \alpha}{y - \alpha} & y < x < \alpha, \\ 0 & x \ge \alpha. \end{cases} \tag{2.52}$$

After the critical time y_c when the solution becomes singular we need to define a weak solution. We seek a solution with a single discontinuity. Formula (2.50)

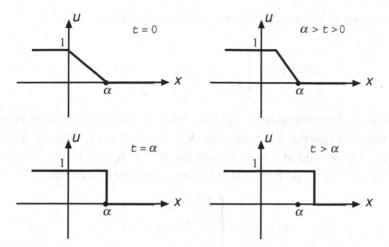

Figure 2.6 Several snapshots in the development of a shock wave.

implies that the discontinuity moves with a speed $\frac{1}{2}$. Therefore the following weak solution is compatible with the integral balance even for $y > \alpha$:

$$u(x, y) = \begin{cases} 1 & x < \alpha + \frac{1}{2}(y - \alpha), \\ 0 & x > \alpha + \frac{1}{2}(y - \alpha). \end{cases} \tag{2.53}$$

The solution thus constructed has the structure of a moving jump discontinuity. It describes a step function moving at a constant speed. Such a solution is called a *shock wave*. Several snapshots of the formation and propagation of a shock wave are depicted in Figure 2.6.

Strictly speaking, the solution is not continuously differentiable even at time $y = 0$; however, this is a minor complication, since it can be shown that the formula for the classical solution is valid even when the derivative of the initial data has finitely many discontinuities as long as it is bounded.

Example 2.15 We now consider the opposite case where the initial data are increasing:

$$u(x, 0) = \begin{cases} 0 & x \leq 0, \\ x/\alpha & 0 < x < \alpha, \\ 1 & x > \alpha. \end{cases} \tag{2.54}$$

Since this time $h' \geq 0$, there is no critical (positive) time where the characteristics intersect. On the contrary, the characteristics diverge. This situation is called an *expansion wave*, in contrast to the wave in the previous example which is called a *compression wave*. We use the classical solution formula to obtain

$$u(x, y) = \begin{cases} 0 & x \leq 0, \\ \dfrac{x}{\alpha + y} & 0 < x < \alpha + y, \\ 1 & x \geq \alpha + y. \end{cases} \tag{2.55}$$

It is useful to consider for both examples the limiting case where $\alpha \to 0$. In Example 2.14 the initial data are the same as the shock weak solution (2.53), and therefore this solution is already valid at $y = 0$. In contrast, in Example 2.15 the characteristics expand, the singularity is smoothed out at once, and the solution is

$$u(x, y) = \begin{cases} 0 & x \leq 0, \\ \dfrac{x}{y} & 0 < x < y, \\ 1 & x \geq y. \end{cases} \tag{2.56}$$

We notice, though, that we could in principle write in the expansion wave case a weak solution that has a shock wave structure:

$$u(x, y) = \begin{cases} 0 & x < \alpha + \frac{1}{2}(y - \alpha), \\ 1 & x > \alpha + \frac{1}{2}(y - \alpha). \end{cases} \tag{2.57}$$

We see that the weak formulation by itself does not have a unique solution! Additional arguments are needed to pick out the correct solution among the several options. In the case we consider here it is intuitively clear that the shock solution (2.57) is not adequate, since slower characteristics starting from the ray $x < 0$ cannot overtake the faster characteristics that start from the ray $x > 0$. Another, more physical, approach to this problem will be described now.

The theory of weak solutions and shock waves is quite difficult. We therefore present essentially the same ideas we developed above from a somewhat different perspective. Instead of looking at the specific canonical equation (2.48) with general initial conditions, we look at a more general PDE with canonical initial conditions. Specifically we consider the following first-order quasilinear PDE

$$u_y + \frac{\partial}{\partial x} F(u) = 0. \tag{2.58}$$

Equations of this kind are called *conservation laws*. To understand the name, let us recall the derivation of the heat equation in Section 1.4.1. The energy balance (1.8) is actually of the form (2.58), where F denotes flux. In the canonical example (2.48), the flux is $F = \frac{1}{2}u^2$, and u is typically interpreted as mass density.

Equation (2.58) is supplemented with the initial condition

$$u(x, 0) = \begin{cases} u^- & x < 0, \\ u^+ & x > 0. \end{cases} \tag{2.59}$$

To write the weak formulation for (2.58), we assume that the solution takes the shape of a shock wave of the form

$$u(x, y) = \begin{cases} u^- & x < \gamma(y), \\ u^+ & x > \gamma(y). \end{cases} \qquad (2.60)$$

It remains therefore to find the shock orbit $x = \gamma(y)$. We find γ by integrating (2.58) with respect to x along the interval (x_1, x_2), with $x_1 < \gamma$, $x_2 > \gamma$. Taking (2.60) into account, we get

$$\frac{\partial}{\partial y}\left\{[x_2 - \gamma(y)]u^+ + [\gamma(y) - x_1]u^-\right\} = F(u^+) - F(u^-). \qquad (2.61)$$

The last equation implies

$$\gamma_y(y) = \frac{F(u^+) - F(u^-)}{u^+ - u^-} := \frac{[F]}{[u]}, \qquad (2.62)$$

where we used the notation $[\cdot]$ to denote the change (jump) of a quantity across the shock.

Conservation laws appear in many areas of continuum mechanics (including hydrodynamics, gas dynamics, combustion, etc.), where the jump equation (2.62) is called the *Rankine–Hugoniot* condition.

Notice that in the special case of $F = \frac{1}{2}u^2$ that we considered earlier, the rule (2.62) reduces to (2.50). We also point out that we integrated (2.58) along an arbitrary finite interval, although the values of x_1 and x_2 do not appear in the final conclusion. The reason for introducing this artificial interval is that the integral of (2.58) over the real line is not bounded. Since we interpret u as a density of a physical quantity (such as mass), our model is really artificial, and a realistic model will have to take into account the effects of finite boundaries.

Our analysis of the general conservation law (2.58) is not yet complete. From our study of Example 2.15 we expect that shock would only occur if the characteristics collide. In the case of general conservation laws, this condition is expressed as:

The entropy condition Characteristics must enter the shock curve, and are not allowed to emanate from it.

The motivation for the entropy condition is rooted in gas dynamics and the second law of thermodynamics. In order not to stray too much away from the theory of PDEs, we shall give a heuristic reasoning, based on the interpretation of entropy as minus the amount of "information" stored in a given physical system. We thus phrase the second law of thermodynamics as stating that in a closed system information is only lost as time y increases, and cannot be created. Now, we have shown that characteristics carry with them information on the solution of a first-order

PDE. Therefore the emergence of a characteristic *from* a shock is interpreted as a *creation* of information which should be forbidden. To give the entropy condition an algebraic form, we write the conservation law (2.58) as $u_y + F_u u_x = 0$. Therefore the characteristic speed is given by F_u, and the entropy condition can be expressed as

$$F_u(u^-) > \gamma_y > F_u(u^+). \tag{2.63}$$

Applying this rule to the special case $F(u) = \frac{1}{2}u^2$ and using (2.62) we obtain that the shock solution is valid only if $u^- > u^+$, a conclusion we reached earlier from different considerations.

The theory of conservation laws has a nice application to the real-world problem of traffic flow. We therefore end this section by a qualitative analysis of this problem. Consider the flow of cars along one direction in a road. Although cars are discrete entities, we model them as a continuum, and denote by $u(x, y)$ the car density at a point x and time y.

A great deal of research has been devoted to the question of how to model the flux term $F(u)$. Clearly the flux is very low in bumper-to-bumper traffic, where each car barely moves. It may be surprising at first sight, but the flux is also low when the traffic is very light. In this case drivers tend to drive fast, and maintain a large distance between each other (at least this is what they ought to do when they drive fast. . .). Therefore the flux, which counts the total number of cars crossing a given point per time, is low. If we assume that a car occupies on average a length of 5 m, then the highest density is $u_b = 200$ cars/km. It was found experimentally that the maximal flux is about $F_{max} = 1500$ cars/hour, and is achieved at a speed of about 35 km/hour (20 miles/hour) (see [22] for a detailed discussion of traffic flow in the current context). Therefore the optimal density (if one wants to maximize the flux) is $u_{max} \sim 43$ cars/km. The concave shape of $F(u)$ is depicted in Figure 2.7.

Let us look at some practical implications of the model. Suppose that at time $y = 0$ there is a traffic jam at some point $x = x_j$. This could be caused by an accident, a red traffic light, a policeman directing the traffic, etc. Assume further that there is a line

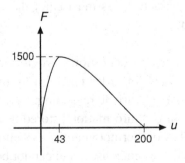

Figure 2.7 The traffic flux F as a function of the density u.

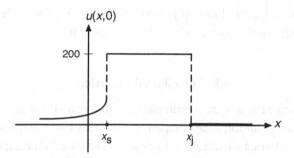

Figure 2.8 The car density at a red traffic light.

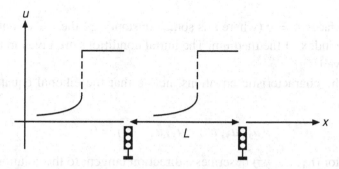

Figure 2.9 Traffic flow through a sequence of traffic lights.

of stationary cars extending from x_s to x_j (with $x_j > x_s$). Cars approach the traffic jam from $x < x_s$. At some point the drivers slow down, as they reach the regime where the car density is greater than u_{max}. Therefore the density u just before x_s is as shown schematically in Figure 2.8. The Rankine–Hugoniot condition (2.62) implies that the shock speed is negative. Although the curve $u(x, 0)$ is increasing, the derivative F_u is now negative, and therefore the entropy condition holds. We conclude that a shock wave will propagate from x_s *backwards*. Indeed as drivers approach a traffic jam there is a stage when they enter the shock and have to decelerate rapidly. The opposite occurs as we leave the point x_j. The density is decreasing and the entropy condition is violated. Therefore we have an expansion wave.

Our analysis can be applied to the design of traffic lights timing. Assume there are several consecutive traffic lights, separated by a distance L (see Figure 2.9). When the traffic approaches one of the traffic lights, a shock propagates backward. To estimate the speed of the shock we assume that behind the shock the density is optimal and so the flux is F_{max}. Then $\gamma_y = -F_{max}/(u_b - u_{max})$. Therefore the time it takes the shock to reach the previous traffic light is

$$T_s = \frac{L(u_b - u_{max})}{F_{max}}.$$

If the red light is maintained over a period exceeding T_s, the high density profile will extend throughout the road, and traffic will come to a complete stop.

2.8 The eikonal equation

Before proceeding to the general nonlinear case, let us analyze in detail the special case of the eikonal equation (see Chapter 1). We shall see that this equation can also be solved by characteristics. The two-dimensional eikonal equation takes the form

$$u_x^2 + u_y^2 = n^2, \tag{2.64}$$

where the surfaces $u = c$ (where c is some constant) are the wavefronts, and n is the refraction index of the medium. The initial conditions are given in the form of an initial curve Γ.

To write the characteristic equations, notice that the eikonal equation can be expressed as

$$(u_x, u_y, n^2) \cdot (u_x, u_y, -1) = 0.$$

Thus the vector (u_x, u_y, n^2) describes a direction tangent to the solution (integral) surface. To verify this argument algebraically, write equations for the x and y components of the characteristic curve, and check that the equation for the u component is consistent with (2.64).

We thus set

$$\frac{dx}{dt} = u_x, \quad \frac{dy}{dt} = u_y, \quad \frac{du}{dt} = n^2. \tag{2.65}$$

Since u_x and u_y are unknown at this stage, we compute

$$\frac{d^2x}{dt^2} = \frac{d}{dt}(u_x) = u_{xx}\frac{dx}{dt} + u_{xy}\frac{dy}{dt} = u_{xx}u_x + u_{xy}u_y$$
$$= \frac{1}{2}(u_x^2 + u_y^2)_x = \frac{1}{2}(n^2(x,y))_x, \tag{2.66}$$

and similarly

$$\frac{d^2y}{dt^2} = \frac{1}{2}(n^2(x,y))_y. \tag{2.67}$$

To write the solution of the eikonal equation, notice that it follows from the definition of the characteristic curves that

$$\frac{du}{dt} = u_x\frac{dx}{dt} + u_y\frac{dy}{dt} = u_x^2 + u_y^2 = n^2. \tag{2.68}$$

Integrating the last equation leads to a formula that determines u at the point $(x(t), y(t))$ in terms of the initial value of u and the values of the refraction index along the integration path:

$$u(x(t), y(t)) = u(x(0), y(0)) + \int_0^t n^2(x(\tau), y(\tau))d\tau, \qquad (2.69)$$

where $(x(t), y(t))$ is a solution of (2.66) and (2.67).

Before solving specific examples, we should clarify an important point regarding the initial conditions for the characteristic equations. Since the original equation (2.65) involves the derivatives of u that are not known at this stage, we eliminated these derivatives by differentiating the characteristic equations once more with respect to the parameter t. Indeed the equations we obtained ((2.66) and (2.67)) no longer depend on u itself; however these are second-order equations! Therefore, it is not enough to provide a single initial condition (such as the initial point of the characteristic curve on the initial curve Γ), but, rather, we must provide the derivatives (x_t, y_t) too. Equivalently, we should provide the vector tangent to the characteristic at the initial point. For this purpose we shall use the fact that the required vector is precisely the gradient (u_x, u_y) of u. From the eikonal equation itself we know that the size of that vector is $n(x, y)$, and from the initial condition we can find its projection at each point of Γ in the direction tangent to Γ. But obviously the size of a planar vector and its projection along a given direction determine the vector uniquely. Hence we obtain the additional initial condition.

Example 2.16 Solve the eikonal equation (2.64) for a medium with a constant refraction index $n = n_0$, and initial condition $u(x, 2x) = 1$.

The physical meaning of the initial condition is that the wavefront is a straight line. The characteristic equations are $d^2x/dt^2 = 0 = d^2y/dt^2$. Thus, the characteristics are straight lines, emanating from the initial line $y = 2x$. Since u is constant on such a line, the gradient of u is orthogonal to it. Hence the second initial condition for the characteristic is

$$\frac{dx}{dt}(0) = \frac{2}{\sqrt{5}}n_0, \quad \frac{dy}{dt}(0) = -\frac{1}{\sqrt{5}}n_0.$$

We thus obtain:

$$x(t, s) = \frac{2}{\sqrt{5}}n_0 t + x_0(s), \quad y(t, s) = -\frac{1}{\sqrt{5}}n_0 t + y_0(s), \quad u(t, s) = n_0^2 t + u_0(s).$$

$$(2.70)$$

In order to find $x_0(s)$ and $y_0(s)$, we write the initial curve parameterically as $(s, 2s, 1)$. Substituting the initial curve into (2.70) leads to the integral surface

$$(x, y, u) = \left(\frac{2}{\sqrt{5}} n_0 t + s, \ -\frac{1}{\sqrt{5}} n_0 t + 2s, \ n_0^2 t + 1 \right). \tag{2.71}$$

Eliminating $t = (2x - y)/\sqrt{5}n_0$, we obtain the explicit solution

$$u(x, y) = 1 + \frac{n_0}{\sqrt{5}} (2x - y).$$

The solution we have obtained has a simple physical interpretation: in a homogeneous medium the characteristic curves are straight lines (classical light rays), and an initial planar wavefront propagates in the direction orthogonal to them. Therefore all wavefronts are planar.

Example 2.17 Compute the function $u(x, y)$ satisfying the eikonal equation $u_x^2 + u_y^2 = n^2$ and the initial condition $u(x, 1) = n\sqrt{1 + x^2}$ (n is a constant parameter).

Write the initial conditions parametrically in the form $(x, y, u) = (s, 1, n\sqrt{1 + s^2})$. This condition implies $x_t(0, s) = u_x = ns/\sqrt{1 + s^2}$. Substituting the last expression into the eikonal equation gives $y_t(0, s) = u_y = n/\sqrt{1 + s^2}$. Integrating the characteristic equations we obtain

$$(x(t, s), y(t, s), u(t, s)) = \left(\frac{ns}{\sqrt{1 + s^2}} t + s, \ \frac{n}{\sqrt{1 + s^2}} t + 1, \ n(nt + \sqrt{1 + s^2}) \right).$$

In order to write an explicit solution, observe the identity

$$x^2 + y^2 = (nt + \sqrt{1 + s^2})^2$$

satisfied by the integral surface. Therefore, the solution is $u = n\sqrt{x^2 + y^2}$. This solution represents a spherical wave starting from a single point at the origin of coordinates.

2.9 General nonlinear equations

The general first-order nonlinear equation takes the form

$$F(x, y, u, u_x, u_y) = 0. \tag{2.72}$$

We shall develop a solution method for such equations. The method is an extension of the method of characteristics. To simplify the presentation we shall use the notation $p = u_x$, $q = u_y$.

Consider a point (x_0, y_0, u_0) on the integral surface. We want to find the slope of a (characteristic) curve on the integral surface passing through this point. In the quasilinear case the equation determined directly the slope of a specific curve on the integral surface. We shall now construct that curve in a somewhat different manner. Let us write for this purpose the equation of the tangent plane to the integral surface through (x_0, y_0, u_0):

$$p(x - x_0) + q(y - y_0) - (u - u_0) = 0. \tag{2.73}$$

Notice that the derivatives p and q at (x_0, y_0, u_0) are not independent. Equation (2.72) imposes the relation

$$F(x_0, y_0, u_0, p, q) = 0. \tag{2.74}$$

The last two equations define a one-parameter family of tangent planes. Such a family spreads out a cone. To honor the French geometer Gaspard Monge (1746–1818) this cone is called the *Monge cone*. The natural candidate for the curve we seek in the tangent plane (defining for us the direction of the characteristic curve) is, therefore, the generator of the Monge cone. To compute the generator we differentiate (2.73) by the parameter p:

$$(x - x_0) + \frac{dq}{dp}(y - y_0) = 0. \tag{2.75}$$

Assume that F is not degenerate, i.e. $F_p^2 + F_q^2$ never vanishes. Without loss of generality assume $F_q \neq 0$. Then it follows from (2.74) and the implicit function theorem that

$$q'(p) = -\frac{F_p}{F_q}. \tag{2.76}$$

Substituting (2.76) into (2.75) we obtain the equation for the Monge cone generator:

$$\frac{x - x_0}{F_p} = \frac{y - y_0}{F_q}. \tag{2.77}$$

Equations (2.73) and (2.77) imply three differential equations for the characteristic curves:

$$\begin{aligned}
x_t &= F_p(x, y, u, p, q), \\
y_t &= F_q(x, y, u, p, q), \\
u_t &= p F_p(x, y, u, p, q) + q F_q(x, y, u, p, q).
\end{aligned} \tag{2.78}$$

It is easy to verify that equations (2.78) coincide in the quasilinear case with the characteristic equations (2.15). However in the fully nonlinear case the characteristic equations (2.78) do not form a closed system. They contain the hitherto unknown functions p and q. In other words, the characteristic curves carry with them a tangent plane that has to be found as part of the solution. To derive equations for p and q we write

$$p_t = u_{xx}x_t + u_{xy}y_t = u_{xx}F_p + u_{xy}F_q, \qquad (2.79)$$

and similarly

$$q_t = u_{yx}F_p + u_{yy}F_q. \qquad (2.80)$$

In order to eliminate u_{xx}, u_{xy} and u_{yy}, recall that the equation $F = 0$ holds along the characteristic curves. We therefore obtain upon differentiation

$$F_x + pF_u + p_xF_p + q_xF_q = 0, \qquad (2.81)$$
$$F_y + qF_u + p_yF_p + q_yF_q = 0. \qquad (2.82)$$

Substituting (2.81) and (2.82) into (2.79) and (2.80) leads to

$$p_t = -F_x - pF_u,$$
$$q_t = -F_y - qF_u.$$

To summarize we write the entire set of characteristic equations:

$$
\begin{aligned}
x_t &= F_p(x, y, u, p, q),\\
y_t &= F_q(x, y, u, p, q),\\
u_t &= pF_p(x, y, u, p, q) + qF_q(x, y, u, p, q),\\
p_t &= -F_x(x, y, u, p, q) - pF_u(x, y, u, p, q),\\
q_t &= -F_y(x, y, u, p, q) - qF_u(x, y, u, p, q).
\end{aligned} \qquad (2.83)
$$

A simple computation of F_t, using (2.83), indeed verifies that the PDE holds at all points along a characteristic curve. The main addition to the theory we presented earlier for the quasilinear case is that the characteristic curves have been replaced by more complex geometric structures. Since each characteristic curve now drags with it a tangent plane, we call these structures *characteristic strips*, and equations (2.83) are called the *strip equations*.

We are now ready to formulate the general Cauchy problem for first-order PDEs. Consider (2.72) with the initial condition given by an initial curve $\Gamma \in C^1$:

$$x = x_0(s), \quad y = y_0(s), \quad u = u_0(s). \qquad (2.84)$$

We shall show in the proof of the next theorem how to derive initial conditions also for p and q in order to obtain a complete initial value problem for the system (2.83). We do not expect every Cauchy problem to be solvable. Clearly some form of transversality condition must be imposed. It turns out, however, that slightly more than that is required:

Definition 2.18 Let a point $P_0 = (x_0(s_0), y_0(s_0), u_0(s_0), p_0(s_0), q_0(s_0))$ satisfy the compatibility conditions

$$F(P_0) = 0, \quad u_0'(s_0) = p_0(s_0)x_0'(s_0) + q_0(s_0)y_0'(s_0). \tag{2.85}$$

If, in addition,

$$x_0'(s_0)F_q(P_0) - y_0'(s_0)F_p(P_0) \neq 0 \tag{2.86}$$

is satisfied, then we say that the Cauchy problem (2.72), (2.84) satisfies the *generalized transversality condition* at the point P_0.

Theorem 2.19 *Consider the Cauchy problem (2.72), (2.84). Assume that the generalized transversality condition (2.85)–(2.86) holds at P_0. Then there exists $\varepsilon > 0$ and a unique solution $(x(t, s), y(t, s), u(t, s), p(t, s), q(t, s))$ for the Cauchy problem which is defined for $|s - s_0| + |t| < \varepsilon$. Moreover, the parametric representation defines a smooth integral surface $u = u(x, y)$.*

Proof We start by deriving full initial conditions for the system (2.83). Actually the Cauchy problem has already provided three conditions for (x, y, u):

$$x(0, s) = x_0(s), \quad y(0, s) = y_0(s), \quad u(0, s) = u_0(s). \tag{2.87}$$

We are left with the task of finding initial conditions

$$p(0, s) = p_0(s), \quad q(0, s) = q_0(s), \tag{2.88}$$

for $p(t, s)$ and $q(t, s)$. Clearly $p_0(s)$ and $q_0(s)$ must satisfy at every point s the differential condition

$$u_0'(s) = p_0(s)x_0'(s) + q_0(s)y_0'(s), \tag{2.89}$$

and the equation itself

$$F(x_0(s), y_0(s), u_0(s), p_0(s), q_0(s)) = 0. \tag{2.90}$$

However (2.85) guarantees that these two requirements indeed hold at $s = s_0$. The transversality condition (2.86) ensures that the Jacobian of the system (2.89)–(2.90) (with respect to the variables p_0 and q_0) does not vanish at s_0. Therefore, the implicit function theorem implies that one can derive from (2.89)–(2.90) the required initial conditions for p_0 and q_0. Hence the characteristic equations have a full set of initial conditions in a neighborhood $|s - s_0| < \delta$. Since the system of ODEs (2.83) is well-posed, the existence of a unique smooth solution

$$(x(t, s), y(t, s), u(t, s), p(t, s), q(t, s))$$

is guaranteed for (t, s) in a neighborhood of $(0, s_0)$. As in the quasilinear case, one can verify that (2.83) and (2.90) imply that

$$F(x(t, s), y(t, s), u(t, s), p(t, s), q(t, s)) = 0 \qquad \forall \, |s - s_0| < \delta, \, |t| < \varepsilon. \quad (2.91)$$

In order that the parametric representation for (x, y, u) will define a smooth surface $u(x, y)$, we must show that the mapping $(x(t, s), y(t, s))$ can be inverted to a smooth mapping $(t(x, y), s(x, y))$. Such an inversion exists if the Jacobian $J = \partial(x, y)/\partial(t, s)$ does not vanish. But the characteristic equations imply

$$J \big|_{(0,s_0)} = \frac{\partial(x, y)}{\partial(s, t)}\bigg|_{(0,s_0)} = x_s(0, s_0)y_t(0, s_0) - x_t(0, s_0)y_s(0, s_0) \neq 0, \quad (2.92)$$

where the last inequality follows from the transversality condition (2.86).

We have thus constructed a smooth function $u(x, y)$. Does it satisfy the Cauchy problem for the nonlinear PDE? This requires that the relations

$$u_t(t, s) = p(t, s)x_t(t, s) + q(t, s)y_t(t, s) \qquad (2.93)$$

and

$$u_s(t, s) = p(t, s)x_s(t, s) + q(t, s)y_s(t, s) \qquad (2.94)$$

would hold. Condition (2.93) is clearly valid since the characteristic equations were, in fact, constructed to satisfy it. The compatibility condition (2.94) holds on the initial curve, i.e. at $t = 0$. It remains, though, to check this condition also for values of t other than zero. We therefore define the auxiliary function

$$R(t, s) = u_s - px_s - qy_s.$$

We have to show that $R(t, s) = 0$. As we have already argued, the initial data for p and q imply

$$R(0, s) = 0.$$

To check that R also vanishes for other values of t we compute

$$R_t = u_{st} - p_t x_s - p x_{st} - q_t y_s - q y_{st}$$
$$= \frac{\partial}{\partial s}(u_t - px_t - qy_t) + p_s x_t + q_s y_t - p_t x_s - q_t y_s.$$

Using (2.93) and the characteristic equations we get

$$R_t = p_s F_p + q_s F_q + x_s(F_x + pF_u) + y_s(F_y + qF_u)$$
$$= F_u(px_s + qy_s) + p_s F_p + q_s F_q + x_s F_x + y_s F_y.$$

Adding and subtracting $F_u u_s$ to the last expression we find

$$R_t = F_s - F_u R = -F_u R,$$

where we used the fact that $F_s = 0$ which follows from (2.91). Since the initial condition for the linear homogeneous ODE $R_t = -F_u R$ is homogeneous too ($R(0, s) = 0$), it follows that $R(t, s) \equiv 0$.

To demonstrate the method we just developed, let us solve again the Cauchy problem from Example 2.16.

Example 2.20 We write the eikonal equation in the form

$$F(x, y, u, p, q) = p^2 + q^2 - n_0^2 = 0. \tag{2.95}$$

Hence the characteristic equations are

$$\begin{aligned} x_t &= 2p, \\ y_t &= 2q, \\ u_t &= 2n_0^2, \\ p_t &= 0, \\ q_t &= 0. \end{aligned} \tag{2.96}$$

The initial conditions for (x, y, u) are given by

$$x(0, s) = s, \quad y(0, s) = 2s, \quad u(0, s) = 1. \tag{2.97}$$

We use (2.89)–(2.90) to derive the initial data for p and q:

$$p(0, s) + 2q(0, s) = 0,$$
$$p^2(0, s) + q^2(0, s) - n_0^2 = 0.$$

Solving these equation we obtain

$$p(0, s) = \frac{2n_0}{\sqrt{5}}, \quad q(0, s) = -\frac{n_0}{\sqrt{5}}. \tag{2.98}$$

It is an easy matter to solve the full characteristic equations:

$$(x, y, u, p, q) = \left(s + \frac{2n_0}{\sqrt{5}}t, \, 2s - \frac{n_0}{\sqrt{5}}t, \, 1 + 2n_0^2 t, \, \frac{2n_0}{\sqrt{5}}, \, -\frac{n_0}{\sqrt{5}} \right), \tag{2.99}$$

After eliminating t we finally obtain

$$u(x, y) = 1 + \frac{n_0}{\sqrt{5}}(2x - y).$$

Notice that the parametric representation obtained in the current example is different from the one we derived in Example 2.16, since the parameter t we used here is half the parameter used in Example 2.16.

2.10 Exercises

2.1 Consider the equation $u_x + u_y = 1$, with the initial condition $u(x, 0) = f(x)$.
   (a) What are the projections of the characteristic curves on the (x, y) plane?
   (b) Solve the equation.
2.2 Solve the equation $x u_x + (x + y)u_y = 1$ with the initial conditions $u(1, y) = y$. Is the solution defined everywhere?
2.3 Let p be a real number. Consider the PDEs

$$x u_x + y u_y = pu \qquad -\infty < x < \infty, \quad -\infty < y < \infty.$$

   (a) Find the characteristic curves for the equations.
   (b) Let $p = 4$. Find an explicit solution that satisfies $u = 1$ on the circle $x^2 + y^2 = 1$.
   (c) Let $p = 2$. Find two solutions that satisfy $u(x, 0) = x^2$, for every $x > 0$.
   (d) Explain why the result in (c) does not contradict the existence–uniqueness theorem.
2.4 Consider the equation $y u_x - x u_y = 0$ $(y > 0)$. Check for each of the following initial conditions whether the problem is solvable. If it is solvable, find a solution. If it is not, explain why.

   (a) $u(x, 0) = x^2$.
   (b) $u(x, 0) = x$.
   (c) $u(x, 0) = x, \quad x > 0$.

2.5 Let $u(x, y)$ be an integral surface of the equation

$$a(x, y)u_x + b(x, y)u_y + u = 0,$$

where $a(x, y)$ and $b(x, y)$ are positive differentiable functions in the entire plane. Define

$$D = \{(x, y), |x| < 1, |y| < 1\}.$$

   (a) Prove that the projection on the (x, y) plane of each characteristic curve passing through a point in D intersects the boundary of D at exactly two points.
   (b) Show that if u is positive on the boundary of D, then it is positive at every point in D.

(c) Suppose that u attains a local minimum (maximum) at a point $(x_0, y_0) \in D$. Evaluate $u(x_0, y_0)$.

(d) Denote by m the minimal value of u on the boundary of D. Assume $m > 0$. Show that $u(x, y) \geq m$ for all $(x, y) \in D$.

Remark This is an atypical example of a first-order PDE for which a maximum principle holds true. Maximum principles are important tools in the study of PDEs, and they are valid typically for second-order elliptic and parabolic PDEs (see Chapter 7).

2.6 The equation $xu_x + (x^2 + y)u_y + (y/x - x)u = 1$ is given along with the initial condition $u(1, y) = 0$.

(a) Solve the problem for $x > 0$. Compute $u(3, 6)$.

(b) Is the solution defined for the entire ray $x > 0$?

2.7 Solve the Cauchy problem $u_x + u_y = u^2$, $u(x, 0) = 1$.

2.8 (a) Solve the equation $xuu_x + yuu_y = u^2 - 1$ for the ray $x > 0$ under the initial condition $u(x, x^2) = x^3$.

(b) Is there a unique solution for the Cauchy problem over the entire real line $-\infty < x < \infty$?

2.9 Consider the equation

$$uu_x + u_y = -\frac{1}{2}u.$$

(a) Show that there is a unique integral surface in a neighborhood of the curve

$$\Gamma_1 = \{(s, 0, \sin s) \mid -\infty < s < \infty\}.$$

(b) Find the parametric representation $x = x(t, s)$, $y = y(t, s)$, $u = u(t, s)$ of the integral surface S for initial condition of part (a).

(c) Find an integral surface S_1 of the same PDE passing through the initial curve

$$\Gamma_1 = \{(s, s, 0) \mid -\infty < s < \infty\}.$$

(d) Find a parametric representation of the intersection curves of the surfaces S and S_1.
Hint Try to characterize that curve relative to the PDE.

2.10 A river is defined by the domain

$$D = \{(x, y) \mid \ |y| < 1, \ -\infty < x < \infty\}.$$

A factory spills a contaminant into the river. The contaminant is further spread and convected by the flow in the river. The velocity field of the fluid in the river is only in the x direction. The concentration of the contaminant at a point (x, y) in the river and at time τ is denoted by $u(x, y, \tau)$. Conservation of matter and momentum implies that u satisfies the first-order PDE

$$u_\tau - (y^2 - 1)u_x = 0.$$

The initial condition is $u(x, y, 0) = e^y e^{-x^2}$.

(a) Find the concentration u for all (x, y, τ).

(b) A fish lives near the point $(x, y) = (2, 0)$ at the river. The fish can tolerate contaminant concentration levels up to 0.5. If the concentration exceeds this level, the fish will die at once. Will the fish survive? If yes, explain why. If no, find the time in which the fish will die.

Hint Notice that y appears in the PDE just as a parameter.

2.11 Solve the equation $(y^2 + u)u_x + yu_y = 0$ in the domain $y > 0$, under the initial condition $u = 0$ on the planar curve $x = y^2/2$.

2.12 Solve the equation $u_y + u^2 u_x = 0$ in the ray $x > 0$ under the initial condition $u(x, 0) = \sqrt{x}$. What is the domain of existence of the solution?

2.13 Consider the equation $uu_x + xu_y = 1$, with the initial condition $\left(\frac{1}{2}s^2 + 1, \frac{1}{6}s^3 + s, s\right)$. Find a solution. Are there other solutions? If not, explain why; if there *are* further solutions, find at least two of them, and explain the lack of uniqueness.

2.14 Consider the equation $xu_x + yu_y = 1/\cos u$.

(a) Find a solution to the equation that satisfies the condition $u(s^2, \sin s) = 0$ (you can write down the solution in the implicit form $F(x, y, u) = 0$).

(b) Find some domain of s values for which there exists a unique solution.

2.15 (a) Find a function $u(x, y)$ that solves the Cauchy problem

$$(x + y^2)u_x + yu_y + \left(\frac{x}{y} - y\right)u = 1, \quad u(x, 1) = 0 \quad x \in \mathbb{R}.$$

(b) Check whether the transversality condition holds.

(c) Draw the projections on the (x, y) plane of the initial condition and the characteristic curves emanating from the points $(2, 1, 0)$ and $(0, 1, 0)$.

(d) Is the solution you obtained in (a) defined at the origin $(x, y) = (0, 0)$? Explain your answer in light of the existence–uniqueness theorem.

2.16 Solve the Cauchy problem

$$xu_x + yu_y = -u, \quad u(\cos s, \sin s) = 1 \quad 0 \le s \le \pi.$$

Is the solution defined everywhere?

2.17 Consider the equation

$$xu_x + u_y = 1.$$

(a) Find a characteristic curve passing through the point $(1, 1, 1)$.

(b) Show that there exists a unique integral surface $u(x, y)$ satisfying $u(x, 0) = \sin x$.

(c) Is the solution defined for all x and y?

2.18 Consider the equation $uu_x + u_y = -\frac{1}{2}u$.

(a) Find a solution satisfying $u(x, 2x) = x^2$.

(b) Is the solution unique?

2.19 (a) Find a function $u(x, y)$ that solves the Cauchy problem

$$x^2 u_x + y^2 u_y = u^2, \quad u(x, 2x) = x^2 \quad x \in \mathbb{R}.$$

(b) Check whether the transversality condition holds.

(c) Draw the projections on the (x, y) plane of the initial curve and the characteristic curves that start at the points $(1, 2, 1)$ and $(0, 0, 0)$.

(d) Is the solution you found in part (a) defined for all x and y?

2.20 Consider the equation

$$yu_x - uu_y = x.$$

(a) Write a parametric representation of the characteristic curves.

(b) Solve the Cauchy problem

$$yu_x - uu_y = x,$$
$$u(s, s) = -2s \qquad -\infty < s < \infty.$$

(c) Is the following Cauchy problem solvable:

$$yu_x - uu_y = x,$$
$$u(s, s) = s \qquad -\infty < s < \infty?$$

(d) Set

$$w_1 = x + y + u, \quad w_2 = x^2 + y^2 + u^2, \quad w_3 = xy + xu + yu.$$

Show that $w_1(w_2 - w_3)$ is constant along each characteristic curve.

2.21 (a) Find a function $u(x, y)$ that solves the Cauchy problem

$$xu_x - yu_y = u + xy, \quad u(x, x) = x^2 \quad 1 \le x \le 2.$$

(b) Check whether the transversality condition holds.

(c) Draw the projections on the (x, y) plane of the initial curve and the characteristic curves emanating from the points $(1, 1, 1)$ and $(2, 2, 4)$.

(d) Is the solution you found in (a) well defined in the entire plane?

2.22 Solve the Cauchy problem $u_x^2 + u_y = 0, u(x, 0) = x$.

2.23 Let $u(x, t)$ be the solution to the Cauchy problem

$$u_t + cu_x + u^2 = 0, \quad u(x, 0) = x,$$

where c is a constant, t denotes time, and x denotes a space coordinate.

(a) Solve the problem.

(b) A person leaves the point x_0 at time $t = 0$, and moves in the positive x direction with a velocity c (i.e. the quantity $x - ct$ is fixed for him). Show that if $x_0 > 0$, then the solution as seen by the person approaches zero as $t \to \infty$.

(c) What will be observed by such a person if $x_0 < 0$, or if $x_0 = 0$?

2.24 (a) Solve the problem

$$xu_x - uu_y = y,$$
$$u(1, y) = y \qquad -\infty < y < \infty.$$

(b) Is the solution unique? What is the maximal domain where it is defined?

2.25 Find at least five solutions for the Cauchy problem

$$u_x + u_y = 1, \quad u(x, x) = x.$$

2.26 (a) Solve the problem

$$xu_y - yu_x + u = 0,$$
$$u(x, 0) = 1 \qquad x > 0.$$

(b) Is the solution unique? What is the maximal domain where it is defined?

2.27 (a) Use the Lagrange method to find a function $u(x, y)$ that solves the problem

$$uu_x + u_y = 1 \qquad\qquad\qquad (2.100)$$
$$u(3x, 0) = -x \quad -\infty < x < \infty. \qquad (2.101)$$

(b) Show that the curve $\{(3x, 2, 4 - 3x)| -\infty < x < \infty\}$ is contained in the solution surface $u(x, y)$.

(c) Solve

$$uu_x + u_y = 1$$
$$u(3x, 2) = 4 - 3x \quad -\infty < x < \infty.$$

2.28 Analyze the following problems using the Lagrange method. For each problem determine whether there exists a unique solution, infinitely many solutions or no solution at all. If there is a unique solution, find it; if there are infinitely many solutions, find at least two of them. Present all solutions explicitly.

(a)

$$xuu_x + yuu_y = x^2 + y^2 \quad x > 0, \ y > 0,$$
$$u(x, 1) = \sqrt{x^2 + 1}.$$

(b)

$$xuu_x + yuu_y = x^2 + y^2 \quad x > 0, \ y > 0,$$
$$u(x, x) = \sqrt{2}x.$$

2.29 Consider the equation

$$xu_x + (1 + y)u_y = x(1 + y) + xu.$$

(a) Find the general solution.

(b) Assume an initial condition of the form $u(x, 6x - 1) = \phi(x)$. Find a necessary and sufficient condition for ϕ that guarantees the existence of a solution to the problem. Solve the problem for the appropriate ϕ that you found.

(c) Assume an initial condition of the form $u(-1, y) = \psi(y)$. Find a necessary and sufficient condition for ψ that guarantees the existence of a solution to the problem. Solve the problem for the appropriate ψ that you found.

(d) Explain the differences between (b) and (c).

2.30 (a) Find a compatibility condition for the Cauchy problem

$$u_x^2 + u_y^2 = 1, \qquad u(\cos s, \sin s) = 0 \quad 0 \le s \le 2\pi.$$

(b) Solve the above Cauchy problem.

(c) Is the solution uniquely defined?

3

Second-order linear equations in two independent variables

3.1 Introduction

In this chapter we classify the family of second-order linear equations for functions in two independent variables into three distinct types: hyperbolic (e.g., the wave equation), parabolic (e.g., the heat equation), and elliptic equations (e.g., the Laplace equation). It turns out that solutions of equations of the same type share many exclusive qualitative properties. We show that by a certain change of variables any equation of a particular type can be transformed into a canonical form which is associated with its type.

3.2 Classification

We concentrate in this chapter on second-order linear equations for functions in two independent variables x, y. Such an equation has the form

$$L[u] = au_{xx} + 2bu_{xy} + cu_{yy} + du_x + eu_y + fu = g, \qquad (3.1)$$

where $a, b, \ldots, f, g$ are given functions of x, y, and $u(x, y)$ is the unknown function. We introduced the factor 2 in front of the coefficient b for convenience. We assume that the coefficients a, b, c do not vanish simultaneously.

The operator

$$L_0[u] = au_{xx} + 2bu_{xy} + cu_{yy}$$

that consists of the second-(highest-)order terms of the operator L is called the *principal part* of L. It turns out that many fundamental properties of the solutions of (3.1) are determined by its principal part, and, more precisely, by the *sign* of the discriminant $\delta(L) := b^2 - ac$ of the equation. We classify the equation according to the sign of $\delta(L)$.

64

Definition 3.1 Equation (3.1) is said to be *hyperbolic* at a point (x, y) if

$$\delta(L)(x, y) = b(x, y)^2 - a(x, y)c(x, y) > 0,$$

it is said to be *parabolic* at (x, y) if $\delta(L)(x, y) = 0$, and it is said to be *elliptic* at (x, y) if $\delta(L)(x, y) < 0$.

Let Ω be a domain in $\mathbb{R}^2$ (i.e. Ω is an open connected set). The equation is hyperbolic (resp., parabolic, elliptic) in Ω, if it is hyperbolic (resp., parabolic, elliptic) at all points $(x, y) \in \Omega$.

Definition 3.2 The transformation $(\xi, \eta) = (\xi(x, y), \eta(x, y))$ is called a *change of coordinates* (or a nonsingular transformation) if the Jacobian $J := \xi_x \eta_y - \xi_y \eta_x$ of the transformation does not vanish at any point (x, y).

Lemma 3.3 *The type of a linear second-order PDE in two variables is invariant under a change of coordinates. In other words, the type of the equation is an intrinsic property of the equation and is independent of the particular coordinate system used.*

Proof Let

$$L[u] = au_{xx} + 2bu_{xy} + cu_{yy} + du_x + eu_y + fu = g, \qquad (3.2)$$

and let $(\xi, \eta) = (\xi(x, y), \eta(x, y))$ be a nonsingular transformation. Write $w(\xi, \eta) = u(x(\xi, \eta), y(\xi, \eta))$. We claim that w is a solution of a second-order equation of the same type. Using the chain rule one finds that

$$u_x = w_\xi \xi_x + w_\eta \eta_x,$$
$$u_y = w_\xi \xi_y + w_\eta \eta_y,$$
$$u_{xx} = w_{\xi\xi} \xi_x^2 + 2w_{\xi\eta} \xi_x \eta_x + w_{\eta\eta} \eta_x^2 + w_\xi \xi_{xx} + w_\eta \eta_{xx},$$
$$u_{xy} = w_{\xi\xi} \xi_x \xi_y + w_{\xi\eta}(\xi_x \eta_y + \xi_y \eta_x) + w_{\eta\eta} \eta_x \eta_y + w_\xi \xi_{xy} + w_\eta \eta_{xy},$$
$$u_{yy} = w_{\xi\xi} \xi_y^2 + 2w_{\xi\eta} \xi_y \eta_y + w_{\eta\eta} \eta_y^2 + w_\xi \xi_{yy} + w_\eta \eta_{yy}.$$

Substituting these formulas into (3.2), we see that w satisfies the following linear equation:

$$\ell[w] := Aw_{\xi\xi} + 2Bw_{\xi\eta} + Cw_{\eta\eta} + Dw_\xi + Ew_\eta + Fw = G,$$

where the coefficients of the principal part of the linear operator ℓ are given by

$$A(\xi, \eta) = a\xi_x^2 + 2b\xi_x \xi_y + c\xi_y^2,$$
$$B(\xi, \eta) = a\xi_x \eta_x + b(\xi_x \eta_y + \xi_y \eta_x) + c\xi_y \eta_y,$$
$$C(\xi, \eta) = a\eta_x^2 + 2b\eta_x \eta_y + c\eta_y^2.$$

Notice that we do not need to compute the coefficients of the lower-order derivatives (D, E, F) since the type of the equation is determined only by its principal part (i.e. by the coefficients of the second-order terms). An elementary calculation shows that these coefficients satisfy the following matrix equation:

$$\begin{pmatrix} A & B \\ B & C \end{pmatrix} = \begin{pmatrix} \xi_x & \xi_y \\ \eta_x & \eta_y \end{pmatrix} \begin{pmatrix} a & b \\ b & c \end{pmatrix} \begin{pmatrix} \xi_x & \eta_x \\ \xi_y & \eta_y \end{pmatrix}.$$

Denote by J the Jacobian of the transformation. Taking the determinant of the two sides of the above matrix equation, we find

$$-\delta(\ell) = AC - B^2 = J^2(ac - b^2) = -J^2 \delta(L).$$

Therefore, the type of the equation is invariant under nonsingular transformations. □

In Chapter 1 we encountered the three (so called) *fundamental equations of mathematical physics*: the heat equation, the wave equations and the Laplace equation. All of them are linear second-order equations. One can easily verify that the wave equation is hyperbolic, the heat equation is parabolic, and the Laplace equation is elliptic.

We shall show in the next sections that if (3.1) is hyperbolic (resp., parabolic, elliptic) in a domain D, then one can find a coordinate system in which the equation has a simpler form that we call the *canonical form* of the equation. Moreover, in such a case the principal part of the canonical form is equal to the principal part of the fundamental equation of mathematical physics of the same type. This is one of the reasons for studying these fundamental equations.

Definition 3.4 The *canonical form of a hyperbolic equation* is

$$\ell[w] = w_{\xi\eta} + \ell_1[w] = G(\xi, \eta),$$

where ℓ_1 is a first-order linear differential operator, and G is a function.

Similarly, the canonical form of a parabolic equation is

$$\ell[w] = w_{\xi\xi} + \ell_1[w] = G(\xi, \eta),$$

and the canonical form of an elliptic equation is

$$\ell[w] = w_{\xi\xi} + w_{\eta\eta} + \ell_1[w] = G(\xi, \eta).$$

Note that the principal part of the canonical form of a hyperbolic equation is not equal to the wave operator. We shall show in Section 4.2 that a simple (linear) change of coordinates transforms the wave equation into the equation $w_{\xi\eta} = 0$.

3.3 Canonical form of hyperbolic equations

Theorem 3.5 *Suppose that (3.1) is hyperbolic in a domain D. There exists a coordinate system (ξ, η) in which the equation has the canonical form*

$$w_{\xi\eta} + \ell_1[w] = G(\xi, \eta),$$

where $w(\xi, \eta) = u(x(\xi, \eta), y(\xi, \eta))$, ℓ_1 is a first-order linear differential operator, and G is a function which depends on (3.1).

Proof Without loss of generality, we may assume that $a(x, y) \neq 0$ for all $(x, y) \in D$. We need to find two functions $\xi = \xi(x, y)$, $\eta = \eta(x, y)$ such that

$$A(\xi, \eta) = a\xi_x^2 + 2b\xi_x\xi_y + c\xi_y^2 = 0,$$
$$C(\xi, \eta) = a\eta_x^2 + 2b\eta_x\eta_y + c\eta_y^2 = 0.$$

The equation that was obtained for the function η is actually the same equation as for ξ; therefore, we need to solve only one equation. It is a first-order equation that is not quasilinear; but as a quadratic form in ξ it is possible to write it as a product of two linear terms

$$\frac{1}{a}\left[a\xi_x + (b - \sqrt{b^2 - ac})\xi_y\right]\left[a\xi_x + (b + \sqrt{b^2 - ac})\xi_y\right] = 0.$$

Therefore, we need to solve the following linear equations:

$$a\xi_x + (b + \sqrt{b^2 - ac})\xi_y = 0, \tag{3.3}$$
$$a\xi_x + (b - \sqrt{b^2 - ac})\xi_y = 0. \tag{3.4}$$

In order to obtain a nonsingular transformation $(\xi(x, y), \eta(x, y))$ we choose ξ to be a solution of (3.3) and η to be a solution of (3.4).

These equations are a special case of Example 2.4. The characteristic equations for (3.3) are

$$\frac{dx}{dt} = a, \quad \frac{dy}{dt} = b + \sqrt{b^2 - ac}, \quad \frac{d\xi}{dt} = 0.$$

Therefore, ξ is constant on each characteristic. The characteristics are solutions of the equation

$$\frac{dy}{dx} = \frac{b + \sqrt{b^2 - ac}}{a}. \tag{3.5}$$

The function η is constant on the characteristic determined by

$$\frac{dy}{dx} = \frac{b - \sqrt{b^2 - ac}}{a}. \tag{3.6}$$

$\square$

Definition 3.6 The solutions of (3.5) and (3.6) are called the two families of the *characteristics* (or *characteristic projections*) of the equation $L[u] = g$.

Example 3.7 Consider the *Tricomi equation*:

$$u_{xx} + x u_{yy} = 0 \qquad x < 0. \tag{3.7}$$

Find a mapping $q = q(x, y), r = r(x, y)$ that transforms the equation into its canonical form, and present the equation in this coordinate system.

The characteristic equations are

$$\frac{dy_\pm}{dx} = \pm\sqrt{-x},$$

and their solutions are $\frac{3}{2} y_\pm \pm (-x)^{3/2} = \text{constant}$. Thus, the new independent variables are

$$q(x, y) = \frac{3}{2} y + (-x)^{3/2}, \qquad r(x, y) = \frac{3}{2} y - (-x)^{3/2}.$$

Clearly,

$$q_x = -r_x = -\frac{3}{2}(-x)^{1/2}, \qquad q_y = r_y = \frac{3}{2}.$$

Define $v(q, r) = u(x, y)$. By the chain rule

$$u_x = \frac{-3}{2}(-x)^{1/2} v_q + \frac{3}{2}(-x)^{1/2} v_r, \qquad u_y = \frac{3}{2}(v_q + v_r),$$

$$u_{xx} = -\frac{9}{4} x v_{qq} - \frac{9}{4} x v_{rr} + 2\frac{9}{4} x v_{qr} + \frac{3}{4}(-x)^{-1/2}(v_q - v_r),$$

$$u_{yy} = -\frac{9}{4}(v_{qq} + 2v_{qr} + v_{rr}).$$

Substituting these expressions into the Tricomi equation we obtain

$$u_{xx} + x u_{yy} = -9(q - r)^{2/3} \left[v_{qr} + \frac{v_q - v_r}{6(q - r)} \right] = 0.$$

Example 3.8 Consider the equation

$$u_{xx} - 2 \sin x \, u_{xy} - \cos^2 x \, u_{yy} - \cos x \, u_y = 0. \tag{3.8}$$

Find a coordinate system $s = s(x, y)$, $t = t(x, y)$ that transforms the equation into its canonical form. Show that in this coordinate system the equation has the form $v_{st} = 0$, and find the general solution.

The characteristic equations are

$$\frac{dy_\pm}{dx} = -\sin x \pm \sqrt{\sin^2 x + \cos^2 x} = -\sin x \pm 1.$$

Consequently, the solutions are $y_\pm = \cos x \pm x + \text{constant}$. The requested transformation is

$$s(x, y) = \cos x + x - y, \qquad t(x, y) = \cos x - x - y.$$

Consider now the function $v(s, t) = u(x, y)$ and substitute it into (3.8). We get

$$\begin{aligned}
&[v_{ss}(-\sin x + 1)^2 + 2v_{st}(-\sin x + 1)(-\sin x - 1) + v_{tt}(-\sin x - 1)^2 \\
&+ v_s(-\cos x) + v_t(-\cos x)] - 2\sin x \, [v_{ss}(\sin x - 1) + v_{st}(\sin x - 1) \\
&+ v_{st}(\sin x + 1) + v_{tt}(\sin x + 1)] - \cos^2 x \, [v_{ss} + 2v_{st} + v_{tt}] \\
&- \cos x(-v_s - v_t) = 0.
\end{aligned}$$

Thus, $-4v_{st} = 0$, and the canonical form is

$$v_{st} = 0.$$

It is easily checked that its general solution is $v(s, t) = F(s) + G(t)$, for every $F, G \in C^2(\mathbb{R})$. Therefore, the general solution of (3.8) is

$$u(x, y) = F(\cos x + x - y) + G(\cos x - x - y).$$

3.4 Canonical form of parabolic equations

Theorem 3.9 *Suppose that (3.1) is parabolic in a domain D. There exists a coordinate system (ξ, η) where the equation has the canonical form*

$$w_{\xi\xi} + \ell_1[w] = G(\xi, \eta),$$

where $w(\xi, \eta) = u(x(\xi, \eta), y(\xi, \eta))$, ℓ_1 is a first-order linear differential operator, and G is a function which depends on (3.1).

Proof Since $b^2 - ac = 0$, we may assume that $a(x, y) \neq 0$ for all $(x, y) \in D$. We need to find two functions $\xi = \xi(x, y)$, $\eta = \eta(x, y)$ such that $B(\xi, \eta) = C(\xi, \eta) = 0$ for all $(x, y) \in D$. It is enough to make $C = 0$, since the parabolicity of the equation will then imply that $B = 0$. Therefore, we need to find a function η that is a solution of the equation

$$C(\xi, \eta) = a\eta_x^2 + 2b\eta_x\eta_y + c\eta_y^2 = \frac{1}{a}(a\eta_x + b\eta_y)^2 = 0.$$

From this it follows that η is a solution of the first-order linear equation

$$a\eta_x + b\eta_y = 0. \tag{3.9}$$

Hence, the solution η is constant on each characteristic, i.e., on a curve that is a solution of the equation

$$\frac{dy}{dx} = \frac{b}{a}. \tag{3.10}$$

Now, the only constraint on the second independent variable ξ, is that the Jacobian of the transformation should not vanish in D, and we may take any such function ξ. Note that a parabolic equation admits only one family of characteristics while for hyperbolic equations we have two families. $\square$

Example 3.10 Prove that the equation

$$x^2 u_{xx} - 2xy u_{xy} + y^2 u_{yy} + x u_x + y u_y = 0 \tag{3.11}$$

is parabolic and find its canonical form; find the general solution on the half-plane $x > 0$.

We identify $a = x^2$, $2b = -2xy$, $c = y^2$; therefore, $b^2 - ac = x^2 y^2 - x^2 y^2 = 0$ and the equation is parabolic. The equation for the characteristics is

$$\frac{dy}{dx} = -\frac{y}{x},$$

and the solution is $xy = $ constant. Therefore, we define $\eta(x, y) = xy$. The second variable can be simply chosen as $\xi(x, y) = x$. Let $v(\xi, \eta) = u(x, y)$. Substituting the new coordinates ξ and η into (3.11), we obtain

$$x^2(y^2 v_{\eta\eta} + 2y v_{\xi\eta} + v_{\xi\xi}) - 2xy(v_\eta + xy v_{\eta\eta} + x v_{\xi\eta})$$
$$+ x^2 v_{\eta\eta} + xy v_\eta + x v_\xi + xy v_\xi = 0.$$

Thus,

$$\xi^2 v_{\xi\xi} + \xi v_\xi = 0,$$

or $v_{\xi\xi} + (1/\xi) v_\xi = 0$, and this is the desired canonical form.

Setting $w = v_\xi$, we arrive at the first-order ODE $w_\xi + (1/\xi) w = 0$. The solution is $\ln w = -\ln \xi + \tilde{f}(\eta)$, or $w = f(\eta)/\xi$. Hence, v satisfies

$$v = \int v_\xi d\xi = \int w d\xi = \int \frac{f(\eta)}{\xi} d\xi = f(\eta) \ln \xi + g(\eta).$$

Therefore, the general solution $u(x, y)$ of (3.11) is $u(x, y) = f(xy) \ln x + g(xy)$, where $f, g \in C^2(\mathbb{R})$ are arbitrary real functions.

3.5 Canonical form of elliptic equations

The computation of a canonical coordinate system for the elliptic case is somewhat more subtle than in the hyperbolic case or in the parabolic case. Nevertheless, under the additional assumption that the coefficients of the principal part of the equation are real analytic functions, the procedure for determining the canonical transformation is quite similar to the one for the hyperbolic case.

Definition 3.11 Let D a planar domain. A function $f : D \to \mathbb{R}$ is said to be *real analytic* in D if for each point $(x_0, y_0) \in D$, we have a convergent power series expansion

$$f(x, y) = \sum_{k=0}^{\infty} \sum_{j=0}^{k} a_{j,k-j}(x - x_0)^j(y - y_0)^{k-j},$$

valid in some neighborhood N of (x_0, y_0).

Theorem 3.12 *Suppose that (3.1) is elliptic in a planar domain D. Assume further that the coefficients a, b, c are real analytic functions in D. Then there exists a coordinate system (ξ, η) in which the equation has the canonical form*

$$w_{\xi\xi} + w_{\eta\eta} + \ell_1[w] = G(\xi, \eta),$$

where ℓ_1 is a first-order linear differential operator, and G is a function which depends on (3.1).

Proof Without loss of generality we may assume that $a(x, y) \neq 0$ for all $(x, y) \in D$. We are looking for two functions $\xi = \xi(x, y)$, $\eta = \eta(x, y)$ that satisfy the equations

$$A(\xi, \eta) = a\xi_x^2 + 2b\xi_x\xi_y + c\xi_y^2 = C(\xi, \eta) = a\eta_x^2 + 2b\eta_x\eta_y + c\eta_y^2, \quad (3.12)$$

$$B(\xi, \eta) = a\xi_x\eta_x + b(\xi_x\eta_y + \xi_y\eta_x) + c\xi_y\eta_y = 0. \quad (3.13)$$

This is a system of two nonlinear first-order equations. The main difficulty in the elliptic case is that (3.12)–(3.13) are coupled. In order to decouple these equations, we shall use the complex plane and the analyticity assumption. We may write the system (3.12)–(3.13) in the following form:

$$a(\xi_x^2 - \eta_x^2) + 2b(\xi_x\xi_y - \eta_x\eta_y) + c(\xi_y^2 - \eta_y^2) = 0, \quad (3.14)$$

$$a\xi_x i\eta_x + b(\xi_x i\eta_y + \xi_y i\eta_x) + c\xi_y i\eta_y = 0, \quad (3.15)$$

where $i = \sqrt{-1}$. Define the complex function $\phi = \xi + i\eta$. The system (3.14)–(3.15) is equivalent to the complex valued equation

$$a\phi_x^2 + 2b\phi_x\phi_y + c\phi_y^2 = 0.$$

Surprisingly, we have arrived at the same equation as in the hyperbolic case. But in the elliptic case the equation does not admit any real solution, or, in other words, elliptic equations do not have characteristics. As in the hyperbolic case, we factor out the above quadratic PDE, and obtain two linear equations, but now these are complex valued differential equations (where x, y are complex variables!). The nontrivial question of the existence and uniqueness of solutions immediately arises. Fortunately, it is known that if the coefficients of these first-order linear equations are real analytic then it is possible to solve them using the same procedure as in the real case. Moreover, the solutions of the two equations are complex conjugates.

So, we need to solve the equations

$$a\phi_x + (b \pm i\sqrt{ac - b^2})\phi_y = 0. \tag{3.16}$$

As before, the solutions ϕ, ψ are constant on the "characteristics" (which are defined on the complex plane):

$$\frac{dy}{dx} = \frac{b \pm i\sqrt{ac - b^2}}{a}. \tag{3.17}$$

As in the hyperbolic case, the equation in the new coordinates system has the form

$$4v_{\phi\psi} + \cdots = 0.$$

This is still not the elliptic canonical form with real coefficients. We return to our real variables ξ and η using the linear transformation

$$\xi = \operatorname{Re}\phi, \quad \eta = \operatorname{Im}\phi.$$

Since ξ and η are solutions of the system (3.12)–(3.13), it follows that in the variables ξ and η the equation has the canonical form. In Exercise 3.9 the reader will be asked to prove that the Jacobian of the canonical transformations in the elliptic case and in the hyperbolic case do not vanish.   □

Example 3.13 Consider the Tricomi equation:

$$u_{xx} + xu_{yy} = 0, \qquad x > 0. \tag{3.18}$$

Find a canonical transformation $q = q(x, y)$, $r = r(x, y)$ and the corresponding canonical form.

The differential equations for the "characteristics" are $dy/dx = \pm\sqrt{-x}$, and their solutions are $\frac{3}{2}y \pm i(x)^{3/2} = $ constant. Therefore, the canonical variables are $q(x, y) = \frac{3}{2}y$ and $r(x, y) = -(x)^{3/2}$. Clearly,

$$q_x = 0, \; q_y = \frac{3}{2} \qquad r_x = -\frac{3}{2}(x)^{1/2}, \; r_y = 0.$$

Set $v(q, r) = u(x, y)$. Hence,

$$u_x = -\frac{3}{2}(x)^{1/2}v_r, \qquad u_y = \frac{3}{2}v_q,$$

$$u_{xx} = \frac{9}{4}xv_{rr} - \frac{3}{4}(x)^{-1/2}v_r, \qquad u_{yy} = \frac{9}{4}v_{qq}.$$

Substituting these into the Tricomi equation we obtain the canonical form

$$\frac{1}{x}u_{xx} + u_{yy} = \frac{9}{4}\left(v_{qq} + v_{rr} + \frac{1}{3r}v_r\right) = 0.$$

3.6 Exercises

3.1 Consider the equation

$$u_{xx} - 6u_{xy} + 9u_{yy} = xy^2.$$

(a) Find a coordinates system (s, t) in which the equation has the form:

$$9v_{tt} = \tfrac{1}{3}(s - t)t^2.$$

(b) Find the general solution $u(x, y)$.
(c) Find a solution of the equation which satisfies the initial conditions $u(x, 0) = \sin x, u_y(x, 0) = \cos x$ for all $x \in \mathbb{R}$.

3.2 (a) Show that the following equation is hyperbolic:

$$u_{xx} + 6u_{xy} - 16u_{yy} = 0.$$

(b) Find the canonical form of the equation.
(c) Find the general solution $u(x, y)$.
(d) Find a solution $u(x, y)$ that satisfies $u(-x, 2x) = x$ and $u(x, 0) = \sin 2x$.

3.3 Consider the equation

$$u_{xx} + 4u_{xy} + u_x = 0.$$

(a) Bring the equation to a canonical form.
(b) Find the general solution $u(x, y)$ and check by substituting back into the equation that your solution is indeed correct.
(c) Find a specific solution satisfying

$$u(x, 8x) = 0, \qquad u_x(x, 8x) = 4e^{-2x}.$$

3.4 Consider the equation

$$y^5 u_{xx} - y u_{yy} + 2u_y = 0, \qquad y > 0.$$

(a) Find the canonical form of the equation.
(b) Find the general solution $u(x, y)$ of the equation.

(c) Find the solution $u(x, y)$ which satisfies $u(0, y) = 8y^3$, and $u_x(0, y) = 6$, for all $y > 0$.

3.5 Consider the equation

$$x u_{xx} - y u_{yy} + \tfrac{1}{2}(u_x - u_y) = 0.$$

(a) Find the domain where the equation is elliptic, and the domain where it is hyperbolic

(b) For each of the above two domains, find the corresponding canonical transformation.

3.6 Consider the equation

$$u_{xx} + (1 + y^2)^2 u_{yy} - 2y(1 + y^2)u_y = 0.$$

(a) Find the canonical form of the equation.

(b) Find the general solution $u(x, y)$ of the equation.

(c) Find the solution $u(x, y)$ which satisfies $u(x, 0) = g(x)$, and $u_y(x, 0) = f(x)$, where $f, g \in C^2(\mathbb{R})$.

(d) Find the solution $u(x, y)$ for $f(x) = -2x$, and $g(x) = x$.

3.7 Consider the equation

$$u_{xx} + 2u_{xy} + [1 - q(y)]u_{yy} = 0,$$

where

$$q(y) = \begin{cases} -1 & y < -1, \\ 0 & |y| \le 1, \\ 1 & y > 1. \end{cases}$$

(a) Find the domains where the equation is hyperbolic, parabolic, and elliptic.

(b) For each of the above three domains, find the corresponding canonical transformation and the canonical form.

(c) Draw the characteristics for the hyperbolic case.

3.8 Consider the equation

$$4y^2 u_{xx} + 2(1 - y^2)u_{xy} - u_{yy} - \frac{2y}{1 + y^2}(2u_x - u_y) = 0.$$

(a) Find the canonical form of the equation.

(b) Find the general solution $u(x, y)$ of the equation.

(c) Find the solution $u(x, y)$ which satisfies $u(x, 0) = g(x)$, and $u_y(x, 0) = f(x)$, where $f, g \in C^2(\mathbb{R})$ are arbitrary functions.

3.9 (a) Prove that in the hyperbolic case the canonical transformation is nonsingular $(J \neq 0)$.

(b) Prove that in the elliptic case the canonical transformation is nonsingular $(J \neq 0)$.

3.10 Consider the equation

$$u_{xx} - 2u_{xy} + 4e^y = 0.$$

(a) Find the canonical form of the equation.

(b) Find the solution $u(x, y)$ which satisfies $u(0, y) = f(y)$, and $u_x(0, y) = g(y)$.

3.11 In continuation of Example 3.8, consider the equation

$$u_{xx} - 2 \sin x \, u_{xy} - \cos^2 x \, u_{yy} - \cos x \, u_y = 0.$$

(a) Find a solution of the equation which satisfies $u(0, y) = f(y)$, $u_x(0, y) = g(y)$, where f, g are given functions.

(b) Find conditions on f and g such that the solution $u(x, y)$ of part (a) is a classical solution.

3.12 Consider the equation

$$u_{xx} + y u_{yy} = 0.$$

Find the canonical forms of the equation for the domain where the equation is hyperbolic, and for the domain where it is elliptic.

4

The one-dimensional wave equation

4.1 Introduction

In this chapter we study the one-dimensional wave equation on the real line. The canonical form of the wave equation will be used to show that the Cauchy problem is well-posed. Moreover, we shall derive simple explicit formulas for the solutions. We also discuss some important properties of the solutions of the wave equation which are typical for more general hyperbolic problems as well.

4.2 Canonical form and general solution

The homogeneous wave equation in one (spatial) dimension has the form

$$u_{tt} - c^2 u_{xx} = 0 \qquad -\infty \leq a < x < b \leq \infty, \; t > 0, \qquad (4.1)$$

where $c \in \mathbb{R}$ is called the *wave speed*, a terminology that will be justified in the discussion below.

To obtain the canonical form of the wave equation, define the new variables

$$\xi = x + ct \qquad \eta = x - ct,$$

and set $w(\xi, \eta) = u(x(\xi, \eta), t(\xi, \eta))$ (see Section 3.3 for the method to obtain this canonical transformation). Using the chain rule for the function $u(x, t) = w(\xi(x, t), \eta(x, t))$, we obtain

$$u_t = w_\xi \xi_t + w_\eta \eta_t = c(w_\xi - w_\eta), \qquad u_x = w_\xi \xi_x + w_\eta \eta_x = w_\xi + w_\eta,$$

and

$$u_{tt} = c^2(w_{\xi\xi} - 2w_{\xi\eta} + w_{\eta\eta}), \qquad u_{xx} = w_{\xi\xi} + 2w_{\xi\eta} + w_{\eta\eta}.$$

Hence,

$$u_{tt} - c^2 u_{xx} = -4c^2 w_{\xi\eta} = 0.$$

This is the canonical form for the wave equation. Since $(w_\xi)_\eta = 0$, it follows that $w_\xi = f(\xi)$, and then $w = \int f(\xi) \, d\xi + G(\eta)$. Therefore, the general solution of the equation $w_{\xi\eta} = 0$ has the form

$$w(\xi, \eta) = F(\xi) + G(\eta),$$

where $F, G \in C^2(\mathbb{R})$ are two arbitrary functions. Thus, in the original variables, the general solution of the wave equation is

$$u(x, t) = F(x + ct) + G(x - ct). \tag{4.2}$$

In other words, if u is a solution of the one-dimensional wave equation, then there exist two real functions $F, G \in C^2$ such that (4.2) holds. Conversely, any two functions $F, G \in C^2$ define a solution of the wave equation via formula (4.2).

For a fixed $t_0 > 0$, the graph of the function $G(x - ct_0)$ has the same shape as the graph of the function $G(x)$, except that it is shifted to the right by a distance ct_0. Therefore, the function $G(x - ct)$ represents a wave moving to the right with velocity c, and it is called a *forward wave*. The function $F(x + ct)$ is a wave traveling to the left with the same speed, and it is called a *backward wave*. Indeed c can be called the *wave speed*.

Equation (4.2) demonstrates that any solution of the wave equation is the sum of two such traveling waves. This observation will enable us to obtain graphical representations of the solutions (the graphical method).

We would like to extend the validity of (4.2). Observe that for any two real piecewise continuous functions F, G, (4.2) defines a piecewise continuous function u that is a superposition of a forward wave and a backward wave traveling in opposite directions with speed c. Moreover, it is possible to find two sequences of smooth functions, $\{F_n(s)\}, \{G_n(s)\}$, converging at any point to F and G, respectively, which converge uniformly to these functions in any bounded and closed interval that does not contain points of discontinuity. The function

$$u_n(x, t) = F_n(x + ct) + G_n(x - ct)$$

is a proper solution of the wave equation, but the limiting function $u(x, t) = F(x + ct) + G(x - ct)$ is not necessarily twice differentiable, and therefore might not be a solution. We call a function $u(x, t)$ that satisfies (4.2) with piecewise continuous functions F, G a *generalized solution* of the wave equation.

Let us further discuss the general solution (4.2). Consider the (x, t) plane. The following two families of lines

$$x - ct = \text{constant}, \quad x + ct = \text{constant},$$

are called the *characteristics* of the wave equation (see Section 3.3). For the wave equation, the characteristics are straight lines in the (x, t) plane with slopes $\pm 1/c$.

It turns out that as for first-order PDEs, the "information" is transferred via these curves.

We arrive now at one of the most important properties of the characteristics. Assume that for a fixed time t_0, the solution u is a smooth function except at one point (x_0, t_0). Clearly, either F is not smooth at $x_0 + ct_0$, and/or the function G is not smooth at $x_0 - ct_0$. There are two characteristics that pass through the point (x_0, t_0); these are the lines

$$x - ct = x_0 - ct_0, \quad x + ct = x_0 + ct_0.$$

Consequently, for any time $t_1 \neq t_0$ the solution u is smooth except at one or two points $x_{\pm}$ that satisfy

$$x_- - ct_1 = x_0 - ct_0, \quad x_+ + ct_1 = x_0 + ct_0.$$

Therefore, the singularities (nonsmoothness) of solutions of the wave equation are traveling only along characteristics. This phenomenon is typical of hyperbolic equations in general: a singularity is not smoothed out; rather it travels at a finite speed. This is in contrast to parabolic and elliptic equations, where, as will be shown in the following chapters, singularities are immediately smoothed out.

Example 4.1 Let $u(x, t)$ be a solution of the wave equation

$$u_{tt} - c^2 u_{xx} = 0,$$

which is defined in the whole plane. Assume that u is constant on the line $x = 2 + ct$. Prove that $u_t + cu_x = 0$.

The solution $u(x, t)$ has the form $u(x, t) = F(x + ct) + G(x - ct)$. Since $u(2 + ct, t) = $ constant, it follows that

$$F(2 + 2ct) + G(2) = \text{constant}.$$

Set $s = 2 + 2ct$, we have $F(s) = $ constant. Consequently $u(x, t) = G(x - ct)$. Computing now the expression $u_t + cu_x$, we obtain

$$u_t + cu_x = -cG'(x - ct) + cG'(x - ct) = 0.$$

4.3 The Cauchy problem and d'Alembert's formula

The *Cauchy problem* for the one-dimensional homogeneous wave equation is given by

$$u_{tt} - c^2 u_{xx} = 0 \qquad -\infty < x < \infty, \ t > 0, \tag{4.3}$$

$$u(x, 0) = f(x), \quad u_t(x, 0) = g(x), \quad -\infty < x < \infty. \tag{4.4}$$

A solution of this problem can be interpreted as the amplitude of a sound wave propagating in a very long and narrow pipe, which in practice can be considered as a one-dimensional infinite medium. This system also represents the vibration of an infinite (ideal) string. The initial conditions f, g are given functions that represent the amplitude u, and the velocity u_t of the string at time $t = 0$.

A *classical (proper) solution* of the Cauchy problem (4.3)–(4.4) is a function u that is continuously twice differentiable for all $t > 0$, such that u and u_t are continuous in the half-space $t \geq 0$, and such that (4.3)–(4.4) are satisfied. Generally speaking, classical solutions should have the minimal smoothness properties in order to satisfy continuously all the given conditions in the classical sense.

Recall that the general solution of the wave equation is of the form

$$u(x, t) = F(x + ct) + G(x - ct). \tag{4.5}$$

Our aim is to find F and G such that the initial conditions of (4.4) are satisfied. Substituting $t = 0$ into (4.5) we obtain

$$u(x, 0) = F(x) + G(x) = f(x). \tag{4.6}$$

Differentiating (4.5) with respect to t and substituting $t = 0$, we have

$$u_t(x, 0) = cF'(x) - cG'(x) = g(x). \tag{4.7}$$

Integration of (4.7) over the integral $[0, x]$ yields

$$F(x) - G(x) = \frac{1}{c} \int_0^x g(s)\, ds + C, \tag{4.8}$$

where $C = F(0) - G(0)$. Equations (4.6) and (4.8) are two linear algebraic equations for $F(x)$ and $G(x)$. The solution of this system of equations is given by

$$F(x) = \frac{1}{2} f(x) + \frac{1}{2c} \int_0^x g(s)\, ds + \frac{C}{2}, \tag{4.9}$$

$$G(x) = \frac{1}{2} f(x) - \frac{1}{2c} \int_0^x g(s)\, ds - \frac{C}{2}. \tag{4.10}$$

By substituting these expressions for F and G into the general solution (4.5), we obtain the formula

$$u(x, t) = \frac{f(x + ct) + f(x - ct)}{2} + \frac{1}{2c} \int_{x-ct}^{x+ct} g(s)\, ds, \tag{4.11}$$

which is called *d'Alembert's formula*. Note that sometimes (4.9)–(4.10) are also useful, as they give us explicit formulas for the forward and the backward waves.

The following examples illustrate the use of d'Alembert's formula.

Example 4.2 Consider the Cauchy problem

$$u_{tt} - u_{xx} = 0 \qquad -\infty < x < \infty, \ t > 0,$$

$$u(x, 0) = f(x) = \begin{cases} 0 & -\infty < x < -1, \\ x + 1 & -1 \le x \le 0, \\ 1 - x & 0 \le x \le 1, \\ 0 & 1 < x < \infty, \end{cases}$$

$$u_t(x, 0) = g(x) = \begin{cases} 0 & -\infty < x < -1, \\ 1 & -1 \le x \le 1, \\ 0 & 1 < x < \infty. \end{cases}$$

(a) Evaluate u at the point $(1, \frac{1}{2})$.
(b) Discuss the smoothness of the solution u.

(a) Using d'Alembert's formula, we find that

$$u(1, \tfrac{1}{2}) = \frac{f(\frac{3}{2}) + f(\frac{1}{2})}{2} + \tfrac{1}{2} \int_{\frac{1}{2}}^{\frac{3}{2}} g(s)\, ds.$$

Since $\frac{3}{2} > 1$ it follows that $f(\frac{3}{2}) = 0$. On the other hand, $0 \le \frac{1}{2} \le 1$; therefore, $f(\frac{1}{2}) = \frac{1}{2}$. Evidently, $\int_{\frac{1}{2}}^{\frac{3}{2}} g(s)ds = \int_{\frac{1}{2}}^{1} 1ds = \frac{1}{2}$. Thus, $u(1, \frac{1}{2}) = \frac{1}{2}$.

(b) The solution is not classical, since $u \notin C^1$. Yet u is a generalized solution of the problem. Note that although g is not continuous, nevertheless the solution u is a continuous function. The singularities of the solution propagate along characteristics that intersect the initial line $t = 0$ at the singularities of the initial conditions. These are exactly the characteristics $x \pm t = -1, 0, 1$. Therefore, the solution is smooth in a neighborhood of the point $(1, \frac{1}{2})$ which does not intersect these characteristics.

Example 4.3 Let $u(x, t)$ be the solution of the Cauchy problem

$$u_{tt} - 9u_{xx} = 0 \qquad -\infty < x < \infty, \ t > 0,$$

$$u(x, 0) = f(x) = \begin{cases} 1 & |x| \le 2, \\ 0 & |x| > 2, \end{cases}$$

$$u_t(x, 0) = g(x) = \begin{cases} 1 & |x| \le 2, \\ 0 & |x| > 2. \end{cases}$$

(a) Find $u(0, \frac{1}{6})$.
(b) Discuss the large time behavior of the solution.

(c) Find the maximal value of $u(x, t)$, and the points where this maximum is achieved.

(d) Find all the points where $u \in C^2$.

(a) Since

$$u(x, t) = \frac{f(x + 3t) + f(x - 3t)}{2} + \frac{1}{6} \int_{x-3t}^{x+3t} g(s)ds,$$

it follows that for $x = 0$ and $t = \frac{1}{6}$, we have

$$u(0, \frac{1}{6}) = \frac{f(\frac{1}{2}) + f(-\frac{1}{2})}{2} + \frac{1}{6} \int_{-\frac{1}{2}}^{\frac{1}{2}} g(s)\,ds = \frac{1+1}{2} + \frac{1}{6} \int_{-\frac{1}{2}}^{\frac{1}{2}} 1\,ds = \frac{7}{6}.$$

(b) Fix $\xi \in \mathbb{R}$ and compute $\lim_{t \to \infty} u(\xi, t)$. Clearly,

$$\lim_{t \to \infty} f(\xi + 3t) = 0, \quad \lim_{t \to \infty} f(\xi - 3t) = 0, \quad \lim_{t \to \infty} \int_{\xi-3t}^{\xi+3t} g(s)\,ds = \int_{-2}^{2} 1\,ds = 4.$$

Therefore, $\lim_{t \to \infty} u(\xi, t) = \frac{2}{3}$.

(c) Recall that for any real functions f, g,

$$\max\{f(x) + g(x)\} \le \max f(x) + \max g(x).$$

It turns out that in our special case there exists a point (x, t), where all the terms in (4.11) attain their maximal value simultaneously, and therefore at such a point the maximum of u is attained.

Indeed, $\max\{f(x + 3t)\} = 1$ which is attained on the strip $-2 \le x + 3t \le 2$. Similarly, $\max\{f(x - 3t)\} = 1$ which is attained on the strip $-2 \le x - 3t \le 2$, while $\max\{\int_{x-3t}^{x+3t} g(s)\,ds = \int_{-2}^{2} 1\,ds = 4$, and it is attained on the intersection of the half-planes $x + 3t \ge 2$ and $x - 3t \le -2$. The intersection of all these sets is the set of all points that satisfy the two equations

$$x + 3t = 2,$$

$$x - 3t = -2.$$

This system has a unique solution at $(x, t) = (0, \frac{2}{3})$. Thus, the solution u achieves its maximum at the point $(0, \frac{2}{3})$, where $u(0, \frac{2}{3}) = \frac{5}{3}$.

(d) The initial conditions are smooth except at the points $x = \pm 2$. Therefore, the solution is smooth at all points that are *not* on the characteristics

$$x \pm 3t = -2, \quad x \pm 3t = 2.$$

The function u is a generalized solution that is piecewise continuous for any fixed time $t > 0$.

The well-posedness of the Cauchy problem follows from the d'Alembert formula.

Theorem 4.4 *Fix* $T > 0$. *The Cauchy problem* (4.3)–(4.4) *in the domain* $-\infty < x < \infty$, $0 \le t \le T$ *is well-posed for* $f \in C^2(\mathbb{R})$, $g \in C^1(\mathbb{R})$.

Proof The existence and uniqueness follow directly from the d'Alembert formula. Indeed, this formula provides us with a solution, and we have shown that any solution of the Cauchy problem is necessarily equal to the d'Alembert solution. Note that from our smoothness assumption ($f \in C^2(\mathbb{R})$, $g \in C^1(\mathbb{R})$), it follows that $u \in C^2(\mathbb{R} \times (0, \infty)) \cap C^1(\mathbb{R} \times [0, \infty))$, and therefore, the d'Alembert solution is a classical solution. On the other hand, for $f \in C(\mathbb{R})$ and g that is locally integrable, the d'Alembert solution is a generalized solution.

It remains to prove the stability of the Cauchy problem, i.e. we need to show that small changes in the initial conditions give rise to a small change in the solution. Let u_i be two solutions of the Cauchy problem with initial conditions f_i, g_i, where $i = 1, 2$. Now, if

$$|f_1(x) - f_2(x)| < \delta, \qquad |g_1(x) - g_2(x)| < \delta,$$

for all $x \in \mathbb{R}$, then for all $x \in \mathbb{R}$ and $0 \le t \le T$ we have

$$|u_1(x, t) - u_2(x, t)| \le \frac{|f_1(x + ct) - f_2(x + ct)|}{2} + \frac{|f_1(x - ct) - f_2(x - ct)|}{2}$$

$$+ \frac{1}{2c} \int_{x-ct}^{x+ct} |g_1(s) - g_2(s)|\, ds < \frac{1}{2}(\delta + \delta) + \frac{1}{2c} 2ct\delta \le (1 + T)\delta.$$

Therefore, for a given $\varepsilon > 0$, we take $\delta < \varepsilon/(1 + T)$. Then for all $x \in \mathbb{R}$ and $0 \le t \le T$ we have

$$|u_1(x, t) - u_2(x, t)| < \varepsilon.$$

$\square$

Remark 4.5 (1) The Cauchy problem is ill-posed on the domain $-\infty < x < \infty$, $t \ge 0$.

(2) The d'Alembert formula is also valid for $-\infty < x < \infty$, $T < t \le 0$, and the Cauchy problem is also well-posed in this domain. The physical interpretation is that the process is reversible.

4.4 Domain of dependence and region of influence

Let us return to the Cauchy problem (4.3)–(4.4), and examine what is the information that actually determines the solution u at a fixed point (x_0, t_0). Consider the (x, t) plane and the two characteristics passing through the point (x_0, t_0):

$$x - ct = x_0 - ct_0, \quad x + ct = x_0 + ct_0.$$

These straight lines intersect the x axis at the points $(x_0 - ct_0, 0)$ and $(x_0 + ct_0, 0)$, respectively. The triangle formed by the these characteristics and the interval $[x_0 - ct_0, x_0 + ct_0]$ is called a *characteristic triangle* (see Figure 4.1).

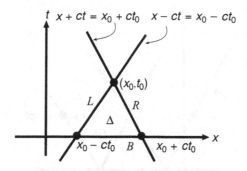

Figure 4.1 Domain of dependence.

By the d'Alembert formula

$$u(x_0, t_0) = \frac{f(x_0 + ct_0) + f(x_0 - ct_0)}{2} + \frac{1}{2c} \int_{x_0 - ct_0}^{x_0 + ct_0} g(s)\, ds. \qquad (4.12)$$

Therefore, the value of u at the point (x_0, t_0) is determined by the values of f at the vertices of the characteristic base and by the values of g along this base. Thus, $u(x_0, t_0)$ depends only on the part of the initial data that is given on the interval $[x_0 - ct_0, x_0 + ct_0]$. Therefore, this interval is called *domain of dependence* of u at the point (x_0, t_0). If we change the initial data at points outside this interval, the value of the solution u at the point (x_0, t_0) will not change. Information on a change in the data travels with speed c along the characteristics, and therefore such information is not available for $t \leq t_0$ at the point x_0. The change will finally influence the solution at the point x_0 at a later time. Hence, for every point (x, t) in a fixed characteristic triangle, $u(x, t)$ is determined only by the initial data that are given on (part of) the characteristic base (see Figure 4.1). Furthermore, if the initial data are smooth on this base, then the solution is smooth in the whole triangle.

We may ask now the opposite question: which are the points on the half-plane $t > 0$ that are influenced by the initial data on a fixed interval $[a, b]$? The set of all such points is called the *region of influence* of the interval $[a, b]$. It follows from the discussion above that the points of this interval influence the value of the solution u at a point (x_0, t_0) if and only if $[x_0 - ct_0, x_0 + ct_0] \cap [a, b] \neq \emptyset$. Hence the initial data along the interval $[a, b]$ influence only points (x, t) satisfying

$$x - ct \leq b, \quad \text{and} \quad x + ct \geq a.$$

These are the points inside the forward (truncated) characteristic cone that is defined by the base $[a, b]$ and the edges $x + ct = a$, $x - ct = b$ (it is the union of the regions I–IV of Figure 4.2).

Assume, for instance, that the initial data f, g vanish outside the interval $[a, b]$. Then the amplitude of the vibrating string is zero at every point outside the influence

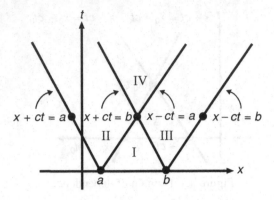

Figure 4.2 Region of influence.

region of this interval. On the other hand, for a fixed point x_0 on the string, the effect of the perturbation (from the zero data) along the interval $[a, b]$ will be felt after a time $t_0 \geq 0$, and eventually, for t large enough, the solution takes the constant value $u(x_0, t) = (1/2c) \int_a^b g(s)\, ds$. This occurs precisely at points (x_0, t) that are inside the cone

$$x_0 - ct \leq a, \quad \text{and} \quad x_0 + ct \geq b,$$

(see region IV in Figure 4.2).

Using these observations, we demonstrate in the following example the so-called *graphical method* for solving the Cauchy problem for the wave equation.

Example 4.6 Consider the Cauchy problem

$$u_{tt} - c^2 u_{xx} = 0 \qquad\qquad -\infty < x < \infty, \ t > 0,$$

$$u(x, 0) = f(x) = \begin{cases} 2 & |x| \leq a, \\ 0 & |x| > a, \end{cases}$$

$$u_t(x, 0) = g(x) = 0 \qquad\qquad -\infty < x < \infty.$$

Draw the graphs of the solution $u(x, t)$ at times $t_i = ia/2c$, where $i = 0, 1, 2, 3$. Using d'Alembert's formula, we write the solution u as a sum of backward and forward waves

$$u(x, t) = \frac{f(x + ct) + f(x - ct)}{2}.$$

Since these waves are piecewise constant functions, it is clear that for each t, the solution u is also a piecewise constant function of x with values $u = 0$, 1, 2.

Consider the (x, t) plane. We draw the characteristic lines that pass through the special points on the initial line $t = 0$ where the initial data are not smooth. In

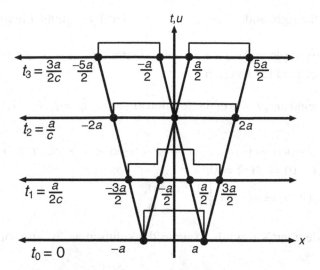

Figure 4.3 The graphical method.

the present problem these are the points $x = \pm a$. We also draw the lines $t = t_i$ that will serve us as the abscissas (x axes) for the graphs of the functions $u(x, t_i)$. Note that the ordinate of the coordinate system is used as the t and the u axes (see Figure 4.3).

Consider the time $t = t_1$. The forward wave has traveled $a/2$ units to the right, and the backward wave has traveled $a/2$ units to the left. The support (the set of points where the function is not zero) of the forward wave at time t_1 is the interval $[-a/2, 3a/2]$, while $[-3a/2, a/2]$ is the support of the backward wave at t_1. Therefore, the support of the solution at t_1 is the interval $[-3a/2, 3a/2]$, i.e. the region of influence of $[-a, a]$ at $t = t_1$. Now, at the intersection of the supports of the two waves (the interval $[-a/2, a/2]$) u takes the value $1 + 1 = 2$, while on the intervals $[-3a/2, -a/2)$, $(a/2, 3a/2]$, where the supports do not intersect, u takes the value 1. Obviously, $u = 0$ at all other points.

Consider the time $t_2 = a/c$. The support of the forward (backward) wave is $[0, 2a]$ ($[-2a, 0]$, respectively). Consequently, the support of the solution u is $[-2a, 2a]$, i.e. the region of influence of $[-a, a]$ at $t = t_2$. The intersection of the supports of the two waves is the point $x = 0$, where u takes the value 2. On the intervals $[-2a, 0)$, $(0, 2a]$, u is 1. Obviously, $u = 0$ at all other points.

At the time $t_3 = 3a/2c$, the support of the forward (backward) wave is $[a/2, 5a/2]$ ($[-5a/2, -a/2]$, respectively), and there is no interaction between the waves. Therefore, the solution at these intervals equals 1, and it equals zero otherwise.

To conclude, the first step of the graphical method is to compute and to draw the graphs of the forward and backward waves. Then, for a given time t, we shift these two shapes to the right and, respectively, to the left by ct units. Finally, we add the two graphs.

In the next example we use the graphical method to investigate the influence of the initial velocity on the solution.

Example 4.7 Find the graphs of the solution $u(x, t_i)$, $t_i = i$, $i = 1, 4$ for the problem

$$
\begin{aligned}
u_{tt} - u_{xx} &= 0 && -\infty < x < \infty,\ t > 0, \\
u(x, 0) &= f(x) = 0, && -\infty < x < \infty, \\
u_t(x, 0) &= g(x) = \begin{cases} 0 & x < 0, \\ 1 & x \geq 0. \end{cases}
\end{aligned}
$$

We apply d'Alembert's formula to write the solution as the sum of forward and backward waves:

$$
u(x, t) = \frac{1}{2} \int_{x-t}^{0} g(s)\, ds + \frac{1}{2} \int_{0}^{x+t} g(s)\, ds = -\frac{\max\{0, x - t\}}{2} + \frac{\max\{0, x + t\}}{2}.
$$

Since both the forward and backward waves are piecewise linear functions, the solution $u(\cdot, t)$ for all times t is a piecewise linear function of x.

We draw in the plane (x, t) the characteristics emanating from the points where the initial condition is nonsmooth. In our case this happens at just one point, namely $x = 0$. We also depict the lines $t = t_i$ that form the abscissas for the graph of $u(x, t_i)$ (see Figure 4.4).

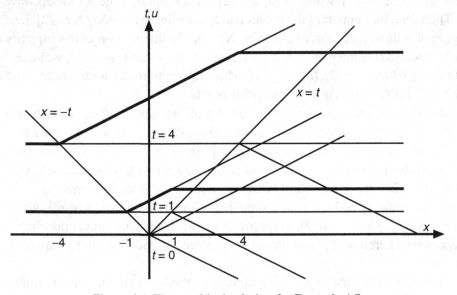

Figure 4.4 The graphical solution for Example 4.7.

At time t_1, the forward wave has moved one unit to the right, and the backward wave has moved one unit to the left. The forward wave is supported on the interval $[1, \infty)$, while the backward wave is supported on $[-1, \infty)$. Therefore the solution is supported on $[-1, \infty)$. It is clear that on the interval $[-1, 1]$ the solution forms a linear function thanks to the backward wave. Specifically, $u = (x + 1)/2$ there. In the interval $[1, \infty)$ the forward wave is a linear function with slope $-1/2$, while the backward wave is a linear function with slope $1/2$. Therefore the solution itself, being a superposition of these two waves, is constant $u = 1$ there.

A similar consideration can be used to draw the graph for $t = 4$. Actually, the solution can be written explicitly as

$$u(x, t) = \begin{cases} 0 & x < -t, \\ \dfrac{x + t}{2} & -t \leq x \leq t, \\ t & x > t. \end{cases}$$

Notice that for each fixed $x_0 \in \mathbb{R}$, the solution $u(x_0, t)$ (considered as a function of t) is not bounded.

4.5 The Cauchy problem for the nonhomogeneous wave equation

Consider the following Cauchy problem

$$u_{tt} - c^2 u_{xx} = F(x, t) \qquad -\infty < x < \infty, \ t > 0, \qquad (4.13)$$
$$u(x, 0) = f(x), \quad u_t(x, 0) = g(x) \qquad -\infty < x < \infty. \qquad (4.14)$$

This problem models, for example, the vibration of a very long string in the presence of an external force F. As in the homogeneous case f, g are given functions that represent the shape and the vertical velocity of the string at time $t = 0$.

As in every linear problem, the uniqueness for the homogeneous problem implies the uniqueness for the nonhomogeneous problem.

Proposition 4.8 *The Cauchy problem (4.13)–(4.14) admits at most one solution.*

Proof Assume that u_1, u_2 are solutions of problem (4.13)–(4.14). We should prove that $u_1 = u_2$. The function $u = u_1 - u_2$ is a solution of the homogeneous problem

$$u_{tt} - c^2 u_{xx} = 0 \qquad -\infty, < x < \infty, \ t > 0, \qquad (4.15)$$
$$u(x, 0) = 0, \quad u_t(x, 0) = 0 \qquad -\infty < x < \infty. \qquad (4.16)$$

On the other hand, $v(x, t) = 0$ is also a solution of the same (homogeneous) problem. By Theorem 4.4, $u = v = 0$, hence, $u_1 = u_2$. $\square$

Next, using an explicit formula, we prove, as in the homogeneous case, the existence of a solution of the Cauchy problem (4.13)–(4.14). For this purpose, recall Green's formula for a pair of functions P, Q in a planar domain Ω with a piecewise smooth boundary Γ:

$$\iint_\Omega [Q(x,t)_x - P(x,t)_t] \, dx \, dt = \oint_\Gamma [P(x,t) \, dx + Q(x,t) \, dt].$$

Let $u(x,t)$ be a solution of problem (4.13)–(4.14). Integrate the two sides of the PDE (4.13) over a characteristic triangle $\triangle$ with a fixed upper vertex (x_0, t_0). The three edges of this triangle (base, right and left edges) will be denoted by B, R, L, respectively (see Figure 4.1). We have

$$-\iint_\triangle F(x,t) \, dx \, dt = \iint_\triangle (c^2 u_{xx} - u_{tt}) \, dx \, dt.$$

Using Green's formula with $Q = c^2 u_x$ and $P = u_t$, we obtain

$$-\iint_\triangle F(x,t) \, dx \, dt = \oint_\Gamma (u_t \, dx + c^2 u_x \, dt) = \int_B + \int_R + \int_L (u_t \, dx + c^2 u_x \, dt).$$

On the base B we have $dt = 0$; therefore, using the initial conditions, we get

$$\int_B (u_t \, dx + c^2 u_x \, dt) = \int_{x_0-ct_0}^{x_0+ct_0} u_t(x,0) \, dx = \int_{x_0-ct_0}^{x_0+ct_0} g(x) \, dx.$$

On the right edge R, $x + ct = x_0 + ct_0$, hence $dx = -c \, dt$. Consequently,

$$\int_R (u_t \, dx + c^2 u_x \, dt) = -c \int_R (u_t \, dt + u_x \, dx) = -c \int_R du$$

$$= -c[u(x_0, t_0) - u(x_0 + ct_0, 0)] = -c[u(x_0, t_0) - f(x_0 + ct_0)].$$

Similarly, on the left edge L, $x - ct = x_0 - ct_0$, implying $dx = c \, dt$, and

$$\int_L (u_t \, dx + c^2 u_x \, dt) = c \int_L (u_t \, dt + u_x \, dx) = c \int_L du$$

$$= c[u(x_0 - ct_0, 0) - u(x_0, t_0)] = c[f(x_0 - ct_0) - u(x_0, t_0)].$$

Therefore,

$$-\iint_\triangle F(x,t) \, dx \, dt = \int_{x_0-ct_0}^{x_0+ct_0} g(x) \, dx + c[f(x_0 - ct_0) + f(x_0 + ct_0) - 2u(x_0, t_0)].$$

Solving for u gives

$$u(x_0, t_0) = \frac{f(x_0 + ct_0) + f(x_0 - ct_0)}{2} + \frac{1}{2c} \int_{x_0-ct_0}^{x_0+ct_0} g(x) \, dx + \frac{1}{2c} \iint_\triangle F(x,t) \, dx \, dt.$$

We finally obtain an explicit formula for the solution at an arbitrary point (x, t):

$$u(x, t) = \frac{f(x + ct) + f(x - ct)}{2} + \frac{1}{2c} \int_{x-ct}^{x+ct} g(s) \, ds + \frac{1}{2c} \iint_{\Delta} F(\xi, \tau) \, d\xi \, d\tau.$$
$$(4.17)$$

This formula is also called d'Alembert's formula.

Remark 4.9 (1) Note that for $F = 0$ the two d'Alembert's formulas coincide, and actually, we have obtained another proof of the d'Alembert formula (4.11).

(2) The value of u at a point (x_0, t_0) is determined by the values of the given data on the *whole* characteristic triangle whose upper vertex is the point (x_0, t_0). This is the domain of dependence for the nonhomogeneous Cauchy problem.

It remains to prove that the function u in (4.17) is indeed a solution of the Cauchy problem. From the superposition principle it follows that u in (4.17) is the desired solution, if and only if the function

$$v(x, t) = \frac{1}{2c} \iint_{\Delta} F(\xi, \tau) \, d\xi \, d\tau = \frac{1}{2c} \int_0^t \int_{x-c(t-\tau)}^{x+c(t-\tau)} F(\xi, \tau) \, d\xi \, d\tau$$

is a solution of the Cauchy problem

$$u_{tt} - c^2 u_{xx} = F(x, t) \qquad -\infty < x < \infty, \ t > 0, \qquad (4.18)$$
$$u(x, 0) = 0, \quad u_t(x, 0) = 0 \quad -\infty < x < \infty. \qquad (4.19)$$

We shall prove that v is a solution of the initial value problem (4.18)–(4.19) under the assumption that F and F_x are continuous. Clearly,

$$v(x, 0) = 0.$$

In order to take derivatives, we shall use the formula

$$\frac{\partial}{\partial t} \int_{a(t)}^{b(t)} G(\xi, t) \, d\xi = G(b(t), t) b'(t) - G(a(t), t) a'(t) + \int_{a(t)}^{b(t)} \frac{\partial}{\partial t} G(\xi, t) \, d\xi.$$

Hence,

$$v_t(x, t) = \frac{1}{2c} \int_x^x F(\xi, t) \, d\xi + \frac{1}{2} \int_0^t [F(x + c(t - \tau), \tau) + F(x - c(t - \tau), \tau)] \, d\tau$$

$$= \frac{1}{2} \int_0^t [F(x + c(t - \tau), \tau) + F(x - c(t - \tau), \tau)] \, d\tau.$$

In particular,

$$v_t(x, 0) = 0.$$

By taking the second derivative with respect to t, we have

$$v_{tt}(x, t) = F(x, t) + \frac{c}{2} \int_0^t [F_x(x + c(t - \tau), \tau) - F_x(x - c(t - \tau), \tau)]\, d\tau.$$

Similarly,

$$v_x(x, t) = \frac{1}{2c} \int_0^t [F(x + c(t - \tau), \tau) - F(x - c(t - \tau), \tau)]\, d\tau,$$

$$v_{xx}(x, t) = \frac{1}{2c} \int_0^t [F_x(x + c(t - \tau), \tau) - F_x(x - c(t - \tau), \tau)]\, d\tau.$$

Therefore, $v(x, t)$ is a solution of the nonhomogeneous wave equation (4.18), and the homogeneous initial conditions (4.19) are satisfied. Note that all the above differentiations are justified provided that $F, F_x \in C(\mathbb{R}^2)$.

Theorem 4.10 *Fix $T > 0$. The Cauchy problem (4.13)–(4.14) in the domain $-\infty < x < \infty$, $0 \le t \le T$ is well-posed for $F, F_x \in C(\mathbb{R}^2)$, $f \in C^2(\mathbb{R})$, $g \in C^1(\mathbb{R})$.*

Proof Recall that the uniqueness has already been proved, and the existence follows from d'Alembert's formula. It remains to prove stability, i.e. we need to show that small changes in the initial conditions and the external force give rise to a small change in the solution. For $i = 1, 2$, let u_i be the solution of the Cauchy problem with the corresponding function F_i, and the initial conditions f_i, g_i. Now, if

$$|F_1(x, t) - F_2(x, t)| < \delta, \quad |f_1(x) - f_2(x)| < \delta, \quad |g_1(x) - g_2(x)| < \delta,$$

for all $x \in \mathbb{R}$, $0 \le t \le T$, then for all $x \in \mathbb{R}$ and $0 \le t \le T$ we have

$$|u_1(x, t) - u_2(x, t)| \le \frac{|f_1(x + ct) - f_2(x + ct)|}{2} + \frac{|f_1(x - ct) - f_2(x - ct)|}{2}$$

$$+ \frac{1}{2c} \int_{x-ct}^{x+ct} |g_1(s) - g_2(s)|\, ds + \frac{1}{2c} \iint_\Delta |F_1(\xi, \tau) - F_2(\xi, \tau)|\, d\xi\, d\tau$$

$$< \frac{1}{2}(\delta + \delta) + \frac{1}{2c} 2ct\delta + \frac{1}{2c} ct^2 \delta \le (1 + T + T^2/2)\delta.$$

Therefore, for a given $\varepsilon > 0$, we choose $\delta < \varepsilon/(1 + T + T^2/2)$. Thus, for all $x \in \mathbb{R}$ and $0 \le t \le T$, we have

$$|u_1(x, t) - u_2(x, t)| < \varepsilon.$$

Note that δ does not depend on the wave speed c. □

Corollary 4.11 *Suppose that f, g are even functions, and for every $t \ge 0$ the function $F(\cdot, t)$ is even too. Then for every $t \ge 0$ the solution $u(\cdot, t)$ of the Cauchy problem (4.13)–(4.14) is also even. Similarly, the solution is an odd function or a* *periodic function with a period L (as a function of x) if the data are odd functions or periodic functions with a period L.*

Proof We prove the first part of the corollary. The other parts can be shown similarly. Let u be the solution of the problem and define the function $v(x, t) = u(-x, t)$. Clearly,

$$v_x(x, t) = -u_x(-x, t), \quad v_t(x, t) = u_t(-x, t)$$

and

$$v_{xx}(x, t) = u_{xx}(-x, t), \quad v_{tt}(x, t) = u_{tt}(-x, t).$$

Therefore,

$$v_{tt}(x, t) - c^2 v_{xx}(x, t) = u_{tt}(-x, t) - c^2 u_{xx}(-x, t)$$
$$= F(-x, t) = F(x, t) \quad -\infty < x < \infty, \ t > 0.$$

Thus, v is a solution of the nonhomogeneous wave equation (4.13). Furthermore,

$$v(x, 0) = u(-x, 0) = f(-x) = f(x), \quad v_t(x, 0) = u_t(-x, 0) = g(-x) = g(x).$$

It means that v is also a solution of the initial value problem (4.13)–(4.14). Since the solution of this problem is unique, we have $v(x, t) = u(x, t)$, which implies $u(-x, t) = u(x, t)$. $\qquad\square$

Example 4.12 Solve the following Cauchy problem

$$u_{tt} - 9u_{xx} = e^x - e^{-x} \quad -\infty < x < \infty, \ t > 0,$$
$$u(x, 0) = x \qquad\qquad -\infty < x < \infty,$$
$$u_t(x, 0) = \sin x \qquad\qquad -\infty < x < \infty.$$

Using the d'Alembert formula, we have

$$u(x, t) = \frac{1}{2}[f(x + ct) + f(x - ct)] + \frac{1}{2c}\int_{x-ct}^{x+ct} g(s)ds$$
$$+ \frac{1}{2c}\int_{\tau=0}^{\tau=t}\int_{\xi=x-c(t-\tau)}^{\xi=x+c(t-\tau)} F(\xi, \tau)\, d\xi\, d\tau.$$

Hence,

$$u(x, t) = \frac{1}{2}[x + 3t + x - 3t] + \frac{1}{6}\int_{x-3t}^{x+3t} \sin s\, ds$$
$$+ \frac{1}{6}\int_{\tau=0}^{\tau=t}\int_{\xi=x-3(t-\tau)}^{\xi=x+3(t-\tau)} (e^\xi - e^{-\xi})\, d\xi\, d\tau$$
$$= x + \frac{1}{3}\sin x \sin 3t - \frac{2}{9}\sinh x + \frac{2}{9}\sinh x \cosh 3t.$$

As expected, for all $t \geq 0$, the solution u is an odd function of x.

Remark 4.13 In many cases it is possible to reduce a nonhomogeneous problem to a homogeneous problem if we can find a particular solution v of the given nonhomogeneous equation. This will eliminate the need to perform the double integration which appears in the d'Alembert formula (4.17). The technique is particularly useful when F has a simple form, for example, when $F = F(x)$ or $F = F(t)$. Suppose that such a particular solution v is found, and consider the function $w = u - v$. By the superposition principle, w should solve the following homogeneous Cauchy problem:

$$
\begin{aligned}
w_{tt} - w_{xx} &= 0 & -\infty < x < \infty, \; t > 0, \\
w(x, 0) &= f(x) - v(x, 0) & -\infty < x < \infty, \\
w_t(x, 0) &= g(x) - v_t(x, 0) & -\infty < x < \infty.
\end{aligned}
$$

Hence, w can be found using the d'Alembert formula for the homogeneous equation. Then $u = v + w$ is the solution of our original problem.

We illustrate this idea through the following example.

Example 4.14 Solve the problem

$$
\begin{aligned}
u_{tt} - u_{xx} &= t^7 & -\infty < x < \infty, \; t > 0, \\
u(x, 0) &= 2x + \sin x & -\infty < x < \infty, \\
u_t(x, 0) &= 0 & -\infty < x < \infty.
\end{aligned}
$$

Because of the special form of the nonhomogeneous equation, we look for a particular solution of the form $v = v(t)$. Indeed it can be easily verified that $v(x, t) = \frac{1}{72}t^9$ is such a solution. Consequently, we need to solve the homogeneous problem

$$
\begin{aligned}
w_{tt} - w_{xx} &= 0 & -\infty < x < \infty, \; t > 0, \\
w(x, 0) &= f(x) - v(x, 0) = 2x + \sin x & -\infty < x < \infty, \\
w_t(x, 0) &= g(x) - v_t(x, 0) = 0 & -\infty < x < \infty.
\end{aligned}
$$

Using d'Alembert's formula for the homogeneous equation, we have

$$
w(x, t) = 2x + \tfrac{1}{2}\sin(x + t) + \tfrac{1}{2}\sin(x - t),
$$

and the solution of the original problem is given by

$$
u(x, t) = 2x + \sin x \cos(t) + \frac{t^9}{72}.
$$

4.6 Exercises

4.1 Complete the proof of Corollary 4.11.

4.2 Solve the problem

$$
\begin{aligned}
u_{tt} - u_{xx} &= 0 & 0 < x < \infty, \; t > 0, \\
u(0, t) &= t^2 & t > 0, \\
u(x, 0) &= x^2 & 0 \le x < \infty, \\
u_t(x, 0) &= 6x & 0 \le x < \infty,
\end{aligned}
$$

and evaluate $u(4, 1)$ and $u(1, 4)$.

4.3 Consider the problem

$$
u_{tt} - 4u_{xx} = 0 \quad -\infty < x < \infty, \; t > 0,
$$

$$
u(x, 0) = \begin{cases} 1 - x^2 & |x| \le 1, \\ 0 & \text{otherwise}, \end{cases}
$$

$$
u_t(x, 0) = \begin{cases} 4 & 1 \le x \le 2, \\ 0 & \text{otherwise}. \end{cases}
$$

(a) Using the graphical method, find $u(x, 1)$.
(b) Find $\lim_{t \to \infty} u(5, t)$.
(c) Find the set of all points where the solution is singular (nonclassical).
(d) Find the set of all points where the solution is not continuous.

4.4 (a) Solve the following initial boundary value problem for a vibrating semi-infinite string which is fixed at $x = 0$:

$$
\begin{aligned}
u_{tt} - u_{xx} &= 0 & 0 < x < \infty, \; t > 0, \\
u(0, t) &= 0 & t > 0, \\
u(x, 0) &= f(x) & 0 \le x < \infty, \\
u_t(x, 0) &= g(x) & 0 \le x < \infty,
\end{aligned}
$$

where $f \in C^2([0, \infty))$ and $g \in C^1([0, \infty))$ satisfy the compatibility conditions $f(0) = f''(0) = g(0) = 0$.

Hint Extend the functions f and g as odd functions $\tilde{f}$ and $\tilde{g}$ over the real line. Solve the Cauchy problem with initial data $\tilde{f}$ and $\tilde{g}$, and show that the restriction of this solution to the half-plane $x \ge 0$ is a solution of the problem. Recall that the solution of the Cauchy problem with odd data is odd. In particular, the solution with odd data is zero for $x = 0$ and all $t \ge 0$.

(b) Solve the problem with $f(x) = x^3 + x^6$, and $g(x) = \sin^2 x$, and evaluate $u(1, i)$ for $i = 1, 2, 3$. Is the solution classical?

4.5 Consider the problem

$$
u_{tt} - u_{xx} = 0 \quad -\infty < x < \infty, \; t > 0,
$$

$$
u(x, 0) = \begin{cases} 8x - 2x^2 & 0 \le x \le 4, \\ 0 & \text{otherwise}, \end{cases}
$$

$$
u_t(x, 0) = \begin{cases} 16 & 0 \le x \le 4, \\ 0 & \text{otherwise}. \end{cases}
$$

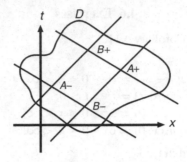

Figure 4.5 A drawing for the parallelogram identity.

(a) Find a formula for the forward and backward waves.
(b) Using the graphical method, draw the graph of $u(x, i)$ for $i = 4, 8, 12$.
(c) Find $u(\pm5, 2), u(\pm3, 4)$.
(d) Find $\lim_{t\to\infty} u(5, t)$.

4.6 (a) Solve the following initial boundary value problem for a vibrating semi-infinite string with a free boundary condition:

$$
\begin{aligned}
u_{tt} - u_{xx} &= 0 & & 0 < x < \infty,\ t > 0, \\
u_x(0, t) &= 0 & & t > 0, \\
u(x, 0) &= f(x) & & 0 \le x < \infty, \\
u_t(x, 0) &= g(x) & & 0 \le x < \infty,
\end{aligned}
$$

where $f \in C^2([0, \infty))$ and $g \in C^1([0, \infty))$ satisfy the compatibility conditions $f'_+(0) = g'_+(0) = 0$.

Hint Extend the functions f and g as even functions $\tilde{f}$ and $\tilde{g}$ on the line. Solve the Cauchy problem with initial data $\tilde{f}$ and $\tilde{g}$, and show that the restriction of this solution to the half-plane $x \ge 0$ is a solution of the problem.

(b) Solve the problem with $f(x) = x^3 + x^6$, $g(x) = \sin^3 x$, and evaluate $u(1, i)$ for $i = 1, 2, 3$. Is the solution classical?

4.7 (a) Let $u(x, t)$ be a solution of the wave equation $u_{tt} - u_{xx} = 0$ in a domain $D \subset \mathbb{R}^2$. Let a, b be real numbers such that the parallelogram with vertices $A_\pm = (x_0 \pm a, t_0 \pm b)$, $B_\pm = (x_0 \pm b, t_0 \pm a)$ is contained in D (see Figure 4.5). Prove the *parallelogram identity*:

$$
u(x_0 - a, t_0 - b) + u(x_0 + a, t_0 + b) = u(x_0 - b, t_0 - a) + u(x_0 + b, t_0 + a).
$$

(b) Derive the corresponding identity when the wave speed $c \ne 1$.

(c) Using the parallelogram identity, solve the following initial boundary value problem for a vibrating semi-infinite string with a nonhomogeneous boundary condition:

$$
\begin{aligned}
u_{tt} - u_{xx} &= 0 & & 0 < x < \infty,\ t > 0, \\
u(0, t) &= h(t) & & t > 0, \\
u(x, 0) &= f(x) & & 0 \le x < \infty, \\
u_t(x, 0) &= g(x) & & 0 \le x < \infty,
\end{aligned}
$$

where $f, g, h \in C^2([0, \infty))$.

Hint Distinguish between the cases $x - t > 0$ and $x - t \leq 0$.

(d) From the explicit formula that was obtained in part (c), derive the corresponding compatibility conditions, and prove that the problem is well-posed.

(e) Derive an explicit formula for the solution and deduce the corresponding compatibility conditions for the case $c \neq 1$.

4.8 Solve the following initial boundary value problem using the parallelogram identity

$$
\begin{aligned}
u_{tt} - u_{xx} &= 0 && 0 < x < \infty, \ 0 < t < 2x, \\
u(x, 0) &= f(x) && 0 \leq x < \infty, \\
u_t(x, 0) &= g(x) && 0 \leq x < \infty, \\
u(x, 2x) &= h(x) && x \geq 0,
\end{aligned}
$$

where $f, g, h \in C^2([0, \infty))$.

4.9 Solve the problem

$$
\begin{aligned}
u_{tt} - u_{xx} &= 1 && -\infty < x < \infty, \ t > 0, \\
u(x, 0) &= x^2 && -\infty < x < \infty, \\
u_t(x, 0) &= 1 && -\infty < x < \infty.
\end{aligned}
$$

4.10 (a) Solve the *Darboux problem*:

$$
\begin{aligned}
u_{tt} - u_{xx} &= 0 && t > \max\{-x, x\}, \ t \geq 0, \\
u(x, t) &= \begin{cases} \phi(t) & x = t, \ t \geq 0, \\ \psi(t) & x = -t, \ t \geq 0, \end{cases}
\end{aligned}
$$

where $\phi, \psi \in C^2([0, \infty)$ satisfies $\phi(0) = \psi(0)$.

(b) Prove that the problem is well posed.

4.11 A pressure wave generated as a result of an explosion satisfies the equation

$$
P_{tt} - 16 P_{xx} = 0
$$

in the domain $\{(x, t) \mid -\infty < x < \infty, \ t > 0\}$, where $P(x, t)$ is the pressure at the point x and time t. The initial conditions at the explosion time $t = 0$ are

$$
P(x, 0) = \begin{cases} 10 & |x| \leq 1, \\ 0 & |x| > 1, \end{cases}
$$

$$
P_t(x, 0) = \begin{cases} 1 & |x| \leq 1, \\ 0 & |x| > 1. \end{cases}
$$

A building is located at the point $x_0 = 10$. The engineer who designed the building determined that it will sustain a pressure up to $P = 6$. Find the time t_0 when the pressure at the building is maximal. Will the building collapse?

4.12 (a) Solve the problem

$$
\begin{aligned}
u_{tt} - u_{xx} &= 0 && 0 < x < \infty, \ 0 < t, \\
u(0, t) &= \frac{t}{1+t} && 0 \leq t, \\
u(x, 0) &= u_t(x, 0) = 0 && 0 \leq x < \infty.
\end{aligned}
$$

(b) Show that the limit

$$\lim_{x \to \infty} u(cx, x) := \phi(c)$$

exists for all $c > 0$. What is the limit?

4.13 Consider the Cauchy problem

$$u_{tt} - 4u_{xx} = F(x, t) \qquad -\infty < x < \infty, \ t > 0,$$
$$u(x, 0) = f(x), \quad u_t(x, 0) = g(x) \qquad -\infty < x < \infty,$$

where

$$f(x) = \begin{cases} x & 0 < x < 1, \\ 1 & 1 < x < 2, \\ 3 - x & 2 < x < 3, \\ 0 & x > 3, x < 0, \end{cases}$$

$$g(x) = \begin{cases} 1 - x^2 & |x| < 1, \\ 0 & |x| > 1, \end{cases}$$

and $F(x, t) = -4e^x$ on $t > 0$, $-\infty < x < \infty$.
(a) Is the d'Alembert solution of the problem a classical solution? If your answer is negative, find all the points where the solution is singular.
(b) Evaluate the solution at $(1, 1)$.

4.14 Solve the problem

$$u_{tt} - 4u_{xx} = e^x + \sin t \qquad -\infty < x < \infty, \ t > 0,$$
$$u(x, 0) \quad = 0 \qquad -\infty < x < \infty,$$
$$u_t(x, 0) \quad = \frac{1}{1 + x^2} \qquad -\infty < x < \infty.$$

4.15 Find the general solution of the problem

$$u_{ttx} - u_{xxx} = 0, \quad u_x(x, 0) = 0, \quad u_{xt}(x, 0) = \sin x,$$

in the domain $\{(x, t) \mid -\infty < x < \infty, \ t > 0\}$.

4.16 Solve the problem

$$u_{tt} - u_{xx} = xt \qquad -\infty < x < \infty, \ t > 0,$$
$$u(x, 0) \quad = 0 \qquad -\infty < x < \infty,$$
$$u_t(x, 0) \quad = e^x \qquad -\infty < x < \infty.$$

4.17 (a) Without using the d'Alembert formula find a solution $u(x, t)$ of the problem

$$u_{tt} - u_{xx} = \cos(x + t) \qquad -\infty < x < \infty, \ t > 0,$$
$$u(x, 0) = x, \quad u_t(x, 0) = \sin x \qquad -\infty < x < \infty.$$

(b) Without using the d'Alembert formula find $v(x, t)$ that is a solution of the problem

$$v_{tt} - v_{xx} = \cos(x + t) \qquad -\infty < x < \infty, \ t > 0,$$
$$v(x, 0) = 0, \quad v_t(x, 0) = 0 \qquad -\infty < x < \infty.$$

(c) Find the PDE and initial conditions that are satisfied by the function $w := u - v$.

(d) Which of the functions u, v, w (as a function of x) is even? Odd? Periodic?

(e) Evaluate $v(2\pi, \pi)$, $w(0, \pi)$.

4.18 Solve the problem

$$
\begin{aligned}
u_{tt} - 4u_{xx} &= 6t & -\infty < x < \infty, \, t > 0, \\
u(x, 0) &= x & -\infty < x < \infty, \\
u_t(x, 0) &= 0 & -\infty < x < \infty,
\end{aligned}
$$

without using the d'Alembert formula.

4.19 Let $u(x, t)$ be a solution of the equation $u_{tt} - u_{xx} = 0$ in the whole plane. Suppose that $u_x(x, t)$ is constant on the line $x = 1 + t$. Assume also that $u(x, 0) = 1$ and $u(1, 1) = 3$. Find such a solution $u(x, t)$. Is this solution uniquely determined?

5

The method of separation of variables

5.1 Introduction

We examined in Chapter 1 Fourier's work on heat conduction. In addition to developing a general theory for heat flow, Fourier discovered a method for solving the initial boundary value problem he derived. His solution led him to propose the bold idea that any real valued function defined on a closed interval can be represented as a series of trigonometric functions. This is known today as the *Fourier expansion*. D'Alembert and the Swiss mathematician Daniel Bernoulli (1700–1782) had actually proposed a similar idea before Fourier. They claimed that the vibrations of a finite string can be formally represented as an infinite series involving sinusoidal functions. They failed, however, to see the generality of their observation.

Fourier's method for solving the heat equation provides a convenient method that can be applied to many other important linear problems. The method also enables us to deduce several properties of the solutions, such as asymptotic behavior, smoothness, and well-posedness. Historically, Fourier's idea was a breakthrough which paved the way for new developments in science and technology. For example, Fourier analysis found many applications in pure mathematics (number theory, approximation theory, etc.). Several fundamental theories in physics (quantum mechanics in particular) are heavily based on Fourier's idea, and the entire theory of signal processing is based on Fourier's method and its generalizations.

Nevertheless, Fourier's method cannot always be applied for solving linear differential problems. The method is applicable only for problems with an appropriate symmetry. Moreover, the equation and the domain should share the same symmetry, and in most cases the domain should be bounded. Another drawback follows from the representation of the solution as an infinite series. In many cases it is not easy to prove that the formal solution given by this method is indeed a proper solution. Finally, even in the case when one can prove that the series converges to a classical solution, it might happen that the rate of convergence is very slow. Therefore, such a representation of the solution may not always be practical.

Fourier's method for solving linear PDEs is based on the technique of *separation of variables*. Let us outline the main steps of this technique. First we search for solutions of the homogeneous PDE that are called *product solutions* (or *separated solutions*). These solutions have the special form

$$u(x, t) = X(x)T(t),$$

and in general they should satisfy certain additional conditions. In many cases, these additional conditions are just homogeneous boundary conditions. It turns out that X and T should be solutions of linear ODEs that are easily derived from the given PDE. In the second step, we use a generalization of the superposition principle to generate out of the separated solutions a more general solution of the PDE, in the form of an infinite series of product solutions. In the last step we compute the coefficients of this series.

Since the separation of variables method relies on several deep ideas and also involves several technical steps, we present in the current chapter the technique for solving several relatively simple problems without much theoretical justification. The theoretical study is postponed to Chapter 6. Since Fourier's method is based on constructing solutions of a specific type, we introduce towards the end of the chapter the energy method, which is used to prove that the solutions we have constructed are indeed unique.

5.2 Heat equation: homogeneous boundary conditions

Consider the following heat conduction problem in a finite interval:

$$u_t - ku_{xx} = 0 \qquad 0 < x < L, \; t > 0, \tag{5.1}$$
$$u(0, t) = u(L, t) = 0 \qquad t \geq 0, \tag{5.2}$$
$$u(x, 0) = f(x) \quad 0 \leq x \leq L, \tag{5.3}$$

where f is a given initial condition, and k is a positive constant. In order to make (5.2) consistent with (5.3), we assume the *compatibility condition*

$$f(0) = f(L) = 0.$$

The equation and the domain are drawn schematically in Figure 5.1

The problem defined above corresponds to the evolution of the temperature $u(x, t)$ in a homogeneous one-dimensional heat conducting rod of length L (i.e. the rod is narrow and is laterally insulated) whose initial temperature (at time $t = 0$) is known and is such that its two ends are immersed in a zero temperature bath.

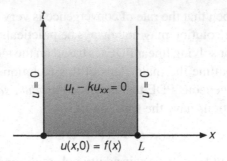

Figure 5.1 The initial boundary value problem for the heat equation together with the domain.

We assume that there is no internal source that heats (or cools) the system. Note that the problem (5.1)–(5.3) is an initial boundary value problem that is linear and homogeneous. Recall also that the boundary condition (5.2) is called the *Dirichlet condition*. At the end of the present section, we shall also discuss other boundary conditions.

We start by looking for solutions of the PDE (5.1) that satisfy the boundary conditions (5.2), and have the special form

$$u(x, t) = X(x)T(t), \tag{5.4}$$

where X and T are functions of the variables x and t, respectively. At this step we do not take into account the initial condition (5.3). Obviously, we are not interested in the zero solution $u(x, t) = 0$. Therefore, we seek functions X and T that do not vanish identically.

Differentiate the separated solution (5.4) once with respect to t and twice with respect to x and substitute these derivatives into the PDE. We then obtain

$$XT_t = kX_{xx}T.$$

Now, we carry out a simple but decisive step – *the separation of variables step*. We move to one side of the PDE all the functions that depend only on x and to the other side the functions that depend only on t. We thus write

$$\frac{T_t}{kT} = \frac{X_{xx}}{X}. \tag{5.5}$$

Since x and t are independent variables, differentiating (5.5) with respect to t implies that there exists a constant denoted by λ (which is called the *separation constant*) such that

$$\frac{T_t}{kT} = \frac{X_{xx}}{X} = -\lambda. \tag{5.6}$$

Equation (5.6) leads to the following system of ODEs:

$$\frac{d^2 X}{dx^2} = -\lambda X \quad 0 < x < L, \tag{5.7}$$

$$\frac{dT}{dt} = -\lambda k T \quad t > 0, \tag{5.8}$$

which are coupled only by the separation constant λ. The function u satisfies the boundary conditions (5.2) if and only if

$$u(0, t) = X(0)T(t) = 0, \quad u(L, t) = X(L)T(t) = 0.$$

Since u is not the trivial solution $u = 0$, it follows that

$$X(0) = X(L) = 0.$$

Therefore, the function X should be a solution of the boundary value problem

$$\frac{d^2 X}{dx^2} + \lambda X = 0 \quad 0 < x < L, \tag{5.9}$$

$$X(0) = X(L) = 0. \tag{5.10}$$

Consider the system (5.9)–(5.10). A nontrivial solution of this system is called an *eigenfunction* of the problem with an *eigenvalue* λ. The problem (5.9)–(5.10) is called an *eigenvalue problem*. The boundary condition (5.10) is called (as in the PDE case) the Dirichlet boundary condition.

Note that the problem (5.9)–(5.10) is not an initial boundary problem for an ODE (for which it is known that there exists a unique solution). Rather, it is a boundary value problem for an ODE. It is not clear a priori that there exists a solution for any value of λ. On the other hand, if we can write the general solution of the ODE for every λ, then we need only to check for which λ there exists a solution that also satisfies the boundary conditions.

Fortunately, (5.9) is quite elementary. It is a second-order linear ODE with constant coefficients, and its general solution (which depends on λ) has the following form:

1. if $\lambda < 0$, then $X(x) = \alpha e^{\sqrt{-\lambda}x} + \beta e^{-\sqrt{-\lambda}x}$,
2. if $\lambda = 0$, then $X(x) = \alpha + \beta x$,
3. if $\lambda > 0$, then $X(x) = \alpha \cos(\sqrt{\lambda}x) + \beta \sin(\sqrt{\lambda}x)$,

where α, β are arbitrary real numbers.

We implicitly assume that λ is real, and we do not consider the complex case (although this case can, in fact, be treated similarly). In Chapter 6, we show that the system (5.9)–(5.10) does not admit a solution with a nonreal λ. In other words, all the eigenvalues of the problem are real numbers.

Negative eigenvalue ($\lambda < 0$) The general solution can be written in a more convenient form: instead of choosing the two exponential functions as the fundamental system of solutions, we use the basis $\{\sinh(\sqrt{-\lambda}x), \cosh(\sqrt{-\lambda}x)\}$. In this basis, the general solution for $\lambda < 0$ has the form

$$X(x) = \tilde{\alpha} \cosh(\sqrt{-\lambda}x) + \tilde{\beta} \sinh(\sqrt{-\lambda}x). \tag{5.11}$$

The function $\sinh s$ has a unique root at $s = 0$, while $\cosh s$ is a strictly positive function. Since $X(x)$ should satisfy $X(0) = 0$, it follows $\tilde{\alpha} = 0$. The second boundary condition $X(L) = 0$ implies that $\tilde{\beta} = 0$. Hence, $X(x) \equiv 0$ is the trivial solution. In other words, the system (5.9)–(5.10) does not admit a negative eigenvalue.

Zero eigenvalue ($\lambda = 0$) We claim that $\lambda = 0$ is also not an eigenvalue. Indeed, in this case the general solution is a linear function $X(x) = \alpha + \beta x$ that (in the nontrivial case $X \neq 0$) vanishes at most at one point; thus it cannot satisfy the boundary conditions (5.10).

Positive eigenvalue ($\lambda > 0$) The general solution for $\lambda > 0$ is

$$X(x) = \alpha \cos(\sqrt{\lambda}x) + \beta \sin(\sqrt{\lambda}x). \tag{5.12}$$

Substituting this solution into the boundary condition $X(0) = 0$, we obtain $\alpha = 0$. The boundary condition $X(L) = 0$ implies $\sin(\sqrt{\lambda}L) = 0$. Therefore, $\sqrt{\lambda}L = n\pi$, where n a positive integer. We do not have to consider the case $n < 0$, since it corresponds to the same set of eigenvalues and eigenfunctions. Hence, λ is an eigenvalue if and only if

$$\lambda = \left(\frac{n\pi}{L}\right)^2 \quad n = 1, 2, 3, \dots.$$

The corresponding eigenfunctions are

$$X(x) = \sin\frac{n\pi x}{L},$$

and they are uniquely defined up to a multiplicative constant.

In conclusion, the set of all solutions of problem (5.9)–(5.10) is an infinite sequence of eigenfunctions, each associated with a positive eigenvalue. It is convenient to use the notation

$$X_n(x) = \sin\frac{n\pi x}{L}, \quad \lambda_n = \left(\frac{n\pi}{L}\right)^2 \quad n = 1, 2, 3, \dots.$$

Recall from linear algebra that an eigenvalue has *multiplicity m* if the space consisting of its eigenvectors is m-dimensional. An eigenvalue with multiplicity 1 is called *simple*. Using the same terminology, we see that the eigenvalues λ_n for the eigenvalue problem (5.9)–(5.10) are all simple.

Let us deal now with the ODE (5.8). The general solution has the form

$$T(t) = Be^{-k\lambda t}.$$

Substituting λ_n, we obtain

$$T_n(t) = B_n e^{-k(\frac{n\pi}{L})^2 t} \qquad n = 1, 2, 3, \dots. \tag{5.13}$$

From the physical point of view it is clear that the solution of (5.8) must decay in time, hence, we must have $\lambda > 0$. Therefore, we could have guessed a priori that the problem (5.9)–(5.10) would admit only positive eigenvalues.

We have thus obtained the following sequence of separated solutions

$$u_n(x, t) = X_n(x)T_n(t) = B_n \sin\frac{n\pi x}{L} e^{-k(\frac{n\pi}{L})^2 t} \qquad n = 1, 2, 3, \dots. \tag{5.14}$$

The superposition principle implies that any linear combination

$$u(x, t) = \sum_{n=1}^{N} B_n \sin\frac{n\pi x}{L} e^{-k(\frac{n\pi}{L})^2 t} \tag{5.15}$$

of separated solutions is also a solution of the heat equation that satisfies the Dirichlet boundary conditions.

Consider now the initial condition. Suppose it has the form

$$f(x) = \sum_{n=1}^{N} B_n \sin\frac{n\pi x}{L},$$

i.e. it is a linear combination of the eigenfunctions. Then a solution of the heat problem (5.1)–(5.3) is given by

$$u(x, t) = \sum_{n=1}^{N} B_n \sin\frac{n\pi x}{L} e^{-k(\frac{n\pi}{L})^2 t}.$$

Hence, we are able to solve the problem for a certain family of initial conditions.

It is natural to ask at this point how to solve for more general initial conditions? The brilliant (although not fully justified at that time) idea of Fourier was that it is possible to represent an arbitrary function f that satisfies the boundary conditions (5.2) as a unique infinite "linear combination" of the eigenfunctions $\sin(n\pi x/L)$. In other words, it is possible to find constants B_n such that

$$f(x) = \sum_{n=1}^{\infty} B_n \sin\frac{n\pi x}{L}, \tag{5.16}$$

Such a series is called a *(generalized) Fourier series* (or expansion) of the function f with respect to the eigenfunctions of the problem, and $B_n, n = 1, 2\dots$ are called the *(generalized) Fourier coefficients* of the series.

The last ingredient that is needed for solving the problem is called the *generalized superposition principle*. We generalize the superposition principle and apply it also to an *infinite* series of separated solutions. We call such a series a *generalized solution* of the PDE if the series is uniformly converging in every subrectangle that is contained in the domain where the solution is defined. This definition is similar to the definition of generalized solutions of the wave equation that was given in Chapter 4.

In our case the generalized superposition principle implies that the formal expression

$$u(x, t) = \sum_{n=1}^{\infty} B_n \sin \frac{n\pi x}{L} e^{-k(\frac{n\pi}{L})^2 t} \tag{5.17}$$

is a natural candidate for a generalized solution of problem (5.1)–(5.3). By a 'formal solution' we mean that if we ignore questions concerning convergence, continuity, and smoothness, and carry out term-by-term differentiations and substitutions, then we see that all the required conditions of the problem (5.1)–(5.3) are satisfied.

Before proving that under certain conditions (5.17) is indeed a solution, we need to explain how to represent an 'arbitrary' function f as a Fourier series. In other words, we need a method of finding the Fourier coefficients of a given function f.

Surprisingly, this question can easily be answered under the assumption that the Fourier series of f converges uniformly. Fix $m \in \mathbb{N}$, multiply the Fourier expansion (5.16) by the eigenfunction $\sin(m\pi x/L)$, and then integrate the equation term-by-term over $[0, L]$. We get

$$\int_0^L \sin \frac{m\pi x}{L} f(x) \, dx = \sum_{n=1}^{\infty} B_n \int_0^L \sin \frac{m\pi x}{L} \sin \frac{n\pi x}{L} \, dx. \tag{5.18}$$

It is easily checked (see Section A.1) that

$$\int_0^L \sin \frac{m\pi x}{L} \sin \frac{n\pi x}{L} \, dx = \begin{cases} 0 & m \neq n, \\ L/2 & m = n. \end{cases} \tag{5.19}$$

Therefore, the Fourier coefficients are given by

$$B_m = \frac{\int_0^L \sin(m\pi x/L) f(x) \, dx}{\int_0^L \sin^2(m\pi x/L) \, dx} = \frac{2}{L} \int_0^L \sin \frac{m\pi x}{L} f(x) \, dx, \quad m = 1, 2, \ldots. \tag{5.20}$$

In particular, it follows that the Fourier coefficients and the Fourier expansion of f are uniquely determined. Therefore, (5.17) together with (5.20) provides an explicit formula for a (formal) solution of the heat problem. Notice that we have developed a powerful tool! For a given initial condition f, one only has to compute the corresponding Fourier coefficients in order to obtain an explicit solution.

Example 5.1 Consider the problem:

$$u_t - u_{xx} = 0 \qquad 0 < x < \pi, \ t > 0, \tag{5.21}$$
$$u(0, t) = u(\pi, t) = 0 \qquad t \geq 0, \tag{5.22}$$
$$u(x, 0) = f(x) = \begin{cases} x & 0 \leq x \leq \pi/2, \\ \pi - x & \pi/2 \leq x \leq \pi. \end{cases} \tag{5.23}$$

The formal solution is

$$u(x, t) = \sum_{m=1}^{\infty} B_m \sin mx \, e^{-m^2 t}, \tag{5.24}$$

where

$$
\begin{aligned}
B_m &= \frac{2}{\pi} \int_0^{\pi} f(x) \sin mx \, dx \\
&= \frac{2}{\pi} \int_0^{\pi/2} x \sin mx \, dx + \frac{2}{\pi} \int_{\pi/2}^{\pi} (\pi - x) \sin mx \, dx \\
&= \frac{2}{\pi} \left[\frac{-x \cos mx}{m} + \frac{\sin mx}{m^2} \right]_0^{\pi/2} + \frac{2}{\pi} \left[\frac{-(\pi - x) \cos mx}{m} - \frac{\sin mx}{m^2} \right]_{\pi/2}^{\pi} \\
&= \frac{4}{\pi m^2} \sin \frac{m\pi}{2}.
\end{aligned}
$$

But

$$\sin \frac{m\pi}{2} = \begin{cases} 0 & m = 2n, \\ (-1)^{n+1} & m = 2n - 1, \end{cases} \tag{5.25}$$

where $n = 1, 2, \dots$. Therefore, the formal solution is

$$u(x, t) = \sum_{n=1}^{\infty} u_n(x, t) = \frac{4}{\pi} \sum_{n=1}^{\infty} \frac{(-1)^{n+1}}{(2n - 1)^2} \sin[(2n - 1)x] e^{-(2n-1)^2 t}. \tag{5.26}$$

We claim that under the assumption that the Fourier expansion converges to f, the series (5.26) is indeed a classical solution. To verify this statement we assume

$$f(x) = \frac{4}{\pi} \sum_{n=1}^{\infty} \frac{(-1)^{n+1}}{(2n - 1)^2} \sin[(2n - 1)x]. \tag{5.27}$$

The functions obtained by summing only finitely many terms in the Fourier series are depicted in Figure 5.2.

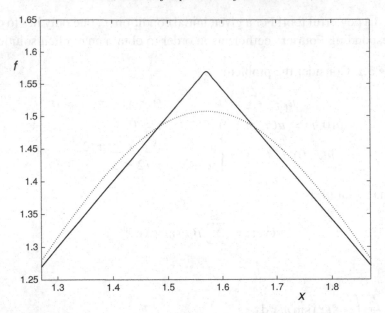

Figure 5.2 The function obtained by summing 100 terms (solid line) and 7 terms (dotted line) for the Fourier expansion of $f(x)$ (5.27). We concentrate the region near the point $x = \pi/2$, where $f(x)$ is not differentiable. If we take just a few terms in the expansion, the actual singularity is smoothed out.

Since

$$|u_n(x, t)| = \frac{4}{\pi} \left| \frac{(-1)^{n+1}}{(2n - 1)^2} \sin[(2n - 1)x] e^{-(2n-1)^2 t} \right| \le \frac{4}{\pi(2n - 1)^2},$$

it follows by the Weierstrass M-test that the series (5.26) converges uniformly to a continuous function in the region

$$\{(x, t) \mid 0 \le x \le \pi, \, t \ge 0\}.$$

Substituting u into the initial and boundary conditions, and using the assumption that the Fourier expansion of f converges to f, we obtain that these conditions are indeed satisfied.

It remains to show that the series (5.26) is differentiable with respect to t, twice differentiable with respect to x, and satisfies the heat equation in the domain

$$D := \{(x, t) \mid 0 < x < \pi, \, t > 0\}.$$

Fix $\varepsilon > 0$. We first show that the series (5.26) is differentiable with respect to t, twice differentiable with respect to x, and satisfies the heat equation in the subdomain

$$D_\varepsilon := \{(x, t) \mid 0 < x < \pi, \, t > \varepsilon\}.$$

For instance, we show that (5.26) can be differentiated with respect to t for $t > \varepsilon$. Indeed, by differentiating $u_n(x, t)$ with respect to t, we obtain that

$$|(u_n(x, t))_t| = \left| \frac{4(2n-1)^2}{\pi(2n-1)^2} \sin[(2n-1)x] e^{-(2n-1)^2 t} \right| \le \frac{4}{\pi} e^{-(2n-1)^2 \varepsilon}.$$

Since the series $(4/\pi) \sum e^{-(2n-1)^2 \varepsilon}$ converges, it follows by the Weierstrass M-test that for every $\varepsilon > 0$ the series $\sum (u_n(x, t))_t$ converges to u_t uniformly in D_ε. Similarly, it can be shown that u has a continuous second-order derivative with respect to x that is obtained by two term-by-term differentiations. Hence,

$$u_t - u_{xx} = \sum_{n=1}^{\infty} (u_n)_t - \sum_{n=1}^{\infty} (u_n)_{xx} = \sum_{n=1}^{\infty} \{(u_n)_t - (u_n)_{xx}\} = 0,$$

where in the last step we used the property that each separated solution $u_n(x, t)$ is a solution of the heat equation. Thus, u is a solution of the PDE in D_ε. Since ε is an arbitrary positive number, it follows that u is a solution of the heat equation in the domain D.

Because the general term u_n decays exponentially in D_ε, it is possible to differentiate (5.26) term-by-term to any order with respect to x and t. The corresponding series converges uniformly in D_ε to the appropriate derivative. Note that k differentiations with respect to x and ℓ differentiations with respect to t contribute to the general term of the series a factor of order $O(n^{k+2\ell})$, but because of the exponential term, the corresponding series is converging.

The important conclusion is that even for nonsmooth initial condition f, the solution has infinitely many derivatives with respect to x and t and it is smooth in the strip D. The nonsmoothness of the initial data disappears immediately (see Figure 5.3). This smoothing effect is known to hold also in more general parabolic problems, in contrast with the hyperbolic case, where singularities propagate along characteristics and in general persist over time.

Another qualitative result that can be deduced from our representation, concerns the large time behavior of the solution (i.e. the behavior in the limit $t \to \infty$). This behavior is directly influenced by the boundary conditions. In particular, it depends on the minimal eigenvalue of the corresponding eigenvalue problem. In our case, all the eigenvalues are strictly positive, and from (5.17) and the uniform convergence in D_ε it follows that

$$\lim_{t \to \infty} u(x, t) = 0 \qquad \forall\, 0 \le x \le L.$$

Hence the temperature along the rod converges to the temperature that is imposed at the end points.

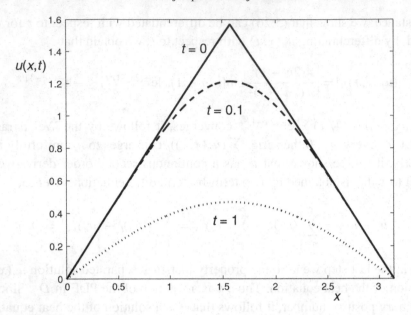

Figure 5.3 The function $u(x, t)$ of (5.26) for $t = 0$, $t = 0.1$, and $t = 1$. Notice that the singularity at $t = 0$ is quickly smoothed out. The graphs were generated with 200 terms in the Fourier expansion. Actually just three or four terms are needed to capture u correctly even for $t = 0.1$.

We conclude this section by mentioning other boundary conditions that appear frequently in heat conduction problems (see Chapter 1). Specifically, we distinguish between two types of boundary conditions:

(a) Separated boundary conditions These boundary conditions can be written as

$$B_0[u] = \alpha u(0, t) + \beta u_x(0, t) = 0, \quad B_L[u] = \gamma u(L, t) + \delta u_x(L, t) = 0 \quad t \geq 0,$$

where

$$\alpha, \beta, \gamma, \delta \in \mathbb{R}, \quad |\alpha| + |\beta| > 0, \quad |\gamma| + |\delta| > 0.$$

This type of boundary condition includes for $\alpha = \gamma = 1$, $\beta = \delta = 0$ the Dirichlet boundary condition

$$u(0, t) = u(L, t) = 0 \quad t \geq 0,$$

which is also called a *boundary condition of the first kind*. Also, for $\alpha = \gamma = 0$, $\beta = \delta = 1$ we obtain

$$u_x(0, t) = u_x(L, t) = 0 \quad t \geq 0,$$

which is called the Neumann condition or a *boundary condition of the second kind*. Recall that the physical interpretation of the Neumann boundary condition for heat problems is that there is no heat flow through the boundary. In our case it means that the rod is insulated. If we impose a Dirichlet condition at one end, and a Neumann condition at the other hand, then the boundary condition is called *mixed*. In the general case, where α, β, γ, δ are nonzero, the boundary condition is called a *boundary condition of the third kind* (or the *Robin condition*). The physical interpretation is that the heat flow at the boundary depends linearly on the temperature.

(b) Periodic boundary condition This boundary condition is imposed for example in the case of heat evolution along a circular wire of length L. Clearly, in this case the temperature $u(x, t)$ and all its derivatives are periodic (as a function of x) with a period L. In addition u satisfies the heat equation on $(0, L)$. The boundary conditions for this problem are

$$u(0, t) = u(L, t), \qquad u_x(0, t) = u_x(L, t) \qquad \forall t \geq 0.$$

The periodicity of all the higher-order derivatives follows from the PDE and the boundary conditions presented above.

5.3 Separation of variables for the wave equation

We now apply the method of separation of variables to solve the problem of a vibrating string without external forces and with two clamped but free ends. Let $u(x, t)$ be the amplitude of the string at the point x and time t, and let f and g be the amplitude and the velocity of the string at time $t = 0$ (see the discussion in Chapter 1 and, in particular, Figure 1.1). We need to solve the problem

$$u_{tt} - c^2 u_{xx} = 0 \qquad 0 < x < L, \quad t > 0, \tag{5.28}$$
$$u_x(0, t) = u_x(L, t) = 0 \qquad t \geq 0, \tag{5.29}$$
$$u(x, 0) = f(x) \quad 0 \leq x \leq L, \tag{5.30}$$
$$u_t(x, 0) = g(x) \quad 0 \leq x \leq L, \tag{5.31}$$

where f, g are given functions and c is a positive constant. The compatibility conditions are given by

$$f'(0) = f'(L) = g'(0) = g'(L) = 0.$$

The problem (5.28)–(5.31) is a linear homogeneous initial boundary value problem. As mentioned above, the conditions (5.29) are called *Neumann boundary conditions*.

Recall that at the first stage of the method, we compute nontrivial separated solutions of the PDE (5.28), i.e. solutions of the form

$$u(x, t) = X(x)T(t), \tag{5.32}$$

that also satisfy the boundary conditions (5.29). Here, as usual, X, T are functions of the variables x and t respectively. At this stage, we do not take into account the initial conditions (5.30)–(5.31).

Differentiating the separated solution (5.32) twice in x and twice in t, and then substituting these derivatives into the wave equation, we infer

$$XT_{tt} = c^2 X_{xx} T.$$

By separating the variables, we see that

$$\frac{T_{tt}}{c^2 T} = \frac{X_{xx}}{X}. \tag{5.33}$$

It follows that there exists a constant λ such that

$$\frac{T_{tt}}{c^2 T} = \frac{X_{xx}}{X} = -\lambda. \tag{5.34}$$

Equation (5.34) implies

$$\frac{d^2 X}{dx^2} = -\lambda X \qquad 0 < x < L, \tag{5.35}$$

$$\frac{d^2 T}{dt^2} = -\lambda c^2 T \quad t > 0. \tag{5.36}$$

The boundary conditions (5.29) for u imply

$$u_x(0, t) = \frac{dX}{dx}(0)T(t) = 0, \quad u_x(L, t) = \frac{dX}{dx}(L)T(t) = 0.$$

Since u is nontrivial it follows that

$$\frac{dX}{dx}(0) = \frac{dX}{dx}(L) = 0.$$

Therefore, the function X should be a solution of the eigenvalue problem

$$\frac{d^2 X}{dx^2} + \lambda X = 0 \qquad 0 < x < L, \tag{5.37}$$

$$\frac{dX}{dx}(0) = \frac{dX}{dx}(L) = 0. \tag{5.38}$$

This eigenvalue problem is also called the Neumann problem.

We have already written the general solution of the ODE (5.37):

1. if $\lambda < 0$, then $X(x) = \alpha \cosh(\sqrt{-\lambda}x) + \beta \sinh(\sqrt{-\lambda}x)$,
2. if $\lambda = 0$, then $X(x) = \alpha + \beta x$,
3. if $\lambda > 0$, then $X(x) = \alpha \cos(\sqrt{\lambda}x) + \beta \sin(\sqrt{\lambda}x)$,

where α, β are arbitrary real numbers.

Negative eigenvalue ($\lambda < 0$) The first boundary condition $(dX/dx)(0) = 0$ implies that $\beta = 0$. Then $(dX/dx)(L) = 0$ implies that $\sinh(\sqrt{-\lambda}L) = 0$. Therefore, $X(x) \equiv 0$ and the eigenvalue problem (5.37)–(5.38) does not admit negative eigenvalues.

Zero eigenvalue ($\lambda = 0$) The general solution is a linear function $X(x) = \alpha + \beta x$. Substituting this solution into the boundary conditions (5.38) implies that $\lambda = 0$ is an eigenvalue with a unique eigenfunction $X_0(x) \equiv 1$ (the eigenfunction is unique up to a multiplicative factor).

Positive eigenvalue ($\lambda > 0$) The general solution for $\lambda > 0$ has the form

$$X(x) = \alpha \cos(\sqrt{\lambda}x) + \beta \sin(\sqrt{\lambda}x). \tag{5.39}$$

Substituting it in $(dX/dx)(0) = 0$, we obtain $\beta = 0$. The boundary condition $(dX/dx)(L) = 0$ implies now that $\sin(\sqrt{\lambda}L) = 0$. Thus $\sqrt{\lambda}L = n\pi$, where $n \in \mathbb{N}$. Consequently, $\lambda > 0$ is an eigenvalue if and only if:

$$\lambda = \left(\frac{n\pi}{L}\right)^2 \quad n = 1, 2, 3, \dots.$$

The associated eigenfunction is

$$X(x) = \cos\frac{n\pi x}{L},$$

and it is uniquely determined up to a multiplicative factor.

Therefore, the solution of the eigenvalue problem (5.37)–(5.38) is an infinite sequence of nonnegative simple eigenvalues and their associated eigenfunctions. We use the convenient notation:

$$X_n(x) = \cos\frac{n\pi x}{L}, \quad \lambda_n = \left(\frac{n\pi}{L}\right)^2 \quad n = 0, 1, 2, \dots.$$

Consider now the ODE (5.36) for $\lambda = \lambda_n$. The solutions are

$$T_0(t) = \gamma_0 + \delta_0 t, \tag{5.40}$$

$$T_n(t) = \gamma_n \cos(\sqrt{\lambda_n c^2}\, t) + \delta_n \sin(\sqrt{\lambda_n c^2}\, t) \quad n = 1, 2, 3, \dots. \tag{5.41}$$

Thus, the product solutions of the initial boundary value problem are given by

$$u_0(x, t) = X_0(x)T_0(t) = \frac{A_0 + B_0 t}{2}, \tag{5.42}$$

$$u_n(x, t) = X_n(x)T_n(t) = \cos\frac{n\pi x}{L}\left(A_n \cos\frac{c\pi nt}{L} + B_n \sin\frac{c\pi nt}{L}\right), \quad n = 1, 2, 3, \dots. \tag{5.43}$$

Applying the (generalized) superposition principle, the expression

$$u(x, t) = \frac{A_0 + B_0 t}{2} + \sum_{n=1}^{\infty} \left(A_n \cos \frac{c\pi nt}{L} + B_n \sin \frac{c\pi nt}{L} \right) \cos \frac{n\pi x}{L} \qquad (5.44)$$

is a (generalized, or at least formal) solution of the problem (5.28)–(5.31). In Exercise 5.2 we show that the solution (5.44) can be represented as a superposition of forward and backward waves. In other words, solution (5.44) is also a generalized solution of the wave equation in the sense defined in Chapter 4.

It remains to find the coefficients A_n, B_n in solution (5.44). Here we use the initial conditions. Assume that the initial data f, g can be expanded into generalized Fourier series with respect to the sequence of the eigenfunctions of the problem, and that these series are uniformly converging. That is,

$$f(x) = \frac{a_0}{2} + \sum_{n=1}^{\infty} a_n \cos \frac{n\pi x}{L}, \qquad (5.45)$$

$$g(x) = \frac{\tilde{a}_0}{2} + \sum_{n=1}^{\infty} \tilde{a}_n \cos \frac{n\pi x}{L}. \qquad (5.46)$$

Again, the (generalized) Fourier coefficients of f and g can easily be determined; for $m \geq 0$, we multiply (5.45) by the eigenfunction $\cos(m\pi x/L)$, and then we integrate over $[0, L]$. We obtain

$$\int_0^L \cos \frac{m\pi x}{L} f(x) \, dx = \frac{a_0}{2} \int_0^L \cos \frac{m\pi x}{L} \, dx + \sum_{n=1}^{\infty} a_n \int_0^L \cos \frac{m\pi x}{L} \cos \frac{n\pi x}{L} \, dx. \qquad (5.47)$$

It is easily checked (see Section A.1) that

$$\int_0^L \cos \frac{m\pi x}{L} \cos \frac{n\pi x}{L} \, dx = \begin{cases} 0 & m \neq n, \\ L/2 & m = n \neq 0, \\ L & m = n = 0. \end{cases} \qquad (5.48)$$

Therefore, the Fourier coefficients of f with respect to the system of eigenfunctions are

$$a_0 = 2 \frac{\int_0^L f(x) \, dx}{\int_0^L 1 \, dx} = \frac{2}{L} \int_0^L f(x) \, dx, \qquad (5.49)$$

$$a_m = \frac{\int_0^L \cos(m\pi x/L) f(x) \, dx}{\int_0^L \cos^2(m\pi x/L) \, dx} = \frac{2}{L} \int_0^L \cos \frac{m\pi x}{L} f(x) \, dx \quad m = 1, 2, \ldots. \qquad (5.50)$$

The Fourier coefficients $\tilde{a}_n$ of g can be computed similarly. Substituting $t = 0$ into (5.44), and assuming that the corresponding series converges uniformly, we obtain

$$u(x, 0) = \frac{A_0}{2} + \sum_{n=1}^{\infty} A_n \cos \frac{n\pi x}{L}$$

$$= f(x) = \frac{a_0}{2} + \sum_{n=1}^{\infty} a_n \cos \frac{n\pi x}{L}.$$

Recall that the (generalized) Fourier coefficients are uniquely determined, and hence $A_n = a_n$ for all $n \geq 0$. In order to compute B_n, we differentiate (5.44) formally (term-by-term) with respect to t and then substitute $t = 0$. We have

$$u_t(x, 0) = \frac{B_0}{2} + \sum_{n=1}^{\infty} B_n \frac{c\pi n}{L} \cos \frac{n\pi x}{L}$$

$$= g(x) = \frac{\tilde{a}_0}{2} + \sum_{n=1}^{\infty} \tilde{a}_n \cos \frac{n\pi x}{L}.$$

Therefore, $B_n = \tilde{a}_n L / c\pi n$ for all $n \geq 1$. Similarly, $B_0 = \tilde{a}_0$. Thus, the problem is formally solved. The uniqueness issue will be discussed at the end of this chapter.

There is a significant difference between the solution (5.17) of the heat problem and the formal solution (5.44). Each term of the solution (5.17) of the heat equation has a decaying exponential factor which is responsible for the smoothing effect for $t > 0$. In (5.44) we have instead a (nondecaying) trigonometric factor. This is related to the fact that hyperbolic equations preserve the singularities of the given data since the rate of the decay of the generalized Fourier coefficients to zero usually depends on the smoothness of the given function (under the assumption that this function satisfies the prescribed boundary conditions). The precise decay rate of the Fourier coefficients is provided for the classical Fourier system by the general theory of Fourier analysis [13].

Example 5.2 Solve the problem

$$\begin{array}{ll} u_{tt} - 4u_{xx} = 0 & 0 < x < 1, \, t > 0, \\ u_x(0, t) = u_x(1, t) = 0 & t \geq 0, \\ u(x, 0) = f(x) = \cos^2 \pi x & 0 \leq x \leq 1, \\ u_t(x, 0) = g(x) = \sin^2 \pi x \cos \pi x & 0 \leq x \leq 1. \end{array} \tag{5.51}$$

The solution of (5.51) was shown to have the form

$$u(x, t) = \frac{A_0 + B_0 t}{2} + \sum_{n=1}^{\infty} (A_n \cos 2n\pi t + B_n \sin 2n\pi t) \cos n\pi x. \tag{5.52}$$

Substituting f into (5.52) implies

$$u(x, 0) = \frac{A_0}{2} + \sum_{n=1}^{\infty} A_n \cos n\pi x = \cos^2 \pi x. \qquad (5.53)$$

The Fourier expansion of f is easily obtained using the trigonometric identity $\cos^2 \pi x = \frac{1}{2} + \frac{1}{2} \cos 2\pi x$. Since the Fourier coefficients are uniquely determined, it follows that

$$A_0 = 1, \quad A_2 = \frac{1}{2}, \quad A_n = 0 \quad \forall n \neq 0, 2. \qquad (5.54)$$

By differentiating the solution with respect to t, and substituting $u_t(x, 0)$ into the second initial condition, we obtain

$$u_t(x, 0) = \frac{B_0}{2} + \sum_{n=1}^{\infty} B_n 2n\pi \cos n\pi x = \sin^2 \pi x \cos \pi x. \qquad (5.55)$$

Similarly, the Fourier expansion of g is obtained using the trigonometric identity $\sin^2 \pi x \cos \pi x = \frac{1}{4} \cos \pi x - \frac{1}{4} \cos 3\pi x$. From the uniqueness of the expansion it follows that

$$B_1 = \frac{1}{8\pi}, \quad B_3 = -\frac{1}{24\pi}, \quad B_n = 0 \quad \forall n \neq 1, 3.$$

Therefore,

$$u(x, t) = \frac{1}{2} + \frac{1}{8\pi} \sin 2\pi t \cos \pi x + \frac{1}{2} \cos 4\pi t \cos 2\pi x - \frac{1}{24\pi} \sin 6\pi t \cos 3\pi x. \qquad (5.56)$$

Since (5.56) contains only a finite number of (smooth) terms, it is verified directly that u is a classical solution of the problem.

5.4 Separation of variables for nonhomogeneous equations

It is possible to upgrade the method of separation of variables to a method for solving nonhomogeneous PDEs. This technique is called also the *method of eigenfunction expansion*. For example, consider the problem

$$\begin{array}{ll} u_{tt} - u_{xx} = \cos 2\pi x \cos 2\pi t & 0 < x < 1, \, t > 0, \\ u_x(0, t) = u_x(1, t) = 0 & t \geq 0, \\ u(x, 0) = f(x) = \cos^2 \pi x & 0 \leq x \leq 1, \\ u_t(x, 0) = g(x) = 2 \cos 2\pi x & 0 \leq x \leq 1. \end{array} \qquad (5.57)$$

In the previous section we found the system of all eigenfunctions and the corresponding eigenvalues of the homogeneous problem. They are

$$X_n(x) = \cos n\pi x, \quad \lambda_n = (n\pi)^2 \qquad n = 0, 1, 2, \dots.$$

Recall Fourier's claim (to be justified in the next chapter) that any reasonable function satisfying the boundary conditions can be uniquely expanded into (generalized) Fourier series with respect to the system of the eigenfunctions of the problem. Since the solution $u(x, t)$ of the problem (5.57) is a twice differentiable function satisfying the boundary conditions, it follows that for a fixed t the solution u can be represented as

$$u(x, t) = \frac{1}{2}T_0(t) + \sum_{n=1}^{\infty} T_n(t) \cos n\pi x, \qquad (5.58)$$

where $T_n(t)$ are the (time dependent) Fourier coefficients of the function $u(\cdot, t)$. Hence, we need to find these coefficients.

Substituting (5.58) into the wave equation (5.57) and differentiating the series term-by-term implies that

$$\frac{1}{2}T_0'' + \sum_{n=1}^{\infty} (T_n'' + n^2\pi^2 T_n) \cos n\pi x = \cos 2\pi t \cos 2\pi x. \qquad (5.59)$$

Note that in the current example, the right hand side of the equation is already given in the form of a Fourier series. The uniqueness of the Fourier expansion implies that the Fourier coefficients of the series of the left hand side of (5.59) are equal to the Fourier coefficients of the series of the right hand side. In particular, for $n = 0$ we obtain the ODE:

$$T_0'' = 0, \qquad (5.60)$$

whose general solution is $T_0(t) = A_0 + B_0 t$. Similarly we obtain for $n = 2$

$$T_2'' + 4\pi^2 T_2 = \cos 2\pi t. \qquad (5.61)$$

The general solution of this linear nonhomogeneous second-order ODE is

$$T_2(t) = A_2 \cos 2\pi t + B_2 \sin 2\pi t + \frac{t}{4\pi} \sin 2\pi t.$$

For $n \neq 0, 2$, we have

$$T_n'' + n^2\pi^2 T_n = 0 \quad \forall n \neq 0, 2. \qquad (5.62)$$

The solution is $T_n(t) = A_n \cos n\pi t + B_n \sin n\pi t$ for all $n \neq 0, 2$. Substituting the solutions of (5.60), (5.61), and (5.62) into (5.58) implies that the solution of the problem is of the form

$$u(x, t) = \frac{A_0 + B_0 t}{2} + \frac{t}{4\pi} \sin 2\pi t \cos 2\pi x$$

$$+ \sum_{n=1}^{\infty} (A_n \cos n\pi t + B_n \sin n\pi t) \cos n\pi x. \tag{5.63}$$

Substituting (5.63) into the first initial condition (5.57), we get

$$u(x, 0) = \frac{A_0}{2} + \sum_{n=1}^{\infty} A_n \cos n\pi x = \cos^2 \pi x = \frac{1}{2} + \frac{1}{2} \cos 2\pi x,$$

therefore,

$$A_0 = 1, \quad A_2 = \frac{1}{2}, \quad A_n = 0 \ \forall n \neq 0, 2.$$

By differentiating (term-by-term) the solution u with respect to t and substituting $u_t(x, 0)$ into the second initial condition of (5.57), we find

$$u_t(x, 0) = \frac{B_0}{2} + \sum_{n=1}^{\infty} n\pi B_n \cos n\pi x = 2 \cos 2\pi x,$$

Hence,

$$B_2 = \frac{1}{\pi}, \quad B_n = 0 \ \forall n \neq 2.$$

Finally

$$u(x, t) = \frac{1}{2} + \left(\frac{1}{2} \cos 2\pi t + \frac{t + 4}{4\pi} \sin 2\pi t \right) \cos 2\pi x.$$

It is clear that this solution is classical, since the (generalized) Fourier series of the solution has only a finite number of nonzero smooth terms, and therefore all the formal operations are justified. Note that the amplitude of the vibrating string grows linearly in t and it is unbounded as $t \to \infty$. This remarkable phenomenon will be discusses further in Chapter 6.

5.5 The energy method and uniqueness

The energy method is a fundamental tool in the theory of PDEs. One of its main applications is in proving the uniqueness of the solution of initial boundary value problems. The method is based on the physical principle of energy conservation, although in some applications the object we refer to mathematically as an 'energy' is not necessarily the actual energy of a physical system.

Recall that in order to prove the uniqueness of solutions for a linear differential problem, it is enough to show that the solution of the corresponding homogeneous PDE with homogeneous initial and boundary conditions is necessarily the zero solution. This basic principle has already been used in Chapter 4 and will be demonstrated again below.

Let us outline the energy method. For certain homogeneous problems it is possible to define an energy integral that is nonnegative and is a nonincreasing function of the time t. In addition, for $t = 0$ the energy is zero and therefore, the energy is zero for all $t \geq 0$. Due to the positivity of the energy, and the zero initial and boundary conditions it will follow that the solution is zero.

We demonstrate the energy method for the problems that have been studied in the present chapter.

Example 5.3 Consider the Neumann problem for the vibrating string

$$u_{tt} - c^2 u_{xx} = F(x, t) \quad 0 < x < L, \ t > 0, \tag{5.64}$$

$$u_x(0, t) = a(t), \quad u_x(L, t) = b(t) \qquad t \geq 0, \tag{5.65}$$

$$u(x, 0) = f(x) \qquad 0 \leq x \leq L, \tag{5.66}$$

$$u_t(x, 0) = g(x) \qquad 0 \leq x \leq L. \tag{5.67}$$

Let u_1, u_2 be two solutions of the problem. By the superposition principle, the function $w := u_1 - u_2$ is a solution of the problem

$$w_{tt} - c^2 w_{xx} = 0 \quad 0 < x < L, \ t > 0, \tag{5.68}$$

$$w_x(0, t) = 0, \quad w_x(L, t) = 0 \quad t \geq 0, \tag{5.69}$$

$$w(x, 0) = 0 \quad 0 \leq x \leq L, \tag{5.70}$$

$$w_t(x, 0) = 0 \quad 0 \leq x \leq L. \tag{5.71}$$

Define the total energy of the solution w at time t as

$$E(t) := \frac{1}{2} \int_0^L (w_t^2 + c^2 w_x^2) \, \mathrm{d}x. \tag{5.72}$$

The first term represents the total kinetic energy of the string, while the second term is the total potential energy. Clearly, E' is given by

$$E'(t) = \frac{\mathrm{d}}{\mathrm{d}t} \left[\frac{1}{2} \int_0^L (w_t^2 + c^2 w_x^2) \, \mathrm{d}x \right] = \int_0^L (w_t w_{tt} + c^2 w_x w_{xt}) \, \mathrm{d}x. \tag{5.73}$$

But

$$c^2 w_x w_{xt} = c^2 \left[\frac{\partial}{\partial x}(w_x w_t) - w_{xx} w_t \right] = c^2 \frac{\partial}{\partial x}(w_x w_t) - w_{tt} w_t.$$

Substituting this identity into (5.73) and using the fundamental theorem of calculus, we have

$$E'(t) = c^2 \int_0^L \frac{\partial}{\partial x}(w_x w_t) \, dx = c^2 (w_x w_t)|_0^L. \qquad (5.74)$$

The boundary condition (5.69) implies that $E'(t) = 0$, hence, $E(t) = $ constant and the energy is conserved.

On the other hand, since for $t = 0$ we have $w(x, 0) = 0$, it follows that $w_x(x, 0) = 0$. Moreover, we have also $w_t(x, 0) = 0$. Therefore, the energy at time $t = 0$ is zero. Thus, $E(t) \equiv 0$.

Since $e(x, t) := w_t^2 + c^2 w_x^2 \geq 0$, and since its integral over $[0, L]$ is zero, it follows that $w_t^2 + c^2 w_x^2 \equiv 0$, which implies that $w_t(x, t) = w_x(x, t) \equiv 0$. Consequently, $w(x, t) \equiv$ constant. By the initial conditions $w(x, 0) = 0$, hence $w(x, t) \equiv 0$. This completes the proof of the uniqueness of the problem (5.64)–(5.67).

Example 5.4 Let us modify the previous problem a little, and instead of the (non-homogeneous) Neumann problem, consider the Dirichlet boundary conditions:

$$u(0, t) = a(t), \quad u(L, t) = b(t) \quad t \geq 0.$$

We use the same energy integral and follow the same steps. We obtain for the function w

$$E'(t) = c^2 (w_x w_t)|_0^L. \qquad (5.75)$$

Since $w(0, t) = w(L, t) = 0$, it follows that $w_t(0, t) = w_t(L, t) = 0$; therefore, $E'(t) = 0$ and in this case too the energy is conserved. The rest of the proof is exactly the same as in the previous example.

Example 5.5 The energy method can also be applied to heat conduction problems. Consider the Dirichlet problem

$$u_t - k u_{xx} = F(x, t) \quad 0 < x < L, \ t > 0, \qquad (5.76)$$

$$u(0, t) = a(t), \quad u(L, t) = b(t) \qquad t \geq 0, \qquad (5.77)$$

$$u(x, 0) = f(x) \qquad 0 \leq x \leq L. \qquad (5.78)$$

As we explained above, we need to prove that if w is a solution of the homogeneous problem with zero initial and boundary conditions, then $w = 0$. In the present case, we define the energy to be:

$$E(t) := \frac{1}{2} \int_0^L w^2 \, dx. \qquad (5.79)$$

The time derivative E' is given by

$$E'(t) = \frac{\mathrm{d}}{\mathrm{d}t}\left(\frac{1}{2}\int_0^L w^2\,dx\right) = \int_0^L w w_t\,dx = \int_0^L k w w_{xx}\,dx. \qquad (5.80)$$

Integrating by parts and substituting the boundary conditions, we have

$$E'(t) = k w w_x |_0^L - \int_0^L k(w_x)^2\,dx = -\int_0^L k(w_x)^2\,dx \le 0,$$

therefore, the energy is not increasing. Since $E(0) = 0$ and $E(t) \ge 0$, it follows that $E \equiv 0$. Consequently, for all $t \ge 0$ we have $w(\cdot, t) \equiv 0$ and the uniqueness is proved. The same proof can also be used for the Neumann problem and even for the boundary condition of the third kind:

$$u(0, t) - \alpha u_x(0, t) = a(t). \quad u(L, t) + \beta u_x(L, t) = b(t) \quad t \ge 0,$$

provided that $\alpha, \beta \ge 0$.

5.6 Further applications of the heat equation

We have seen that the underlying property of the wave equation is to propagate waves, while the heat equation smoothes out oscillations and discontinuities. In this section we shall consider two specific applications of the heat equation that concern signal propagation. In the first application we shall show that a diffusion mechanism can still transmit (to some extent) oscillatory data. In fact, diffusion effects play an important role in one of the most important communication systems. In the second example the goal will be to use the smoothing property of the heat equation to dampen oscillations in the data.

5.6.1 The cable equation

The great success of the telegraph prompted businessmen and governments to lay an underwater cable between France and Britain in 1850. It was realized, however, that the transmission rate through this cable was very low. The British scientist William Thomson (1824–1907) sought to explain this phenomenon. His mathematical model showed that the cable's electrical capacity has a major effect on signal transmission. We shall derive the equation for signal transmission in a cable, solve it, and then explain Thomson's analysis.

A cross section of the cable is shown in Figure 5.4. The cable is modeled as a system of outer and inner conductors separated by an insulating layer. To simplify the analysis we shall consider a two-dimensional model, using x to denote the

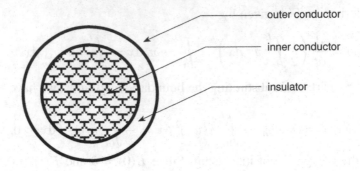

Figure 5.4 The cross section of the cable.

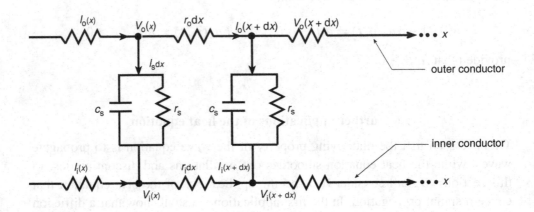

Figure 5.5 A longitudinal cross section of the cable.

longitudinal direction. A small segment of the longitudinal cross section is shown in Figure 5.5. In this segment we see the local resistivity (r_i in the inner conductor, and r_o in the outer conductor), while the insulator is modeled by a capacitor C_s and a resistor r_s in parallel. The transversal current in a horizontal element of length dx is $I_s dx$. Ohm's law for the segment $(x, x + dx)$ implies

$$V_i(x + dx) - V_i(x) = -I_i(x)r_i dx, \quad V_o(x + dx) - V_o(x) = -I_o(x)r_o dx. \quad (5.81)$$

In the limit $dx \to 0$ this becomes

$$\frac{\partial V_i}{\partial x} = -r_i I_i(x), \quad \frac{\partial V_o}{\partial x} = -r_o I_o(x). \quad (5.82)$$

Having balanced the voltage drop in the longitudinal direction, we proceed to write the current conservation equation (Kirchhoff's law). We have

$$I_i(x + dx) = I_i(x) + I_s(x)dx, \quad I_o(x + dx) = I_o(x) - I_s(x)dx. \quad (5.83)$$

Again letting $dx \to 0$ we obtain

$$I_s = \frac{\partial I_i}{\partial x} = -\frac{\partial I_o}{\partial x}. \tag{5.84}$$

Introducing the transinsulator potential $V = V_i - V_o$, we conclude from (5.82) that

$$-\frac{\partial V}{\partial x} = r_i I_i - r_o I_o. \tag{5.85}$$

Differentiating (5.85) by x, and using (5.84) we get

$$-I_s = \frac{1}{r_i + r_o} \frac{\partial^2 V}{\partial x^2}. \tag{5.86}$$

It remains to understand the current I_s. The contribution of the resistor r_s is $-(1/r_s)V$. The current through a capacitor is given by [10] $-C_s \partial V / \partial t$, where C_s denotes the capacitance. Therefore we finally obtain the (passive) *cable equation*

$$\frac{\partial V}{\partial t} = D \frac{\partial^2 V}{\partial x^2} - \beta V \quad D = \frac{1}{C_s(r_i + r_e)}, \quad \beta = \frac{1}{r_s C_s}. \tag{5.87}$$

Note that the capacitor gave rise to a diffusion-like term in the transport equation.

Equation (5.87) can be solved in a finite x interval by the separation of variables method (see, for example, Exercise 5.10). In order to understand its use in communication, we shall assume that the transmitter is located at $x = 0$, and the receiver is at an arbitrary location x up the line. Therefore we solve the cable equation for a semi-infinite interval. To fix ideas, we formulate the following problem:

$$V_t = DV_{xx} - \beta V \quad 0 < x < \infty, \; -\infty < t < \infty, \tag{5.88}$$
$$V(0, t) = A \cos \omega t \quad -\infty < t < \infty, \tag{5.89}$$
$$V(x, t) \to 0 \quad x \to \infty. \tag{5.90}$$

The problem (5.88)–(5.90) can be solved by a variant of the separation of variables method. Our motivation is to seek a solution that will have propagation and oscillation properties as in a wave equation, but also decay properties that are typical of a heat equation. Therefore we seek a solution of the form

$$V(x, t) = Av(x)\cos(\omega t - kx).$$
(5.91)

Substituting (5.91) into (5.88), and defining $\phi := \omega t - kx$, we get

$$-\omega v \sin \phi = D \left(v_{xx} \cos \phi + 2k v_x \sin \phi - k^2 v \cos \phi \right) - \beta v \cos \phi.$$

We first equate the coefficients of the $\cos \phi$ term. This implies

$$v_{xx} - \left(k^2 + \frac{\beta}{D} \right) v = 0.$$

The boundary conditions (5.89)–(5.90) imply $v(0) = 1, v(\infty) = 0$. Therefore $v(x) = \exp[-\sqrt{k^2 + (\beta/D)}\, x]$. Equating now the coefficients of the term $\sin \phi$, using the solution that was found for v, we find that (5.91) is indeed the desired solution if k, ω, and D satisfy the *dispersion relation*

$$\omega = 2Dk\sqrt{k^2 + \frac{\beta}{D}}\,.$$
(5.92)

We now analyze this solution in light of the cable transmission issue. The parameter β represents the loss of energy due to the transinsulator resistivity. Increasing the resistivity will decrease β. We therefore proceed to consider an ideal situation where $\beta = 0$. In this case the solution (5.91) and the dispersion relation (5.92) become

$$V(x, t) = Ae^{-kx} \cos(\omega t - kx), \quad \omega = 2Dk^2.$$
(5.93)

The frequency ω can be interpreted as the rate of transmission. Similarly, we interpret $1/k$ as the distance L between the transmitter and the receiver. Therefore $\omega = 2DL^{-2}$. This formula enabled Thomson to predict that with the parameters of the materials used for the cable, i.e. C_s, r_i, r_o that determine D, and in light of the distance L, the transmission rate would be far below the expected rate. His prediction was indeed fulfilled. Following the great success of his mathematical analysis, Thomson was asked to consult in the next major attempt to lay an underwater communication cable, this time in 1865 between Britain and the USA. The great improvement in production control allowed the manufacture of a high quality cable, and the enterprise met with high technical and financial success. To honor him for his contributions to the transatlantic cable, Thomson was created Lord Kelvin in 1866.

Interest in the cable equation was renewed in the middle of the twentieth century when it was discovered to be an adequate model for signal transmission in biological cells in general, and for neurons in particular. The insulating layer in this case is the cell's membrane. The currents consist of ions, mainly potassium and sodium ions.

In the biological applications, however, one needs to replace the passive resistor r_s with a nonlinear electrical element. The reason is that the current through the cell's membrane flows in special channels with a complex gate mechanism that was deciphered by Hodgkin and Huxley [8]. Moreover, we need to supplement in this case the resulting active cable equation with further dynamical rules for the channel gates.

As another example of the energy method we shall prove that the solution we found for the cable equation is unique. More precisely, we shall prove that the problem (5.88)–(5.90) has a unique solution in the class of functions for which the energy

$$E_w(t) = E(t) := \frac{1}{2} \int_0^\infty [w(x, t)]^2 \, dx \qquad (5.94)$$

is bounded. Namely, we assume that for each solution w of the problem there exists a constant $M_w > 0$ such that $E_w(t) \leq M_w$. We need to prove that if w is a solution of the homogeneous problem with zero boundary conditions, then $w = 0$. We obtain as for the heat equation that

$$E'(t) = \frac{d}{dt}\left(\frac{1}{2}\int_0^\infty w^2 \, dx\right) = \int_0^\infty ww_t \, dx = \int_0^\infty (Dww_{xx} - \beta w^2) \, dx, \quad (5.95)$$

where we have assumed that all the above integrals are finite. Integrating by parts and substituting the boundary conditions, we have

$$E'(t) \leq -\int_0^\infty D(w_x)^2 \, dx - \beta E(t) \leq -\beta E(t).$$

Fixing $T \in \mathbb{R}$ and integrating the above differential inequality from t to $T > t$, we obtain the estimate

$$E(T) \leq E(t)e^{-\beta(T-t)} \leq Me^{-\beta T}e^{\beta t}.$$

Letting $t \to -\infty$ it follows that $E(T) = 0$. Therefore, $E \equiv 0$ which implies that $w = 0$.

5.6.2 Wine cellars

Most types of foodstuff require good temperature control. A well-known example is wine, which is stored in underground wine cellars. The idea is that a good layer of soil will shield the wine from temperature fluctuations with the seasons (and even daily fluctuations). Clearly very deep cellars will do this, but such cellars are costly to build, and inconvenient to use and maintain. Therefore we shall use the solution we found in the previous section for the heat equation in a semi-infinite strip to estimate an adequate depth for a wine cellar. We consider the following model:

$$u_t = Du_{xx} \qquad 0 < x < \infty, \ -\infty < t < \infty, \qquad (5.96)$$

$$u(0, t) = T_0 + A \cos \omega t \qquad -\infty < t < \infty, \qquad (5.97)$$

$$u(x, t) \to T_0 \qquad x \to \infty. \qquad (5.98)$$

Here the x coordinate measures the distance towards the earth center, where $x = 0$ is the earth's surface, D is the earth's diffusion coefficient, and ω represents the ground temperature fluctuations about a fixed temperature T_0. For example, one can take one year as the basic period, which implies $\omega = 0.19 \times 10^{-6} \ \text{s}^{-1}$. Thanks to the superposition principle and formula (5.93), we obtain the solution:

$$V(x, t) = T_0 + A e^{-kx} \cos(\omega t - kx), \quad \omega = 2Dk^2. \qquad (5.99)$$

How should formula (5.99) be used to choose the depth of the cellar? We have already determined ω. The diffusion coefficient D depends on the nature of the soil. It can vary by a factor of 5 or more between dry soil and wet soil and rocks. For the purpose of our model we shall assume an average value of $0.0025 \ \text{cm}^2 \ \text{s}^{-1}$. The ground temperature can fluctuate by $20 \,^\circ\text{C}$. If we want to minimize the fluctuation in the cellar to less than $2 \,^\circ\text{C}$, say, we need to use a depth L such that $e^{-kL} = 0.1$, i.e. $L = 3.7$ m. A smarter choice for the depth L would be the criterion $kL = \pi$, i.e. $L = 5$ m. This will provide two advantages. First, it gives a reduction in the amplitude by a factor of 23, i.e. the fluctuation will be less than $1 \,^\circ\text{C}$. Second, the phase at this depth would be exactly opposite to the phase at zero ground level (with respect to the fixed temperature T_0). This effect is desirable, since other mechanisms of heat transfer, such as opening the door to the cellar, convection of heat by water, etc. would then drive the temperature in the cellar further towards T_0.

5.7 Exercises

5.1 Solve the equation

$$u_t = 17u_{xx} \qquad 0 < x < \pi, \ t > 0,$$

with the boundary conditions

$$u(0, t) = u(\pi, t) = 0 \qquad t \geq 0,$$

and the initial conditions

$$u(x, 0) = \begin{cases} 0 & 0 \leq x \leq \pi/2, \\ 2 & \pi/2 < x \leq \pi. \end{cases}$$

5.2 Prove that the solution we found by separation of variables for the vibration of a free string can be represented as a superposition of a forward and a backward wave.

5.3 (a) Using the separation of variables method find a (formal) solution of a vibrating string with fixed ends:

$$u_{tt} - c^2 u_{xx} = 0 \qquad 0 < x < L, \ 0 < t,$$
$$u(0, t) = u(L, t) = 0 \qquad t \geq 0,$$
$$u(x, 0) = f(x) \qquad 0 \leq x \leq L,$$
$$u_t(x, 0) = g(x) \qquad 0 \leq x \leq L.$$

(b) Prove that the above solution can be represented as a superposition of a forward and a backward wave.

5.4 (a) Find a formal solution of the problem

$$u_{tt} = u_{xx} \qquad 0 < x < \pi, \ t > 0,$$
$$u(0, t) = u(\pi, t) = 0 \qquad t \geq 0,$$
$$u(x, 0) = \sin^3 x \qquad 0 \leq x \leq \pi,$$
$$u_t(x, 0) = \sin 2x \qquad 0 \leq x \leq \pi.$$

(b) Show that the above solution is classical.

5.5 (a) Using the method of separation of variables, find a (formal) solution of the problem

$$u_t - k u_{xx} = 0 \qquad 0 < x < L, \ t > 0,$$
$$u_x(0, t) = u_x(L, t) = 0 \qquad t \geq 0,$$
$$u(x, 0) = f(x) \qquad 0 \leq x \leq L,$$

describing the heat evolution of an insulated one-dimensional rod (Neumann problem).

(b) Solve the heat equation $u_t = 12 u_{xx}$ in $0 < x < \pi, t > 0$ subject to the following boundary and initial conditions:

$$u_x(0, t) = u_x(\pi, t) = 0 \qquad t \geq 0,$$
$$u(x, 0) = 1 + \sin^3 x \qquad 0 \leq x \leq \pi.$$

(c) Find $\lim_{t \to \infty} u(x, t)$ for all $0 < x < \pi$, and explain the physical interpretation of your result.

5.6 (a) Using the separation of variables method find a (formal) solution of the following periodic heat problem:

$$u_t - k u_{xx} = 0 \qquad 0 < x < 2\pi, \ t > 0,$$
$$u(0, t) = u(2\pi, t), \ u_x(0, t) = u_x(2\pi, t) \qquad t \geq 0,$$
$$u(x, 0) = f(x) \qquad 0 \leq x \leq 2\pi,$$

where f is a smooth periodic function. This system describes the heat evolution on a circular insulated wire of length 2π.

(b) Find $\lim_{t \to \infty} u(x, t)$ for all $0 < x < 2\pi$, and explain the physical interpretation of your result.

(c) Show that if v is an arbitrary partial derivative of the solution u, then $v(0, t) = v(2\pi, t)$ for all $t \geq 0$.

5.7 Solve the following heat problem:

$$u_t - k u_{xx} = A \cos \alpha t \qquad 0 < x < 1, \ t > 0,$$
$$u_x(0, t) = u_x(1, t) = 0 \quad t \geq 0,$$
$$u(x, 0) = 1 + \cos^2 \pi x \quad 0 \leq x \leq 1.$$

5.8 Consider the problem

$$u_t - u_{xx} = e^{-t} \sin 3x \qquad 0 < x < \pi, \ t > 0,$$
$$u(0, t) = u(\pi, t) = 0 \qquad t \geq 0,$$
$$u(x, 0) = f(x) \qquad 0 \leq x \leq \pi.$$

(a) Solve the problem using the method of eigenfunction expansion.
(b) Find $u(x, t)$ for $f(x) = x \sin x$.
(c) Show that the solution $u(x, t)$ is indeed a solution of the equation

$$u_t - u_{xx} = e^{-t} \sin 3x \qquad 0 < x < \pi, \ t > 0.$$

5.9 Consider the problem

$$u_t - u_{xx} - hu = 0 \qquad 0 < x < \pi, \ t > 0,$$
$$u(0, t) = u(\pi, t) = 0 \qquad t \geq 0,$$
$$u(x, 0) = x(\pi - x) \quad 0 \leq x \leq \pi,$$

where h is a real constant.
(a) Solve the problem using the method of eigenfunction expansion.
(b) Does $\lim_{t \to \infty} u(x, t)$ exist for all $0 < x < \pi$?
Hint Distinguish between the following cases:

$$\text{(i)} \quad h < 1, \qquad \text{(ii)} \quad h = 1, \qquad \text{(iii)} \quad h > 1.$$

5.10 Consider the problem

$$u_t = u_{xx} + \alpha u \qquad 0 < x < 1, \ t > 0,$$
$$u(0, t) = u(1, t) = 0 \qquad t \geq 0,$$
$$u(x, 0) = f(x) \qquad 0 \leq x \leq 1, \ f \in C([0, 1]).$$

(a) Assume that $\alpha = -1$ and $f(x) = x$ and solve the problem.
(b) Prove that for all $\alpha \leq 0$ and all f, the solution u satisfies $\lim_{t \to \infty} u(x, t) = 0$.
(c) Assume now that $\pi^2 < \alpha < 4\pi^2$. Does $\lim_{t \to \infty} u(x, t)$ exist for all f? If your answer is no, find a necessary and sufficient condition on f which ensures the existence of this limit.

5.11 Consider the following problem:

$$u_{tt} - u_{xx} = 0 \qquad 0 < x < 1,\ t > 0,$$
$$u_x(0, t) = u_x(1, t) = 0 \qquad t \geq 0,$$
$$u(x, 0) = f(x) \qquad 0 \leq x \leq 1,$$
$$u_t(x, 0) = 0 \qquad 0 \leq x \leq 1.$$

(a) Draw (on the (x, t) plane) the domain of dependence of the point $(\frac{1}{3}, \frac{1}{10})$.
(b) Suppose that $f(x) = (x - \frac{1}{2})^3$. Evaluate $u(\frac{1}{3}, \frac{1}{10})$.
(c) Solve the problem with $f(x) = 2 \sin^2 2\pi x$.

5.12 (a) Solve the problem

$$u_t - u_{xx} - \frac{9u}{4} = 0 \qquad 0 < x < \pi,\ t > 0,$$
$$u(0, t) = u_x(\pi, t) = 0 \qquad t \geq 0,$$
$$u(x, 0) = \sin(3x/2) + \sin(9x/2) \qquad 0 \leq x \leq \pi.$$

(b) Compute $\phi(x) := \lim_{t \to \infty} u(x, t)$ for $x \in [0, \pi]$.

5.13 Solve the problem

$$u_t = u_{xx} - u \qquad 0 < x < 1,\ t > 0,$$
$$u(0, t) = u_x(1, t) = 0 \qquad t \geq 0,$$
$$u(x, 0) = x(2 - x) \qquad 0 \leq x \leq 1.$$

5.14 Prove *Duhamel's principle:* for $s \geq 0$, let $v(x, t, s)$ be the solution of the following initial-boundary problem (which depends on the parameter s):

$$v_t - v_{xx} = 0 \qquad 0 < x < L,\ t > s,$$
$$v(0, t, s) = v(L, t, s) = 0 \qquad t \geq s,$$
$$v(x, s, s) = F(x, s) \qquad 0 \leq x \leq L.$$

Prove that the function

$$u(x, t) = \int_0^t v(x, t, s)\, ds$$

is a solution of the nonhomogeneous problem

$$u_t - u_{xx} = F(x, t) \qquad 0 < x < L,\ t > 0,$$
$$u(0, t) = u(L, t) = 0 \qquad t \geq 0,$$
$$u(x, 0) = 0 \qquad 0 \leq x \leq L.$$

5.15 Using the energy method, prove the uniqueness for the problem

$$u_{tt} - c^2 u_{xx} = F(x, t) \qquad 0 < x < L, \ t > 0,$$
$$u_x(0, t) = t^2, \quad u(L, t) = -t \qquad t \geq 0,$$
$$u(x, 0) = x^2 - L^2 \qquad 0 \leq x \leq L,$$
$$u_t(x, 0) = \sin^2 \frac{\pi x}{L} \qquad 0 \leq x \leq L.$$

5.16 Consider the following telegraph problem:

$$\begin{aligned}
u_{tt} + u_t - c^2 u_{xx} &= 0 & a < x < b, \ t > 0, \\
u(a, t) = u_x(b, t) &= 0 & t \geq 0, \\
u(x, 0) &= f(x), & a \leq x \leq b, \\
u_t(x, 0) &= g(x), & a \leq x \leq b.
\end{aligned} \qquad (5.100)$$

Use the energy method to prove that the problem has a unique solution.

5.17 Using the energy method, prove uniqueness for the problem

$$u_{tt} - c^2 u_{xx} + hu = F(x, t) \qquad -\infty < x < \infty, \ t > 0,$$
$$\lim_{x \to \pm\infty} u(x, t) = \lim_{x \to \pm\infty} u_x(x, t) = \lim_{x \to \pm\infty} u_t(x, t) = 0 \qquad t \geq 0,$$
$$\int_{-\infty}^{\infty} (u_t^2 + c^2 u_x^2 + hu^2) \, dx < \infty \qquad t \geq 0,$$
$$u(x, 0) = f(x) \qquad -\infty < x < \infty,$$
$$u_t(x, 0) = g(x) \qquad -\infty < x < \infty,$$

where h is a positive constant.

Hint Use the energy integral

$$E(t) = \frac{1}{2} \int_{-\infty}^{\infty} (w_t^2 + c^2 w_x^2 + hw^2) \, dx.$$

5.18 Let $\alpha, \beta \geq 0, k > 0$. Using the energy method, prove uniqueness for the problem

$$u_t - k u_{xx} = F(x, t) \qquad 0 < x < L, \ t > 0,$$
$$u(0, t) - \alpha u_x(0, t) = a(t), \quad u(L, t) + \beta u_x(L, t) = b(t) \qquad t \geq 0,$$
$$u(x, 0) = f(x) \qquad 0 < x < L.$$

5.19 (a) Prove the following identity:

$$u \left[(y^2 u_x)_x + (x^2 u_y)_y \right] = \operatorname{div} \left(y^2 uu_x, x^2 uu_y \right) - \left[(yu_x)^2 + (xu_y)^2 \right]. \quad (5.101)$$

(b) Let D be a planar bounded domain with a smooth boundary Γ which does not intersect the lines $x = 0$ and $y = 0$. Using the energy method, prove uniqueness for the elliptic problem

$$(y^2 u_x)_x + (x^2 u_y)_y = F(x, t) \quad (x, y) \in D,$$
$$u(x, y) = f(x, y) \quad (x, y) \in \Gamma.$$

Hint Use the divergence theorem $\iint_D \operatorname{div} w \, dxdy = \int_{\partial D} w \cdot n \, d\sigma$ and (5.101).

5.20 *Similarity variables for the heat equation*: the purpose of this exercise is to derive an important canonical solution for the heat equation *and* to introduce the method of similarity variables.

(a) Consider the heat equation

$$u_t - u_{xx} = 0 \qquad x \in \mathbb{R}, \ t \geq 0. \tag{5.102}$$

Set

$$u(x, t) = \phi(\lambda(x, t)),$$

where

$$\lambda(x, t) = \frac{x}{2\sqrt{t}}.$$

Show that u is a solution of (5.102) if and only if $\phi(\lambda)$ is a solution of the ODE $\phi'' + 2\lambda\phi' = 0$, where $' = d/d\lambda$.

(b) Integrate the ODE and show that the function

$$u(x, t) = \mathrm{erf}\left(\frac{x}{2\sqrt{t}}\right)$$

is a solution of (5.102), where $\mathrm{erf}(s)$ is the *error function* defined by

$$\mathrm{erf}(s) := \frac{2}{\sqrt{\pi}} \int_0^s e^{-r^2} dr.$$

(c) The *complementary error function* is defined by

$$\mathrm{erfc}(s) := \frac{2}{\sqrt{\pi}} \int_s^\infty e^{-r^2} dr = 1 - \mathrm{erf}(s).$$

Show that

$$u(x, t) = \mathrm{erfc}\left(\frac{x}{2\sqrt{t}}\right)$$

is a solution of (5.102).

(d) Differentiating $\mathrm{erf}\,(x/2\sqrt{t})$, show that

$$K(x, t) = \frac{1}{\sqrt{4\pi t}} \exp\left(-\frac{x^2}{4t}\right)$$

is a solution of (5.102). K is called the *heat kernel*. We shall consider heat kernels in detail in Chapter 8.

6

Sturm–Liouville problems and eigenfunction expansion

6.1 Introduction

In the preceding chapter we presented several examples of initial boundary value problems that can be solved by the method of separation of variables. In this chapter we shall discuss the theoretical foundation of this method.

We consider two basic initial boundary value problems for which the method of separation of variables is applicable. The first problem is parabolic and concerns heat flow in a nonhomogeneous rod. The corresponding PDE is a generalization of the heat equation. We seek a function $u(x, t)$ that is a solution of the problem

$$u_t - \frac{1}{r(x)m(t)}[(p(x)u_x)_x + q(x)u] = 0 \qquad a < x < b, \ t > 0, \qquad (6.1)$$

$$B_a[u] = \alpha u(a, t) + \beta u_x(a, t) = 0 \qquad t \geq 0, \qquad (6.2)$$

$$B_b[u] = \gamma u(b, t) + \delta u_x(b, t) = 0 \qquad t \geq 0, \qquad (6.3)$$

$$u(x, 0) = f(x) \qquad a \leq x \leq b. \qquad (6.4)$$

The second problem is hyperbolic. It models the vibrations of a nonhomogeneous string. The corresponding PDE is a generalization of the wave equation:

$$u_{tt} - \frac{1}{r(x)m(t)}[(p(x)u_x)_x + q(x)u] = 0 \qquad a < x < b, \ t > 0, \qquad (6.5)$$

$$B_a[u] = \alpha u(a, t) + \beta u_x(a, t) = 0 \qquad t \geq 0, \qquad (6.6)$$

$$B_b[u] = \gamma u(b, t) + \delta u_x(b, t) = 0 \qquad t \geq 0, \qquad (6.7)$$

$$u(x, 0) = f(x), \quad u_t(x, 0) = g(x) \qquad a \leq x \leq b. \qquad (6.8)$$

We assume that the coefficients of these PDEs are real functions that satisfy

$$p, p', q, r \in C([a, b]), \ p(x), r(x) > 0, \ \forall x \in [a, b],$$
$$m \in C([0, \infty)), \ m(t) > 0, \ \forall t \geq 0.$$

We also assume that

$$\alpha, \beta, \gamma, \delta \in \mathbb{R}, \quad |\alpha| + |\beta| > 0, \quad |\gamma| + |\delta| > 0.$$

Note that these boundary conditions include in particular the Dirichlet boundary condition ($\alpha = \gamma = 1$, $\beta = \delta = 0$) and the Neumann boundary condition ($\alpha = \gamma = 0$, $\beta = \delta = 1$).

We concentrate on the parabolic problem; the hyperbolic problem can be dealt with similarly. To apply the method of separation of variables we seek nontrivial separated solutions of (6.1) that satisfy the boundary conditions (6.2)–(6.3) and have the form

$$u(x, t) = X(x)T(t), \tag{6.9}$$

where X and T are functions of one variable, x and t, respectively. Substituting such a product solution into the PDE and separating the variables we obtain

$$\frac{mT_t}{T} = \frac{(pX_x)_x + qX}{rX}. \tag{6.10}$$

The left hand side depends solely on t, while the right hand side is a function of x. Therefore, there exists a constant λ such that

$$\frac{mT_t}{T} = \frac{(pX_x)_x + qX}{rX} = -\lambda. \tag{6.11}$$

Thus, (6.11) is equivalent to the following system of ODEs

$$(pX')' + qX + \lambda rX = 0 \quad a < x < b, \tag{6.12}$$

$$m\frac{dT}{dt} = -\lambda T \quad t > 0. \tag{6.13}$$

By our assumption $u \neq 0$. Since u must satisfy the boundary conditions (6.2)–(6.3), it follows that

$$B_a[X] = 0, \quad B_b[X] = 0.$$

In other words, the function X should be a solution of the boundary value problem

$$(pv')' + qv + \lambda rv = 0 \quad a < x < b, \tag{6.14}$$

$$B_a[v] = B_b[v] = 0. \tag{6.15}$$

The main part of the present chapter is devoted to the solution of the system (6.14)–(6.15). A nontrivial solution of this system is called an *eigenfunction* of the problem associated with the *eigenvalue* λ. The problem (6.14)–(6.15) is called a *Sturm–Liouville eigenvalue problem* in honor of the French mathematicians Jacques Charles Sturm (1803–1855) and Joseph Liouville (1809–1882). The differential operator $L[v] := (pv')' + qv$ is said to be a *Sturm–Liouville operator*. The function r is called a *weight function*.

The notions *eigenfunction* and *eigenvalue* are familiar to the reader from a basic course in linear algebra. Let A be a linear operator acting on a vector space V, and let $\lambda \in \mathbb{C}$. A vector $v \neq 0$ is an *eigenvector* of the operator A with an *eigenvalue* λ, if $A[v] = \lambda v$. The set of all vectors satisfying $A[v] = \lambda v$ is a linear subspace of V, and its dimension is the *multiplicity* of λ. An eigenvalue with multiplicity 1 is called *simple*.

In our (Sturm–Liouville) eigenvalue problem, the corresponding linear operator is the differential operator $-L$, which acts on the space of twice differentiable functions satisfying the corresponding boundary conditions.

Example 6.1 In Chapter 5 we solved the following Sturm–Liouville problem:

$$\frac{d^2 v}{dx^2} + \lambda v = 0 \quad 0 < x < L, \tag{6.16}$$

$$v(0) = v(L) = 0. \tag{6.17}$$

Here $p = r = 1$, $q = 0$, and the boundary condition is of the first kind (Dirichlet). The eigenfunctions and eigenvalues of the problem are:

$$v_n(x) = \sin\frac{n\pi x}{L}, \quad \lambda_n = \left(\frac{n\pi}{L}\right)^2 \quad n = 1, 2, 3, \dots.$$

Example 6.2 We also solved the Sturm–Liouville problem

$$\frac{d^2 v}{dx^2} + \lambda v = 0 \quad 0 < x < L, \tag{6.18}$$

$$v'(0) = v'(L) = 0. \tag{6.19}$$

Here we are dealing with the Neumann boundary condition. The eigenfunctions and eigenvalues of the problem are:

$$v_n(x) = \cos\frac{n\pi x}{L}, \quad \lambda_n = \left(\frac{n\pi}{L}\right)^2 \quad n = 0, 1, 2, \dots.$$

In the following sections we show that the essential properties of the eigenfunctions and eigenvalues of these simple problems are also satisfied in the case of a general Sturm–Liouville problem. We then use these properties to solve the general initial boundary value problems that were presented at the beginning of the current section.

6.2 The Sturm–Liouville problem

Consider the Sturm–Liouville eigenvalue problem

$$(p(x)v')' + q(x)v + \lambda r(x)v = 0 \qquad a < x < b, \tag{6.20}$$

$$B_a[v] := \alpha v(a) + \beta v'(a) = 0, \qquad B_b[v] := \gamma v(b) + \delta v'(b) = 0. \tag{6.21}$$

The first equation is a linear second-order ODE. We assume that the coefficients of this ODE are real functions satisfying

$$p, p', q, r \in C([a, b]), \qquad p(x), r(x) > 0, \quad \forall x \in [a, b].$$

We also assume that

$$\alpha, \beta, \gamma, \delta \in \mathbb{R}, \quad |\alpha| + |\beta| > 0, \quad |\gamma| + |\delta| > 0.$$

Under these assumptions the eigenvalue problem (6.20)–(6.21) is called a *regular Sturm–Liouville problem*. If either of the functions p or r vanishes at least at one end point, or is discontinuous there, or if the problem is defined on an infinite interval, then the Sturm–Liouville problem is said to be *singular*.

Remark 6.3 It is always possible to transform a general linear second–order ODE into an ODE of the Sturm–Liouville form:

$$L[v] := (p(x)v')' + q(x)v = f.$$

Indeed, suppose that

$$M[v] := A(x)v'' + B(x)v' + C(x)v = F(x) \tag{6.22}$$

is an arbitrary linear second-order ODE such that A is a positive continuous function. We denote by p the integration factor $p(x) := \exp\{\int [B(x)/A(x)] \, dx\}$. Multiplying (6.22) by $p(x)/A(x)$ we obtain

$$L[v] := \frac{p(x)}{A(x)} M[v] = p(x)v'' + p'(x)v' + \frac{p(x)}{A(x)} C(x)v$$
$$= (p(x)v')' + q(x)v = f,$$

and we see that the operator M is equivalent to a Sturm–Liouville operator L, where $q(x) = [p(x)/A(x)]C(x)$, and $f(x) = [p(x)/A(x)]F(x)$.

Example 6.4 Let $v \in \mathbb{R}, a > 0$. The equation

$$r^2 w''(r) + r w'(r) + (r^2 - v^2)w(r) = 0 \qquad r > 0 \tag{6.23}$$

is called a *Bessel equation* of order v. Dividing (6.23) by r, using the transformation $x = r/\sqrt{\lambda}$, and limiting our attention to a finite interval, we obtain the following singular Sturm–Liouville problem:

$$(xv'(x))' + \left(\lambda x - \frac{v^2}{x}\right) v(x) = 0 \quad 0 < x < a,$$

$$v(a) = 0, \quad |v(0)| < \infty.$$

Here $p(x) = r(x) = x, q(x) = -v^2/x$. We shall study this equation in some detail in Chapter 9.

In our study of the Sturm–Liouville theory, we shall also deal with the *periodic Sturm–Liouville problem*:

$$(p(x)v')' + q(x)v + \lambda r(x)v = 0 \qquad a < x < b, \tag{6.24}$$

$$v(a) = v(b), \quad v'(a) = v'(b), \tag{6.25}$$

where the coefficients p, q, r are periodic functions of a period $(b - a)$, and

$$p, p', q, r \in C(\mathbb{R}), \qquad p(x), r(x) > 0 \qquad \forall x \in \mathbb{R}.$$

The periodic boundary conditions (6.25) and the ODE (6.24) imply that an eigenfunction can be extended to a periodic function on the real line. This periodic function is a twice differentiable (periodic) function, except possibly at the points $a + k(b - a), k \in \mathbb{Z}$, where a singularity of the second derivative may occur.

Example 6.5 Consider the following periodic Sturm–Liouville problem:

$$\frac{d^2v}{dx^2} + \lambda v = 0 \quad 0 < x < L, \tag{6.26}$$

$$v(0) = v(L), \quad v'(0) = v'(L). \tag{6.27}$$

Here $p = r = 1, \; q = 0$. Recall that the general solution of the ODE (6.26) is of the form:

1. if $\lambda < 0$, then $v(x) = \alpha \cosh(\sqrt{-\lambda}x) + \beta \sinh(\sqrt{-\lambda}x)$,
2. if $\lambda = 0$, then $v(x) = \alpha + \beta x$,
3. if $\lambda > 0$, then $v(x) = \alpha \cos(\sqrt{\lambda}x) + \beta \sin(\sqrt{\lambda}x)$,

where α, β are arbitrary real numbers. Note that we assume again that λ is real. We shall prove later that all the eigenvalues of regular or periodic Sturm–Liouville problems are real.

Negative eigenvalues ($\lambda < 0$) In this case any nontrivial solution of the corresponding ODE is an unbounded function on $\mathbb{R}$. In particular, there is no periodic nontrivial solution for this equation. In other words, the system (6.26)–(6.27) does not admit negative eigenvalues.

Zero eigenvalue ($\lambda = 0$) A linear function is periodic if and only if it is a constant. Therefore, $\lambda = 0$ is an eigenvalue with an eigenfunction **1**.

Positive eigenfunctions ($\lambda > 0$) The general solution for the case $\lambda > 0$ is of the form

$$v(x) = \alpha \cos(\sqrt{\lambda}x) + \beta \sin(\sqrt{\lambda}x). \tag{6.28}$$

Substituting the boundary conditions (6.27) into (6.28) we arrive at a system of algebraic linear equations

$$\alpha \cos(\sqrt{\lambda}L) + \beta \sin(\sqrt{\lambda}L) = \alpha, \tag{6.29}$$
$$\sqrt{\lambda}[-\alpha \sin(\sqrt{\lambda}L) + \beta \cos(\sqrt{\lambda}L)] = \sqrt{\lambda}\beta. \tag{6.30}$$

If α or β equals zero, but $|\alpha| + |\beta| \neq 0$, then obviously $\lambda = (2n\pi/L)^2$, where $n \in \mathbb{N}$. Otherwise, multiplying (6.29) by β, and (6.30) by $\alpha/\sqrt{\lambda}$ implies again that $\lambda = (2n\pi/L)^2$.

Therefore, the system (6.29)–(6.30) has a nontrivial solution if and only if

$$\lambda_n = \left(\frac{2n\pi}{L}\right)^2 \qquad n = 1, 2, 3, \ldots.$$

These eigenvalues have eigenfunctions of the form

$$v_n(x) = \alpha_n \cos\left(\frac{2n\pi x}{L}\right) + \beta_n \sin\left(\frac{2n\pi x}{L}\right). \tag{6.31}$$

It is convenient to select $\{\cos(2n\pi x/L), \sin(2n\pi x/L)\}$ as a basis for the eigenspace corresponding to λ_n.

Therefore, positive eigenvalues of the periodic problem (6.26)–(6.27) are of multiplicity 2. Recall that in the other examples of the Sturm–Liouville problem that we have encountered so far *all* the eigenvalues are simple (i.e. of multiplicity 1). In the sequel, we prove that this is a general property of *regular* Sturm–Liouville problems. Moreover, it turns out that this is the only essential property of a regular Sturm–Liouville problem that does not hold in the periodic case. Note that the maximal multiplicity of an eigenvalue of a Sturm–Liouville problem is 2, since the space of *all* solutions of the ODE (6.20) (without imposing any boundary conditions) is two-dimensional.

In conclusion, the solution of the periodic Sturm–Liouville eigenvalue problem (6.26)–(6.27) is the following infinite sequence of eigenvalues and eigenfunctions

$$\lambda_0 = 0, \qquad u_0(x) = 1, \tag{6.32}$$

$$\lambda_n = \left(\frac{2n\pi}{L}\right)^2, \quad u_n(x) = \cos\frac{2n\pi x}{L}, \quad v_n(x) = \sin\frac{2n\pi x}{L} \qquad n = 1, 2, \ldots.$$

$$\tag{6.33}$$

This system is called the *classical Fourier system* on the interval $[0, L]$.

6.3 Inner product spaces and orthonormal systems

To prepare the ground for the Sturm–Liouville theory we survey basic notions and properties of real inner product spaces. We omit proofs which can be found in standard textbooks on Fourier analysis [13].

Definition 6.6 A real linear space V is said to be a (*real*) *inner product space* if for any two vectors $u, v \in V$ there is a real number $\langle u, v \rangle \in \mathbb{R}$, which is called the inner product of u and v, such that the following properties are satisfied:

1. $\langle u, v \rangle = \langle v, u \rangle$ for all $u, v \in V$.
2. $\langle u + v, w \rangle = \langle u, w \rangle + \langle v, w \rangle$ for all $u, v, w \in V$.
3. $\langle \alpha u, v \rangle = \alpha \langle u, v \rangle$ for all $u, v \in V$, and $\alpha \in \mathbb{R}$.
4. $\langle v, v \rangle \geq 0$ for all $v \in V$, moreover, $\langle v, v \rangle > 0$ for all $v \neq 0$.

In the context of Sturm–Liouville problems the following natural inner product plays an important role:

Definition 6.7 (a) Let f be a real function defined on $[a, b]$ except, possibly, for finitely many points. f is called *piecewise continuous* on $[a, b]$ if it has at most finitely many points of discontinuity, and if at any such point f admits left and right limits (such a discontinuity is called a *jump (or step) discontinuity*).

(b) Two piecewise continuous functions which take the same values at all points in $[a, b]$ except, possibly, for finitely many points are called equivalent. The space of all (equivalent classes of) piecewise continuous functions on $[a, b]$ will be denoted by $E(a, b)$.

(c) If f and f' are piecewise continuous functions, we say that f is *piecewise differentiable*.

(d) Let $r(x)$ be a positive continuous weight function on $[a, b]$. We define the following inner product on the space $E(a, b)$:

$$\langle u, v \rangle_r = \int_a^b u(x)v(x)r(x)\,\mathrm{d}x, \qquad u, v \in E(a, b).$$

The corresponding inner product space is denoted by $E_r(a, b)$. To simplify the notation we shall use $E(a, b)$ for $E_1(a, b)$.

Each inner product induces a *norm* defined by $\|v\| := \langle v, v \rangle^{1/2}$, which satisfies the usual norm properties:

(1) $\|\alpha u\| = |\alpha|\, \|u\|$ for all $u \in V$, and $\alpha \in \mathbb{R}$.
(2) *The triangle inequality*: $\|u + v\| \leq \|u\| + \|v\|$ for all $u, v \in V$.
(3) $\|v\| \geq 0$ for all $v \in V$, moreover, $\|v\| > 0$ for all $v \neq 0$.

In addition this induced norm satisfies the *Cauchy–Schwartz inequality*

$$|\langle u, v \rangle| \le \|u\| \, \|v\|.$$

Definition 6.8 Let $(V, \langle \cdot, \cdot \rangle)$ be an inner product space.

(1) A sequence $\{v_n\}_{n=1}^{\infty}$ *converges to v in the mean* (or in norm), if

$$\lim_{n \to \infty} \|v_n - v\| = 0.$$

(2) Two vectors $u, v \in V$ are called *orthogonal* if $\langle u, v \rangle = 0$.
(3) The sequence $\{v_n\} \subset V$ is said to be *orthogonal* if $v_n \ne 0$ for all $n \in \mathbb{N}$, and $\langle v_n, v_m \rangle = 0$ for all $n \ne m$.
(4) The sequence $\{v_n\} \subset V$ is said to be *orthonormal* if

$$\langle v_n, v_m \rangle = \begin{cases} 0 & m \ne n, \\ 1 & m = n. \end{cases} \tag{6.34}$$

Remark 6.9 Consider the inner product space $E_r(a, b)$. Then convergence in the mean does not imply pointwise convergence on $[a, b]$, and vice versa, a pointwise convergence does not imply convergence in the mean. If, however, $[a, b]$ is a bounded closed interval, then uniform convergence on $[a, b]$ implies convergence in the mean.

As an example, consider the interval $[0, \infty)$ and the weight function $r(x) = 1$. The function

$$\chi_{[\alpha, \beta]}(x) = \begin{cases} 1 & x \in [\alpha, \beta], \\ 0 & x \notin [\alpha, \beta] \end{cases} \tag{6.35}$$

is called the *characteristic function* of the interval $[\alpha, \beta]$.

The sequence of functions $v_n = \chi_{[n, n+1]}$, $n = 1, 2, \ldots$ converges pointwise to zero on $[0, \infty)$, but since $\|v_n\| = 1$, this sequence does not converge in norm to zero.

On the other hand, consider the interval $[0, 1]$, and let $\{[a_n, b_n]\}$ be a sequence of intervals such that each $x \in [0, 1]$ belongs, and also does not belong, to infinitely many intervals $[a_n, b_n]$, and such that $b_n - a_n = 2^{-k(n)}$, where $\{k(n)\}$ is a nondecreasing sequence satisfying $\lim_{n \to \infty} k(n) = \infty$. Since

$$\|\chi_{[a_n, b_n]}\|^2 = \int_0^1 [\chi_{[a_n, b_n]}(x)]^2 \, dx = 2^{-k(n)} \to 0,$$

it follows that the sequence $\{\chi_{[a_n, b_n]}\}$ tends to the zero function in the mean. On the other hand $\{\chi_{[a_n, b_n]}\}$ does not converge at any point of $[0, 1]$ since for a fixed $0 \le x_0 \le 1$, the sequence $\{\chi_{[a_n, b_n]}(x_0)\}$ attains infinitely many times the value 0 and also infinitely many times the value 1.

Remark 6.10 One can easily modify any orthogonal sequence $\{v_n\}$ to obtain an orthonormal sequence $\{\tilde{v}_n\}$, using the normalization process $\tilde{v}_n := (1/\|v_n\|)v_n$.

Using an orthonormal sequence, one can find the *orthogonal projection* of a vector $v \in V$ into a subspace V_N of V, which is the closest vector to v in V_N.

Theorem 6.11 *(a) Let $\{v_n\}_{n=1}^{N}$ be a finite orthonormal sequence, and set $V_N := \text{span}\{v_1, \ldots, v_N\}$. Let $v \in V$, and define*

$$u := \sum_{n=1}^{N} \langle v, v_n \rangle v_n .$$

Then

$$\|v - u\| = \min_{w \in V_N} \{\|v - w\|\} = \sqrt{\|v\|^2 - \sum_{n=1}^{N} \langle v, v_n \rangle^2}. \tag{6.36}$$

In other words, u is the orthogonal projection of v into V_N.

(b) Let $\{v_n\}_{n=1}^{N}$ ($N \leq \infty$) be a finite or infinite orthonormal sequence, and let $v \in V$. Then the following inequality holds:

$$\sum_{n=1}^{N} \langle v, v_n \rangle^2 \leq \|v\|^2 . \tag{6.37}$$

In particular,

$$\lim_{n \to \infty} \langle v, v_n \rangle = 0. \tag{6.38}$$

Definition 6.12 (1) The last claim of Theorem 6.11 (i.e. (6.38)) is called the *Riemann–Lebesgue lemma*.

(2) The coefficients $\langle v, v_n \rangle$ are called *generalized Fourier coefficients* (or simply, Fourier coefficients) of the function v with respect the orthonormal sequence $\{v_n\}_{n=1}^{N}$, where $N \leq \infty$.

(3) The inequality (6.37) is called the *Bessel inequality*. Note that the Bessel inequality (6.37) follows easily from (6.36).

(4) The orthonormal sequence $\{v_n\}_{n=1}^{N}$ is said to be *complete* in V, if for every $v \in V$ we have equality in the Bessel inequality. In this case the equality is call the *Parseval identity*.

The following proposition follows from (6.36).

Proposition 6.13 *Let $\{v_n\}_{n=1}^{\infty}$ be an infinite orthonormal sequence. The following propositions are equivalent:*

(1) $\{v_n\}_{n=1}^{\infty}$ is a complete orthonormal sequence.
(2)

$$\lim_{k \to \infty} \left\| v - \sum_{n=1}^{k} \langle v, v_n \rangle v_n \right\| = 0 ,$$

for all $v \in V$.

Definition 6.14 If $\lim_{k \to \infty} \| v - \sum_{n=1}^{k} \langle v, v_n \rangle v_n \| = 0$ exists, we write

$$v = \sum_{n=1}^{\infty} \langle v, v_n \rangle v_n,$$

and we say that the *Fourier expansion of v converges in norm* (or on average, or in the mean) to v. More generally, the series

$$\sum_{n=1}^{\infty} \langle v, v_n \rangle v_n$$

is called the *generalized Fourier expansion* (or for short, Fourier expansion) of v with respect to the orthonormal system $\{v_n\}_{n=1}^{\infty}$.

Remark 6.15 The notion of convergence in the mean may seem initially to be an abstract mathematical idea. We shall see, however, that in fact it provides the right framework for Fourier's theory of representing a function as a series of an orthonormal sequence.

We end this section with two examples of Fourier expansion.

Example 6.16 Let $E(0, \pi)$ be the inner product space (of equivalent classes) of all piecewise continuous functions in the interval $[0, \pi]$ equipped with the inner product $\langle u, v \rangle = \int_0^\pi u(x)v(x)\,dx$. Consider the sequence

$$u_n(x) = \cos nx, \qquad n = 0, 1, 2, 3, \ldots$$

and recall that in Example 6.2 we computed directly for $m, n = 0, 1, 2, \ldots$

$$\int_0^\pi \cos mx \cos nx \, dx = \begin{cases} 0 & m \neq n, \\ \pi/2 & m = n \neq 0, \\ \pi & m = n = 0. \end{cases} \qquad (6.39)$$

Consequently, the sequence $\{\sqrt{1/\pi}\} \cup \{\sqrt{2/\pi} \cos nx\}_{n=1}^{\infty}$ is orthonormal in the space $E(0, \pi)$. We shall see in the next section that it is, in fact, a *complete* orthonormal sequence in $E(0, \pi)$.

We proceed to compute the Fourier expansion of $u(x) = x$ with respect to that orthonormal sequence. We write the expansion as

$$A_0 \sqrt{\frac{1}{\pi}} + \sum_{n=1}^{\infty} A_n \sqrt{\frac{2}{\pi}} \cos nx,$$

where

$$A_0 = \sqrt{\frac{1}{\pi}} \int_0^\pi u(x)\,dx, \qquad A_n = \sqrt{\frac{2}{\pi}} \int_0^\pi u(x) \cos nx\,dx \quad n \geq 1.$$

Therefore,

$$A_0 = \sqrt{\frac{1}{\pi}} \int_0^\pi x \, dx = \sqrt{\pi}\frac{\pi}{2},$$

$$A_n = \sqrt{\frac{2}{\pi}} \int_0^\pi x \cos nx \, dx = \sqrt{\frac{\pi}{2}} \frac{2\pi}{n^2\pi^2}[(-1)^n - 1].$$

It follows that the Fourier expansion of u in this orthonormal sequence is given by the series

$$\frac{\pi}{2} - \frac{4}{\pi} \sum_{m=1}^{\infty} \frac{1}{(2m-1)^2} \cos(2m-1)x,$$

which converges uniformly on $[0, \pi]$.

Example 6.17 Let $E^0(0, \pi)$ be the subspace of $E(0, \pi)$ (of equivalent classes) of all piecewise continuous functions in the interval $[0, \pi]$ that vanish at the interval's end points. In particular, $E^0(0, \pi)$ is an inner product space with respect to $\langle u, v \rangle = \int_0^\pi u(x)v(x) \, dx$. Consider the sequence of functions

$$v_n(x) = \sin nx \qquad n = 1, 2, 3, \ldots$$

in this space.

The orthogonality of the sequence $\{v_n(x)\}_{n=1}^\infty$ in the space $E^0(0, \pi)$ has already been established in Example 6.1. Specifically, we found that for $m, n = 1, 2, 3, \ldots$

$$\int_0^\pi \sin mx \sin nx \, dx = \begin{cases} 0 & m \neq n, \\ \pi/2 & m = n. \end{cases} \tag{6.40}$$

Therefore, $\{\sqrt{2/\pi} \sin nx\}_{n=1}^\infty$ is indeed an orthonormal (and, as will be shown soon, even a complete orthonormal) sequence in $E^0(0, \pi)$.

The Fourier expansion of $v(x) = x \sin x$ in the current sequence is given by

$$\sum_{n=1}^{\infty} B_n \sqrt{\frac{2}{\pi}} \sin nx, \qquad \text{where } B_n = \sqrt{\frac{2}{\pi}} \int_0^\pi v(x) \sin nx \, dx.$$

We use the identity

$$\sin x \sin nx = \frac{1}{2}[\cos(n-1)x - \cos(n+1)x]$$

to find

$$B_n = \sqrt{\frac{1}{2\pi}} \int_0^\pi x[\cos(n-1)x - \cos(n+1)x] \, dx.$$

An integration by parts leads to

$$B_1 = \left(\frac{\pi}{2}\right)^{3/2}, \qquad B_n = \sqrt{\frac{\pi}{2}} \frac{4n[(-1)^{n+1} - 1]}{\pi(n+1)^2(n-1)^2} \qquad n > 1.$$

We therefore obtain that the Fourier expansion for v in this orthonormal sequence is given by the series

$$\frac{\pi}{2} \sin x + \sum_{n=2}^{\infty} \frac{4n[(-1)^{n+1} - 1]}{\pi(n+1)^2(n-1)^2} \sin nx,$$

which converges uniformly on $[0, \pi]$.

6.4 The basic properties of Sturm–Liouville eigenfunctions and eigenvalues

We now present the essential properties of the eigenvalues and eigenfunctions of regular and periodic Sturm–Liouville problems. We shall point out some properties which are still valid in the irregular case.

We start with an algebraic characterization of λ as an eigenvalue.

Proposition 6.18 *Consider the following regular Sturm–Liouville problem*

$$L[v] + \lambda r v = 0 \quad a < x < b, \tag{6.41}$$
$$B_a[v] = B_b[v] = 0. \tag{6.42}$$

Assume that the pair of functions u_λ, v_λ is a basis of the linear space of all solutions of the ODE (6.41). Then λ is an eigenvalue of the Sturm–Liouville problem if and only if

$$\begin{vmatrix} B_a[u_\lambda] & B_a[v_\lambda] \\ B_b[u_\lambda] & B_b[v_\lambda] \end{vmatrix} = 0. \tag{6.43}$$

Proof A function w is a nontrivial solution of (6.41) if and only if there exist $c, d \in \mathbb{R}$ such that $|c| + |d| > 0$ and such that

$$w(x) = cu_\lambda(x) + dv_\lambda(x).$$

The function w is an eigenfunction with eigenvalue λ if and only if w also satisfies the boundary conditions

$$B_a[w] = cB_a[u_\lambda] + dB_a[v_\lambda] = 0,$$
$$B_b[w] = cB_b[u_\lambda] + dB_b[v_\lambda] = 0.$$

In other words, the vector $(c, d) \neq 0$ is a nontrivial solution of a 2×2 linear homogeneous algebraic system with the coefficients matrix

$$\begin{pmatrix} B_a[u_\lambda] & B_a[v_\lambda] \\ B_b[u_\lambda] & B_b[v_\lambda] \end{pmatrix}.$$

This system has a nontrivial solution if and only if condition (6.43) is satisfied. □

Example 6.19 Let us check the criterion that we just derived for the Sturm–Liouville problem

$$v'' + \lambda v = 0 \qquad 0 < x < L, \tag{6.44}$$

$$v(0) = v'(L) = 0. \tag{6.45}$$

For $\lambda > 0$, the pair of functions

$$u_\lambda(x) = \sin \sqrt{\lambda} x, \quad v_\lambda(x) = \cos \sqrt{\lambda} x$$

forms a basis for the linear space of all solutions of the corresponding ODE. Therefore, λ is an eigenvalue of the problem if and only if

$$\begin{vmatrix} \sin 0 & \cos 0 \\ \sqrt{\lambda} \cos \sqrt{\lambda} L & -\sqrt{\lambda} \sin \sqrt{\lambda} L \end{vmatrix} = -\sqrt{\lambda} \cos \sqrt{\lambda} L = 0. \tag{6.46}$$

Hence,

$$\lambda = \left(\frac{(2n-1)\pi}{2L} \right)^2 \qquad n = 1, 2, \dots.$$

We proceed to list general properties of Sturm–Liouville problems.

1 Symmetry Let L be a Sturm–Liouville operator of the form

$$L[u] = (p(x)u')' + q(x)u,$$

and consider the expression $uL[v] - vL[u]$ for $u, v \in C^2([a, b])$. Using the Leibnitz product rule we have

$$uL[v] - vL[u] = u(pv')' + uqv - v(pu')' - vqu$$
$$= (upv')' - u'pv' - (vpu')' + u'pv'.$$

We thus obtain the *Lagrange identity*:

$$uL[v] - vL[u] = \left[p \left(uv' - vu' \right) \right]'. \tag{6.47}$$

Integrating the Lagrange identity over the interval $[a, b]$ implies the identity

$$\int_a^b (uL[v] - vL[u]) \, dx = p \left(uv' - vu' \right) \Big|_a^b, \tag{6.48}$$

which is called *Green's formula*. Assume that u and v satisfy the boundary conditions (6.21) in the regular case, or (6.25) in the periodic case. Then it can be seen that

$$p \left(uv' - vu' \right) \big|_a^b = 0. \tag{6.49}$$

Therefore, for such u and v we have

$$\int_a^b \left(uL[v] - vL[u] \right) \, \mathrm{d}x = 0. \tag{6.50}$$

The algebraic interpretation of the above formula is that the operator L is a *symmetric operator* on the space of twice differentiable functions that satisfy either the regular boundary conditions (6.21), or the periodic boundary conditions (6.25), with respect to the inner product

$$\langle u, v \rangle = \int_a^b u(x)v(x) \, \mathrm{d}x.$$

Although the formal definition of a symmetric operator will not be given here, the analogy with the case of symmetric matrices acting on the vector space $\mathbb{R}^k$ (equipped with the standard inner product) is evident.

We point out that the operator L is symmetric in many singular cases. For example, if $a = -\infty$, $b = \infty$ and $\lim_{x \to \pm\infty} p(x) = 0$, then L is symmetric on the space of smooth bounded functions with bounded derivatives.

2 Orthogonality The following property also has a well-known analog in the case of symmetric matrices.

Proposition 6.20 *Eigenfunctions which belong to distinct eigenvalues of a regular Sturm–Liouville problem are orthogonal relative to the inner product*

$$\langle u, v \rangle_r = \int_a^b u(x)v(x)r(x) \, \mathrm{d}x.$$

Moreover, this property also holds in the periodic case and in fact also in many singular cases.

Proof Let v_n, v_m be two eigenfunctions belonging to the eigenvalues $\lambda_n \neq \lambda_m$, respectively. Hence,

$$-L[v_n] = \lambda_n r v_n, \tag{6.51}$$
$$-L[v_m] = \lambda_m r v_m. \tag{6.52}$$

Moreover, v_n, v_m satisfy the boundary conditions (6.21).

Multiplying (6.51) by v_m, and (6.52) by v_n, integrating over $[a, b]$, and then taking the difference between the two equations thus obtained, we find

$$-\int_a^b (v_m L[v_n] - v_n L[v_m]) \, dx = (\lambda_n - \lambda_m) \int_a^b v_n v_m r \, dx. \qquad (6.53)$$

Since v_n, v_m satisfy the boundary conditions (6.21), we may use Green's formula (6.50) to infer that

$$(\lambda_n - \lambda_m) \int_a^b v_n v_m r \, dx = 0.$$

But $\lambda_n \neq \lambda_m$, thus $\langle v_n, v_m \rangle_r = 0$.   □

Recall that for the Sturm–Liouville problem of Example 6.1, the orthogonality of the corresponding eigenfunctions was already shown, since we checked that for $m, n = 1, 2, 3, \ldots$

$$\int_0^L \sin \frac{m\pi x}{L} \sin \frac{n\pi x}{L} \, dx = \begin{cases} 0 & m \neq n, \\ L/2 & m = n. \end{cases} \qquad (6.54)$$

In other words, the sequence $\{\sqrt{2/L} \sin(n\pi x/L)\}_{n=1}^\infty$ is an orthonormal system of all the eigenfunctions of this problem.

Similarly, in Example 6.2 we found that for $m, n = 0, 1, 2, \ldots$

$$\int_0^L \cos \frac{m\pi x}{L} \cos \frac{n\pi x}{L} \, dx = \begin{cases} 0 & m \neq n, \\ L/2 & m = n \neq 0, \\ L & m = n = 0. \end{cases} \qquad (6.55)$$

Therefore, $\{\sqrt{1/L}\} \cup \{\sqrt{2/L} \cos(n\pi x/L)\}_{n=1}^\infty$ is an orthonormal system of all the eigenfunctions of the corresponding problem.

Consider now the periodic problem of Example 6.5. From (6.54) we have for $m, n = 1, 2, 3, \ldots$

$$\int_0^L \sin \frac{2m\pi x}{L} \sin \frac{2n\pi x}{L} \, dx = \begin{cases} 0 & m \neq n, \\ L/2 & m = n. \end{cases} \qquad (6.56)$$

From (6.55) we see that for $m, n = 0, 1, 2, \ldots$

$$\int_0^L \cos \frac{2m\pi x}{L} \cos \frac{2n\pi x}{L} \, dx = \begin{cases} 0 & m \neq n, \\ L/2 & m = n \neq 0, \\ L & m = n = 0. \end{cases} \qquad (6.57)$$

In addition, for $m = 1, 2, 3, \ldots, n = 0, 1, 2, \ldots$

$$\int_0^L \sin \frac{2m\pi x}{L} \cos \frac{2n\pi x}{L} \, dx = 0. \qquad (6.58)$$

It follows that our system of all eigenfunctions of the periodic problem is indeed orthogonal, including the orthogonality of eigenfunctions with the same eigenvalue. Moreover, the system

$$\left\{\sqrt{\frac{1}{L}}\right\} \cup \left\{\sqrt{\frac{2}{L}}\cos\frac{2n\pi x}{L}\right\}_{n=1}^{\infty} \cup \left\{\sqrt{\frac{2}{L}}\sin\frac{2n\pi x}{L}\right\}_{n=1}^{\infty}$$

is an *orthonormal* system of all the eigenfunctions of this periodic problem.

Note that the functions

$$u_n = \sin\frac{2n\pi x}{L}, \qquad w_n = \sin\frac{2n\pi x}{L} + \cos\frac{2n\pi x}{L}$$

are two linearly independent eigenfunctions belonging to the same eigenvalue; yet they are not orthogonal. But in such a case of nonsimple eigenvalue, one can carry out the Gram–Schmidt orthogonalization process to obtain an orthonormal system of all the eigenfunctions of the problem.

3 Real eigenvalues

Proposition 6.21 *The eigenvalues of a regular Sturm–Liouville problem are all real. Moreover, this property holds in the periodic case and also in many singular cases.*

Proof Assume that $\lambda \in \mathbb{C}$ is a nonreal eigenvalue with an eigenfunction v. Then

$$L[v] + \lambda r v = (pv')' + qv + \lambda r v = 0, \qquad (6.59)$$
$$B_a[v] = \alpha v(a) + \beta v'(a) = 0, \qquad B_b[v] = \gamma v(b) + \delta v'(b) = 0. \qquad (6.60)$$

Recall that the coefficients of (6.59)–(6.60) are all real. By forming the complex conjugate of (6.59)–(6.60), and interchanging the order of conjugation and differentiation, we obtain

$$\overline{L[v] + \lambda r v} = L[\overline{v}] + \overline{\lambda} r \overline{v} = 0, \qquad (6.61)$$
$$\overline{B_a[v]} = \alpha \overline{v}(a) + \beta \overline{v}'(a) = 0, \qquad \overline{B_b[v]} = \gamma \overline{v}(b) + \delta \overline{v}'(b) = 0. \qquad (6.62)$$

Therefore, $\overline{v}$ is an eigenfunction with eigenvalue $\overline{\lambda}$. By our assumption $\lambda \neq \overline{\lambda}$, and by Proposition 6.20 we have

$$0 = \langle \overline{v}, v \rangle_r = \int_a^b \overline{v(x)} v(x) r(x)\, dx = \int_a^b |v(x)|^2 r(x)\, dx.$$

On the other hand, since $v \neq 0$ and $r(x) > 0$ on $[a, b]$, it follows that $\int_a^b |v(x)|^2 r(x)dx > 0$, which leads to a contradiction. $\qquad \square$

4 Real eigenfunctions Let λ be an eigenvalue with eigenfunction v. Since for every complex number $C \neq 0$, the function Cv is also an eigenfunction with the same eigenvalue λ, it is not true that all the eigenfunctions are real. Moreover, for $n = 1, 2, \ldots$, the complex valued functions $\exp(\pm 2n\pi i x/L)$, which are not scalar multiples of real eigenfunctions, *are* eigenfunctions of the periodic problem of Example 6.5. We can prove, however, the following result.

Proposition 6.22 *Let λ be an eigenvalue of a regular or a periodic Sturm–Liouville problem, and denote by V_λ the subspace spanned by all the eigenfunctions with eigenvalue λ. Then V_λ admits an orthonormal basis of real valued functions.*

Proof Let v be an eigenfunction with eigenvalue λ. Recall that λ is a real number. By separating the real and the imaginary parts of (6.59)–(6.60), it can be checked that both $\operatorname{Re} v$ and $\operatorname{Im} v$ are solutions of the ODE (6.59) that satisfy the boundary conditions (6.60). Since at least one of these two functions is not zero, it follows that at least one of them is an eigenfunction. If λ is simple, then we now have a real basis for V_λ. On the other hand, if the multiplicity of λ is 2, we can consider the real and imaginary parts of two linearly independent eigenfunctions in V_λ. By a simple dimensional consideration, it follows that out of these four real functions, one can extract at least one pair of linearly independent functions. Then one applies the Gram–Schmidt process on such a pair of real eigenfunctions to obtain an orthonormal basis for V_λ. $\qquad\square$

5 Simple eigenvalues

Proposition 6.23 *The eigenvalues of a regular Sturm–Liouville problem are all simple.*

Proof Let v_1, v_2 be two eigenfunctions belonging to the same eigenvalue λ. Then

$$L[v_1] = -\lambda r v_1, \tag{6.63}$$

$$L[v_2] = -\lambda r v_2. \tag{6.64}$$

Therefore,

$$v_2 L[v_1] - v_1 L[v_2] = 0.$$

Recall that by the Lagrange identity

$$v_2 L[v_1] - v_1 L[v_2] = \left[p \left(v_2 v_1' - v_1 v_2' \right) \right]'. \tag{6.65}$$

Hence,

$$Q(x) := p \left(v_2 v_1' - v_1 v_2' \right) = \text{constant}.$$

On the other hand, we have shown that two functions that satisfy the same regular boundary conditions also satisfy $Q(a) = Q(b) = 0$. Since p is a positive function on the entire closed interval $[a, b]$, it follows that the Wronskian

$$W := v_2 v_1' - v_1 v_2'$$

vanishes at the end points. Recall that v_1, v_2 are solutions of the same linear ODE, and therefore, the Wronskian is identically zero. Consequently, the functions v_1, v_2 are linearly dependent. □

Remark 6.24 For the periodic eigenvalue problem of Example 6.5, we have shown that except for the first eigenvalue all the other eigenvalues are *not* simple.

6 Existence of an infinite sequence of eigenvalues The standard proof of the existence of an eigenvalue for matrices uses the characteristic polynomial and therefore cannot be generalized to the Sturm–Liouville case. Actually, it is not clear at all that a Sturm–Liouville problem admits *even one* eigenvalue; in fact, in 1836 both Sturm and Liouville published papers in the same journal where they independently asked exactly this particular question.

Example 6.25 It can be checked that the following singular Sturm–Liouville problem does not admit an eigenvalue.

$$v'' + \lambda v = 0 \quad x \in \mathbb{R},$$

$$\lim_{x \to -\infty} v(x) = \lim_{x \to \infty} v(x) = 0. \tag{6.66}$$

On the other hand, if we change the boundary conditions slightly:

$$v'' + \lambda v = 0 \quad x \in \mathbb{R},$$

$$\sup_{x \in \mathbb{R}} |v(x)| < \infty, \tag{6.67}$$

then the set of all eigenvalues of the problem is the half-line $[0, \infty)$. Indeed, for $\lambda > 0$ the eigenfunctions are $\sin \sqrt{\lambda} x$, $\cos \sqrt{\lambda} x$, while for $\lambda = 0$ the corresponding eigenfunction equals **1**. This set of eigenfunctions is not an orthogonal system with respect to the natural inner product $\langle u, v \rangle = \int_{-\infty}^{\infty} u(x) v(x) \, dx$, since for such a function v we have $\|v\|^2 = \infty$, and hence v does not belong to the corresponding inner product space.

The following proposition demonstrates that for regular problems the picture is simpler (the proof is beyond the scope of this book; see for example [6]).

Proposition 6.26 *The set of all eigenvalues of a regular Sturm–Liouville problem forms an unbounded strictly monotone sequence. We denote this sequence by*

$$\lambda_0 < \lambda_1 < \lambda_2 < \cdots < \lambda_n < \lambda_{n+1} < \cdots.$$

In particular, there are infinitely many eigenvalues, and $\lim_{n \to \infty} \lambda_n = \infty$.

*Moreover, the above statements are also valid in the periodic case, ex-
cept that the sequence $\{\lambda_n\}_{n=0}^{\infty}$ is only nondecreasing (repeated eigenvalues are
allowed).*

Corollary 6.27 *(1) A regular or periodic Sturm–Liouville problem admits an infinite orthonormal sequence of real eigenfunctions in $E_r(a, b)$.*
*(2) The sequence of all eigenvalues is an unbounded subset of the real line that is
bounded from below.*

7 Completeness, and convergence of the Fourier expansion The separation of
variables method (and the justification of Fourier's idea) relies on the following
convergence theorems; the proofs will not be given here (see for example [6]).

Proposition 6.28 *The orthonormal system $\{v_n\}_{n=0}^{\infty}$ of all eigenfunctions of a regular (or periodic) Sturm–Liouville problem is complete in the inner product space
$E_r(a, b)$.*

Definition 6.29 The generalized Fourier expansion of a function v with respect to
the orthonormal system $\{v_n\}_{n=0}^{\infty}$ of all eigenfunctions of a Sturm–Liouville problem
is called the *eigenfunction expansion* of v.

Proposition 6.28 implies that the eigenfunction expansion is converging in the mean
(in norm). In fact, for every function such that $\int_a^b u^2(x)r(x)\,dx < \infty$ the eigenfunction expansion of u converges in norm. If we assume further that the function u is
smoother we arrive at a stronger convergence result.

Proposition 6.30 *Let $\{v_n\}_{n=0}^{\infty}$ be an orthonormal system of all eigenfunctions of a
regular (or periodic) Sturm–Liouville problem.*

*(1) Let f be a piecewise differentiable function on $[a, b]$. Then for all $x \in (a, b)$
    the eigenfunction expansion of f with respect to the system $\{v_n\}_{n=0}^{\infty}$ converges to
    $[f(x_+) + f(x_-)]/2$ (i.e. the average of the two one-side limits of f at x).*
*(2) If f is a continuous and piecewise differentiable function that satisfies the boundary
    conditions of the given Sturm–Liouville problem, then the eigenfunction expansion of
    f with respect to the system $\{v_n\}_{n=0}^{\infty}$ converges uniformly to f on the interval $[a, b]$.*

In the following three examples we demonstrate Proposition 6.28 and Proposition
6.30 for three different eigenfunctions systems.

Example 6.31 Find the eigenfunction expansion of the function $f = 1$ with respect
to the orthonormal system $\{\sqrt{2/L}\sin(n\pi x/L)\}_{n=1}^{\infty}$ of the eigenfunctions of the
Sturm–Liouville problem of Example 6.1.

The Fourier coefficients are given by

$$b_n = \left\langle f, \sqrt{\frac{2}{L}} \sin \frac{n\pi x}{L} \right\rangle = \sqrt{\frac{2}{L}} \int_0^L \sin \frac{n\pi x}{L} \, dx = -\sqrt{\frac{2}{L}} \frac{L}{n\pi} \cos \frac{n\pi x}{L} \Big|_0^L$$

$$= \frac{\sqrt{2L}}{n\pi} [1 - (-1)^n].$$

Therefore, the series

$$\frac{2}{\pi} \sum_{n=1}^{\infty} \frac{[1 - (-1)^n]}{n} \sin \frac{n\pi x}{L} = \frac{4}{\pi} \sum_{k=0}^{\infty} \frac{1}{2k+1} \sin \frac{(2k+1)\pi x}{L} \tag{6.68}$$

is the eigenfunction expansion of f. While it converges to 1 for all $x \in (0, L)$, it does not converges uniformly on $[0, L]$ since f does not satisfy the Dirichlet boundary conditions at the end points.

Example 6.32 Find the eigenfunction expansion of the function $f(x) = x$ with respect to the orthonormal system $\{\sqrt{1/L}\} \cup \{\sqrt{2/L} \cos(n\pi x/L)\}_{n=1}^{\infty}$ of all the eigenfunctions of the Sturm–Liouville problem of Example 6.2.

For $n = 0$, we have

$$a_0 = \left\langle f, \sqrt{\frac{1}{L}} \right\rangle = \sqrt{\frac{1}{L}} \int_0^L x \, dx = \frac{(L)^{3/2}}{2}.$$

For $n \neq 0$, we have

$$a_n = \left\langle f, \sqrt{\frac{2}{L}} \cos \frac{n\pi x}{L} \right\rangle = \sqrt{\frac{2}{L}} \int_0^L x \cos \frac{n\pi x}{L} \, dx = -L \frac{\sqrt{2L}}{(n\pi)^2} [1 - (-1)^n].$$

Therefore, the series

$$\frac{L}{2} - \frac{2L}{\pi^2} \sum_{n=1}^{\infty} \frac{[1 - (-1)^n]}{n^2} \cos \frac{n\pi x}{L} = \frac{L}{2} - \frac{4L}{\pi^2} \sum_{k=0}^{\infty} \frac{1}{(2k+1)^2} \cos \frac{(2k+1)\pi x}{L}$$

is the eigenfunction expansion of f that converges to x for all $x \in (0, L)$. This expansion converges uniformly on $[0, L]$, although the expansion theorem does not ensure this.

Example 6.33 Find the eigenfunction expansion of the function

$$f(x) = \begin{cases} x & 0 \leq x \leq 1, \\ 1 & 1 \leq x \leq 2 \end{cases} \tag{6.69}$$

with respect to the (classical Fourier) orthonormal system

$$\left\{ \sqrt{\frac{1}{2}} \right\} \cup \{\cos n\pi x\}_{n=1}^{\infty} \cup \{\sin n\pi x\}_{n=1}^{\infty},$$

the eigenfunctions of the periodic Sturm–Liouville problem of Example 6.5 on $[0, 2]$.

For $n = 0$, we obtain

$$a_0 = \left\langle f, \sqrt{\frac{1}{2}} \right\rangle = \frac{3}{2\sqrt{2}}.$$

For $n \neq 0$, we have

$$a_n = \langle f, \cos n\pi x \rangle = \int_0^1 x \cos n\pi x \, dx + \int_1^2 \cos n\pi x \, dx = -\frac{[1 - (-1)^n]}{(n\pi)^2}.$$

In addition,

$$b_n = \langle f, \sin n\pi x \rangle = \int_0^1 x \sin n\pi x \, dx + \int_1^2 \sin n\pi x \, dx = -\frac{1}{n\pi}.$$

Therefore, the series

$$\frac{3}{4} + \sum_{n=1}^{\infty} \left[\frac{[(-1)^n - 1]}{n^2\pi^2} \cos n\pi x - \frac{1}{n\pi} \sin n\pi x \right]$$

is the corresponding eigenfunction expansion of f that converges to f for all $x \in (0, 2)$. This expansion does not converge uniformly on $[0, 2]$, since f does not satisfy the periodic boundary conditions.

Although the eigenfunction expansion for a piecewise differentiable function may not converge uniformly, it frequently happens that the expansion converges uniformly on any subinterval that does not contain the end points and jump discontinuities. Recall that at a jump discontinuity, the eigenfunction expansion converges to the average of the two one-sided limits of f. When one draws the graphs of the sums of the first N terms of this eigenfunction expansion, one notices oscillations that appear near the jump points. The oscillations persist even as the number of terms in the expansion is increased. These oscillations (which appear only for *finite* sums) are called the *Gibbs phenomenon* after the American scientist Josiah Willard Gibbs (1839–1903) who discovered them. We demonstrate the Gibbs phenomenon in the following example.

Example 6.34 Consider the following 2π-periodic function:

$$f(x) = \begin{cases} -1 & -\pi < x < 0, \\ 1 & 0 < x < \pi, \end{cases}$$

and $f(x + 2\pi) = f(x)$, which is sometimes called a *square wave*. This function is discontinuous at integer multiples of π. The eigenfunction expansion with respect

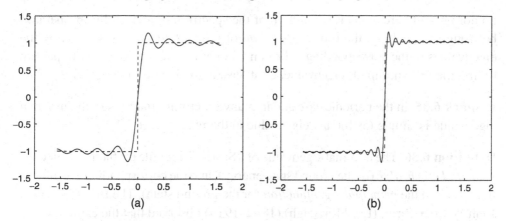

Figure 6.1 The Gibbs phenomenon for the square wave function: the partial sums for (a) $N = 8$ and (b) $N = 24$.

to the classical Fourier series is given by

$$f(x) = \frac{4}{\pi} \sum_{k=0}^{\infty} \frac{\sin(2k+1)x}{2k+1}. \tag{6.70}$$

Note that the eigenfunction expansions (6.68) for $L = \pi$ and (6.70) look the same. Clearly, the series (6.70) does not converge uniformly on $\mathbb{R}$. Consider the partial sum

$$f_N(x) := \frac{4}{\pi} \sum_{k=0}^{N} \frac{\sin(2k+1)x}{2k+1}.$$

In Figure 6.1 the graphs of f_8 and f_{24} are illustrated. It can be seen that while adding terms improves the approximation, no matter how many terms are added, there is always a fluctuation near the jump at $x = 0$ (overshoot before the jump and undershoot after it). To see the oscillation better, we concentrate the graphs on the interval $(-\pi/2, \pi/2)$. The graph of f is drawn (dashed line) in the background for comparison.

8 Rayleigh quotients An important problem that arises frequently in chemistry and physics is how to compute the spectrum of a quantum system. The system is modeled by a Schrödinger operator. In the one-dimensional case such operators are of the Sturm–Liouville type. For instance, the information from the spectrum of the Schrödinger operator enables us to determine the discrete frequencies of the radiation from excited atoms (We shall present an explicit computation of the spectral lines of the hydrogen atom in Chapter 9.) In addition, using the information from the spectrum, one can understand the stability of atoms and molecules. We do not present here a precise definition of the spectrum of a given linear operator, but roughly speaking, the (point) *spectrum* of a quantum system is given by the eigenvalues of the corresponding Schrödinger operator. It is particularly important to find the first (minimal) eigenvalue, or at least a good approximation of it.

Remark 6.35 In the periodic case and in many other important cases, the minimal eigenvalue is simple (as for any eigenvalue in the regular case).

Definition 6.36 The minimal eigenvalue of a Sturm–Liouville problem is called the *principal eigenvalue* (or the ground state energy), and the corresponding eigenfunction is called the *principal eigenfunction* (or the ground state). The British scientist John William Strutt (Lord Rayleigh) (1842–1919) observed that the expression

$$R(u) = -\frac{\int_a^b uL[u]\,dx}{\int_a^b u^2 r\,dx}$$

plays an important role in this context. Therefore $R(u)$ is called the *Rayleigh quotient* of u.

Most of the numerical methods for computing the eigenvalues of a symmetric operator are based on the following variational principle, which is called the *Rayleigh–Ritz formula*.

Proposition 6.37 *The principal eigenvalue λ_0 of a regular Sturm–Liouville problem satisfies the following variational principle:*

$$\lambda_0 = \inf_{u \in V} R(u) = \inf_{u \in V} -\frac{\int_a^b uL[u]\,dx}{\int_a^b u^2 r\,dx}, \tag{6.71}$$

where

$$V = \{u \in C^2([a, b]) \mid B_a[u] = B_b[u] = 0, \quad u \neq 0\}.$$

Moreover, the infimum of the Rayleigh quotient is attained only by the principal eigenfunction.
    For a periodic Sturm–Liouville problem, (6.71) holds true with

$$V = \{u \in C^2([a, b]) \mid v(a) = v(b), \ v'(a) = v'(b), \ v \neq 0\}.$$

Proof The following proof is not complete since it relies on some auxiliary lemmas which we do not prove here. Let $\{\lambda_n\}_{n=0}^\infty$ be the increasing sequence of all eigenvalues of the given problem, and let $\{v_n\}_{n=0}^\infty$ be the orthonormal system of the corresponding eigenfunctions.

If $u \in V$, then the eigenfunction expansion of u converges uniformly to u, i.e.

$$u(x) = \sum_{n=0}^{\infty} a_n v_n(x).$$

Without a rigorous justification, let us exchange the order of summation and differentiation. This implies that

$$L[u] = \sum_{n=0}^{\infty} a_n L[v_n(x)] = -\sum_{n=0}^{\infty} a_n \lambda_n r(x) v_n(x).$$

We substitute the above expression into the numerator of the Rayleigh quotient, and integrate term by term (again, without a rigorous justification), using the orthogonality relations. For the denominator of the Rayleigh quotient, we use the Parseval identity. We obtain

$$R(u) = -\frac{\int_a^b u L[u]\, dx}{\int_a^b u^2 r\, dx} = \frac{\int_a^b \left[\sum_{m=0}^{\infty} \sum_{n=0}^{\infty} a_m a_n \lambda_n r(x) v_m(x) v_n(x) \right] dx}{\sum_{m=0}^{\infty} a_n^2}$$

$$= \frac{\sum_{m=0}^{\infty} \sum_{n=0}^{\infty} a_m a_n \lambda_n \int_a^b r(x) v_m(x) v_n(x)\, dx}{\sum_{m=0}^{\infty} a_n^2}$$

$$= \frac{\sum_{n=0}^{\infty} a_n^2 \lambda_n}{\sum_{m=0}^{\infty} a_n^2} \geq \frac{\sum_{n=0}^{\infty} a_n^2 \lambda_0}{\sum_{m=0}^{\infty} a_n^2} = \lambda_0.$$

Therefore, $R(u) \geq \lambda_0$ for all $u \in V$, and thus, $\inf_{u \in V} R(u) \geq \lambda_0$. It is easily verified that equality holds if and only if $u = C v_0$ (recall that λ_0 is always a simple eigenvalue), and the proposition is proved. □

Remark 6.38 The following alternative method for computing the principal eigenvalue can be derived from the Rayleigh–Ritz formula through an integration by parts of (6.71)

$$\lambda_0 = \inf_{u \in V} \frac{\int_a^b \left(p(u')^2 - qu^2 \right) dx - puu'|_a^b}{\int_a^b u^2 r\, dx}, \tag{6.72}$$

where

$$V = \{ u \in C^2([a, b]) \mid B_a[u] = B_b[u] = 0, \ u \neq 0 \}.$$

Actually, (6.72) is more useful than (6.71) since it does not involve second derivatives. In particular, for the Dirichlet (or Neumann, or periodic) problem, we have

$$\lambda_0 = \inf_{u \in V} \frac{\int_a^b \left(p(u')^2 - qu^2 \right) dx}{\int_a^b u^2 r\, dx}. \tag{6.73}$$

Corollary 6.39 *If $q \leq 0$, and if $puu'|_a^b \leq 0$ for all functions $u \in V$, then all the eigenvalues of the Sturm–Liouville problem are nonnegative. In particular, for the Dirichlet (or Neumann, or periodic) problem, if $q \leq 0$, then all the eigenvalues of the problem are nonnegative.*

Example 6.40 Consider the following Sturm–Liouville problem:

$$\frac{d^2 v}{dx^2} + \lambda v = 0 \quad 0 < x < 1, \tag{6.74}$$

$$v(0) = v(1) = 0. \tag{6.75}$$

We already know that the principal eigenfunction is $v_0(x) = \sin \pi x$, with the principal eigenvalue $\lambda_0 = \pi^2$. If we use the test function $u(x) = x - x^2$ in the Rayleigh quotient, we obtain the bound

$$R(u) = 10 \geq \pi^2 \approx 9.86.$$

This bound is a surprisingly good approximation for λ_0.

In general, it is not possible to explicitly compute the eigenfunctions and the eigenvalues λ_n. But the Rayleigh–Ritz formula has a useful generalization for λ_n with $n \geq 1$. In fact, using Rayleigh quotients with appropriate test functions, one can obtain good approximations for the eigenvalues of the problem.

9 Zeros of eigenfunctions The following beautiful result holds (for a proof see Volume 1 of [4]).

Proposition 6.41 *Consider a regular Sturm–Liouville problem on the interval (a, b). Let $\{\lambda_n\}_{n=0}^\infty$ be the increasing sequence of all eigenvalues, and $\{v_n\}_{n=0}^\infty$ be the corresponding complete orthonormal sequence of eigenfunctions. Then v_n admits exactly n roots on the interval (a, b), for $n = 0, 1, 2 \ldots$. In particular, the principal eigenfunction v_0 does not change its sign in (a, b).*

The reader can check the proposition for the orthonormal systems of Examples 6.1 and 6.2.

10 Asymptotic behavior of high eigenvalues and eigenfunctions As was mentioned above, the eigenvalues λ_n of a regular Sturm–Liouville problem cannot, in general, be computed precisely; using the (generalized) Rayleigh–Ritz formula, however, we can obtain a good approximation for λ_n. It turns out that for large n there is no need to use a numerical method, since the asymptotic behavior of large eigenvalues is given by the following formula discovered by the German–American mathematician Herman Weyl (1885–1955). Write

$$\ell := \int_a^b \sqrt{\frac{r(x)}{p(x)}} \, dx,$$

then

$$\lambda_n \sim \left(\frac{n\pi}{\ell}\right)^2. \tag{6.76}$$

The symbol $\sim$ in the preceding formula means an asymptotic relation, i.e.

$$\lim_{n \to \infty} \frac{\ell^2 \lambda_n}{(n\pi)^2} = 1.$$

Furthermore, it is known that the general solution of the equation $L[u] + \lambda r u = 0$ for large λ behaves as a linear combination of cos and sin. More precisely, the general solution of the above equation for large λ takes the form

$$u(x) \sim [r(x)p(x)]^{-1/4} \left\{ \alpha \cos\left[\sqrt{\lambda} \int_a^x \sqrt{\frac{r(s)}{p(s)}} \, ds\right] + \beta \sin\left[\sqrt{\lambda} \int_a^x \sqrt{\frac{r(s)}{p(s)}} \, ds\right] \right\}.$$

It follows that the orthonormal sequence $\{v_n(x)\}$ of all eigenfunctions is uniformly bounded. Moreover, for large n, we have the asymptotic estimates (Vol I of [4], [9]):

$$|v_n(x)| \leq C_0, \quad \left|\frac{dv_n(x)}{dx}\right| \leq C_1\sqrt{\lambda_n}, \quad \left|\frac{d^2v_n(x)}{d^2x}\right| \leq C_2\lambda_n \leq C_3n^2. \tag{6.77}$$

Example 6.42 Let $h > 0$ be a fixed number. Consider the following mixed eigenvalue problem

$$u'' + \lambda u = 0 \quad 0 < x < 1, \quad u(0) = 0, \quad hu(1) + u'(1) = 0. \tag{6.78}$$

If $\lambda < 0$, then a solution of the ODE above that satisfies the first boundary condition is a function of the form $u_\lambda(x) = C \sinh\sqrt{-\lambda}x$. The boundary condition at $x = 1$ implies that $0 < \tanh\sqrt{-\lambda} = -\sqrt{-\lambda}/h < 0$ which is impossible. This means that there are no negative eigenvalues.

If $\lambda = 0$, then a nontrivial solution of the corresponding ODE is a linear function of the form $u_0(x) = cx + d$ where $|c| + |d| > 0$. But such a function cannot satisfy both boundary conditions, since our boundary conditions clearly imply that $d = 0$ and $c(1 + h) = 0$.

If $\lambda > 0$, then an eigenfunction has the form $u_\lambda(x) = C \sin\sqrt{\lambda}x$, where λ is a solution of the *transcendental* equation

$$\tan\sqrt{\lambda} = -\sqrt{\lambda}/h. \tag{6.79}$$

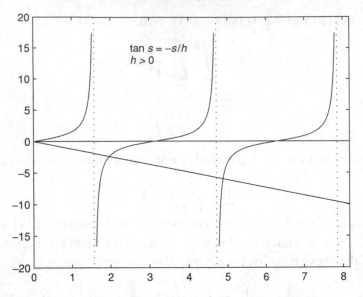

Figure 6.2 The graphical solution of (6.79).

Equation (6.79) cannot be solved analytically. Using the intermediate value theorem, however, we can verify that this transcendental equation has infinitely many roots λ_n that satisfy $(n - 1/2)\pi < \sqrt{\lambda_n} < n\pi$ (see Figure 6.2). Therefore, all the eigenvalues are positive and simple. Note that using Corollary 6.39, we could have concluded directly that there are no negative eigenvalues, since in our case $q = 0$, and for all $u \in V$ we have

$$puu'|_0^1 = -h(u'(1))^2 \le 0.$$

Let us check the asymptotic behavior of the eigenvalues as a function of n, and also as a function of h. Denote the sequence of the eigenvalues of the above Sturm–Liouville problem by $\{\lambda_n^{(h)}\}_{n=1}^\infty$. The nth eigenvalue satisfies

$$(n - 1/2)\pi < \sqrt{\lambda_n^{(h)}} < n\pi \qquad n = 1, 2, \dots .$$

Using our graphical solution (Figure 6.2), we verify that as $n \to \infty$ the asymptotic formula (6.76) is satisfied, namely,

$$\lim_{n \to \infty} \frac{\lambda_n^{(h)}}{n^2 \pi^2} = 1.$$

Let $h \to 0^+$. The slope of the straight line $-s/h$ tends to $-\infty$. Therefore, this line intersects the graph of the function $\tan s$ closer and closer to the negative asymptotes of $\tan s$. Hence,

$$\lim_{h \to 0} \lambda_n^{(h)} = \left[\frac{(2n-1)\pi}{2}\right]^2.$$

Indeed, for $h = 0$ we have the Sturm–Liouville problem

$$u'' + \lambda u = 0 \qquad 0 < x < 1,$$
$$u(0) = 0, \quad u'(1) = 0,$$

(see Example 6.19), and the eigenvalues and eigenfunctions of this problem are

$$\lambda_n = \left[\frac{(2n-1)\pi}{2}\right]^2, \quad u_n(x) = C_n \sin \frac{(2n-1)\pi x}{2}, \quad n = 1, 2, \dots.$$

Similarly, we have

$$\lim_{h \to \infty} \lambda_n^{(h)} = (n\pi)^2,$$

which are the eigenvalues of the limit problem:

$$u'' + \lambda u = 0 \qquad 0 < x < 1,$$
$$u(0) = 0, \quad u(1) = 0,$$

(see Example 6.1). The reader should check as an exercise that for $h > 0$, the nth eigenfunction admits exactly $(n-1)$ zeros in $(0, 1)$, where $n = 1, 2 \dots$.

Example 6.43 Assume now that h is a fixed *negative* number. Consider again the mixed eigenvalue problem

$$u'' + \lambda u = 0 \qquad 0 < x < 1,$$
$$u(0) = 0, \quad hu(1) + u'(1) = 0.$$

If $\lambda < 0$, then a solution of the ODE that satisfies the boundary condition at $x = 0$ is of the form $u_\lambda(x) = C \sinh \sqrt{-\lambda} x$. The boundary condition at $x = 1$ implies $\tanh \sqrt{-\lambda} = -\sqrt{-\lambda}/h$. Since the function $\tanh s$ is a concave increasing function on $[0, \infty)$ that satisfies

$$\tanh(0) = 0, \quad (\tanh)'(0) = 1, \quad \lim_{s \to \infty} \tanh s = 1,$$

it follows that the equation $\tanh s = -s/h$ has a positive solution if and only if $h < -1$. Moreover, under this condition there is exactly one solution (see Figure 6.3). This is a necessary and sufficient condition for the existence of a negative eigenvalue. The corresponding eigenfunction of the unique negative eigenvalue λ_0 is of the form $u_0(x) = \sinh \sqrt{-\lambda_0} x$ that indeed does not vanish on $(0, 1)$.

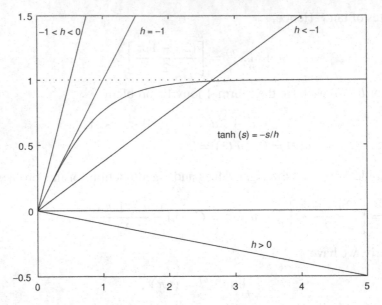

Figure 6.3 The graphical solution method for negative eigenvalues in Example 6.43.

If $\lambda = 0$, then a solution of the corresponding ODE is of the form $u_0(x) = cx + d$. The first boundary condition implies $d = 0$, and from the second boundary condition we have $c(1 + h) = 0$. Consequently, $\lambda = 0$ is an eigenvalue if and only if $h = -1$. If $h = -1$, then $\lambda_0 = 0$ is the minimal eigenvalue, and corresponding eigenfunction is $u_0(x) = x$ (notice that this function does not vanish on $(0, 1)$).

If $\lambda > 0$, then an eigenfunction has the form $u_\lambda(x) = C \sin \sqrt{\lambda} x$, where λ is a solution of the transcendental equation

$$\tan \sqrt{\lambda} = -\sqrt{\lambda}/h.$$

This equation has infinitely many solutions: If $h \leq -1$, then the solutions of this equation satisfy $n\pi < \sqrt{\lambda_n} < (n + 1/2)\pi$, where $n \geq 1$. On the other hand, if $-1 < h < 0$, then the minimal eigenvalue satisfies $0 < \sqrt{\lambda_0} < \pi/2$, while $n\pi < \sqrt{\lambda_n} < (n + 1/2)\pi$ for all $n \geq 1$ (see Figure 6.4). Note that for $n \geq 0$, the function $\lambda_n^{(h)}$ is an increasing function of h.

The asymptotic behavior of the eigenvalues of the present example as a function of h or n is similar to the behavior in the preceding example ($h > 0$).

The reader should check as an exercise the number of roots of the nth eigenfunction for $h < 0$. Here one should distinguish between the following three cases:

(1) The problem admits a negative eigenvalue ($h < -1$).
(2) The principal eigenvalue is zero ($h = -1$).
(3) All the eigenvalues are positive ($h > -1$).

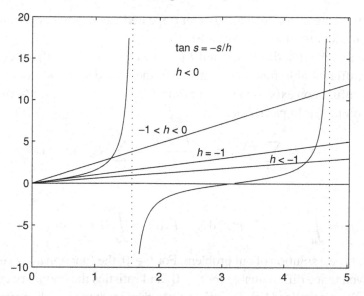

Figure 6.4 The graphical solution method for positive eigenvalues in Example 6.43.

6.5 Nonhomogeneous equations

We turn our attention to the case of nonhomogeneous problems in the general context of the Sturm–Liouville theory.

Consider first a general nonhomogeneous parabolic initial boundary value problem with homogeneous boundary conditions. One can solve a general hyperbolic problem similarly, and we leave this to the reader as an exercise.

We seek a function $u(x, t)$ that is a solution of the problem

$$
\begin{aligned}
&r(x)m(t)u_t - [(p(x)u_x)_x + q(x)u] = F(x, t) && a < x < b, \ t > 0, \\
&B_a[u] = \alpha u(a, t) + \beta u_x(a, t) = 0 && t \geq 0, \\
&B_b[u] = \gamma u(b, t) + \delta u_x(b, t) = 0 && t \geq 0, \\
&u(x, 0) = f(x) && a \leq x \leq b.
\end{aligned}
\tag{6.80}
$$

One can also deal with periodic boundary conditions by the same method.

The related Sturm–Liouville eigenvalue problem that is derived from the *homogeneous* problem using the method of separation of variables is of the form

$$
L[v] + \lambda r v = (pv')' + qv + \lambda r v = 0 \quad a < x < b, \tag{6.81}
$$

$$
B_a[v] = B_b[v] = 0. \tag{6.82}
$$

Let $\{v_n\}_{n=0}^{\infty}$ be the complete orthonormal sequence of eigenfunctions of the problem, and let the corresponding sequence of eigenvalues be denoted by $\{\lambda_n\}_{n=0}^{\infty}$, where the eigenvalues are in nondecreasing order (repeated eigenvalues are allowed, since we may consider also the periodic case).

Suppose that the functions $f(x)$ and $F(x, t)/r(x)$ (for all $t \geq 0$) are continuous, piecewise differentiable functions that satisfy the boundary conditions. It follows that the eigenfunction expansions of $f(x)$ and $F(x, t)/r(x)$ (for all $t \geq 0$) converge uniformly on $[a, b]$. In particular,

$$f(x) = \sum_{n=0}^{\infty} f_n v_n(x), \qquad \frac{F(x, t)}{r(x)} = \sum_{n=0}^{\infty} F_n(t) v_n(x),$$

where

$$f_n = \int_a^b f(x) v_n(x) r(x) \, dx, \qquad F_n(t) = \int_a^b F(x, t) v_n(x) \, dx.$$

Let $u(x, t)$ be a solution of our problem. For $t \geq 0$, the function $u(\cdot, t)$ is continuous (and even twice differentiable for $t > 0$) and satisfies the boundary conditions. Therefore, the generalized Fourier series of u with respect to the orthonormal system $\{v_n\}_{n=0}^{\infty}$ converges (and uniformly so for $t > 0$) and has the form

$$u(x, t) = \sum_{n=0}^{\infty} a_n(t) v_n(x),$$

where

$$a_n(t) = \int_a^b u(x, t) v_n(x) r(x) \, dx.$$

Let us fix $n \geq 0$. Substituting the time derivative of a_n into the PDE leads to

$$m(t) a_n'(t) = \int_a^b m(t) r(x) \frac{\partial u(x, t)}{\partial t} v_n(x) \, dx$$

$$= \int_a^b L[u(x, t)] v_n(x) \, dx + \int_a^b F(x, t) v_n(x) \, dx.$$

Green's formula with respect to the functions $v_n(x)$ and $u(x, t)$ implies that

$$m(t) a_n'(t) = \int_a^b u(x, t) L[v_n(x)] \, dx + \int_a^b F(x, t) v_n(x) \, dx$$

$$= -\lambda_n \int_a^b u(x, t) v_n(x) r(x) \, dx + \int_a^b F(x, t) v_n(x) \, dx$$

$$= -\lambda_n a_n(t) + F_n(t).$$

Therefore, the function a_n is a solution of the ODE

$$m(t) a_n' + \lambda_n a_n = F_n(t).$$

The solution of this first-order linear ODE is given by

$$a_n(t) = a_n(0)e^{-\lambda_n \int_0^t \frac{1}{m(s)}\,ds} + e^{-\lambda_n \int_0^t \frac{1}{m(s)}\,ds}\int_0^t \frac{F_n(\tau)}{m(\tau)}e^{\lambda_n \int_0^\tau \frac{1}{m(s)}\,ds}\,d\tau$$

(see Formula (1) of Section A.3). The continuity of u at $t = 0$, and the initial condition $u(x, 0) = f(x)$ imply that

$$a_n(0) = f_n.$$

Thus, we propose a solution of the form

$$u(x, t) = \sum_{n=0}^{\infty} f_n v_n(x)e^{-\lambda_n \int_0^t \frac{1}{m(s)}\,ds} + \sum_{n=0}^{\infty} v_n(x)e^{-\lambda_n \int_0^t \frac{1}{m(s)}\,ds}\int_0^t \frac{F_n(\tau)}{m(\tau)}e^{\lambda_n \int_0^\tau \frac{1}{m(s)}\,ds}\,d\tau.$$

We need to show that u is indeed a classical solution. For this purpose we estimate the general term of the series and its derivatives. Since m is a positive continuous function on $[0, T]$, there exist constants $0 < c_1 \le c_2$ such that

$$c_2^{-1} \le m(t) \le c_1^{-1},$$

and hence,

$$c_1 t \le \int_0^t \frac{1}{m(s)}\,ds \le c_2 t.$$

Consequently, for all $0 \le t \le T$ we have

$$e^{-c_2 \lambda_n t} \le e^{-\lambda_n \int_0^t \frac{1}{m(s)}\,ds} \le e^{-c_1 \lambda_n t}.$$

Furthermore,

$$\int \frac{e^{\lambda_n \int_0^\tau \frac{1}{m(s)}\,ds}}{m(\tau)}\,d\tau = \frac{e^{\lambda_n \int_0^\tau \frac{1}{m(s)}\,ds}}{\lambda_n} + C.$$

Thus

$$\left| e^{-\lambda_n \int_0^t \frac{1}{m(s)}\,ds}\int_0^t F_n(\tau)\frac{e^{\lambda_n \int_0^\tau \frac{1}{m(s)}\,ds}}{m(\tau)}\,d\tau \right|$$

$$\le e^{-\lambda_n \int_0^t \frac{1}{m(s)}\,ds} \max_{0 \le t \le T} |F_n(t)| \int_0^t \frac{e^{\lambda_n \int_0^t \frac{1}{m(s)}\,ds}}{m(\tau)}\,d\tau$$

$$= \frac{1}{\lambda_n} \max_{0 \le t \le T} |F_n(t)|(1 - e^{-\lambda_n \int_0^t \frac{1}{m(s)}\,ds}) \le \frac{1}{\lambda_n} \max_{0 \le t \le T} |F_n(t)|.$$

Moreover, by the asymptotic estimates (6.77):

$$|v_n(x)| \le C_0, \quad \left|\frac{dv_n(x)}{dx}\right| \le C_1\sqrt{\lambda_n}, \quad \left|\frac{d^2 v_n(x)}{d^2 x}\right| \le C_2\lambda_n \le C_3 n^2.$$

Since $|v_n(x)| \leq C_0$, it follows that $|f_n|$ and $|F_n(t)|$ are uniformly bounded. We assume further that the series $\sum f_n$ and $\sum F_n(t)$ converge absolutely and uniformly (on $0 \leq t \leq T$). This assumption can be verified directly in many cases where f and F are twice differentiable and satisfy the boundary conditions. In other words, we assume that

$$\sum_{n=0}^{\infty} [|f_n| + \max_{0 \leq t \leq T} |F_n(t)|] < \infty.$$

We are now ready to prove that the proposed series is a classical solution. First we show that the eigenfunction expansion of $u(x, t)$ satisfies the parabolic PDE

$$r(x)m(t)u_t - [(p(x)u_x)_x + q(x)u] = F(x, t) \quad a < x < b, \ t > 0.$$

Differentiate (term by term) the series of u, twice with respect to x and once with respect to t for $0 < \varepsilon \leq t \leq T$. We claim that the obtained series is uniformly converging. Indeed, using the asymptotic estimates, $a_n(t)v_n(x)$ (the general term of the series of u), its first- and second-order derivatives with respect to x, and its first derivative with respect to t are all bounded by

$$C\lambda_n e^{-c_1 \lambda_n \varepsilon} + C \max_{0 \leq t \leq T} |F_n(t)| \leq C_1 n^2 e^{-C_2 n^2 \varepsilon} + C \max_{0 \leq t \leq T} |F_n(t)|.$$

By the Weierstrass M-test for uniform convergence, the corresponding series of the function u and its derivatives (up to second order in x and first order in t) converge uniformly in the rectangle

$$\{(x, t) \mid a \leq x \leq b, \ \varepsilon \leq t \leq T\}.$$

Consequently, we may differentiate the series of u term by term, twice with respect to x or once with respect to t, and therefore, u is evidently a solution of the nonhomogeneous PDE and satisfies the boundary conditions for $t > 0$.

For $0 \leq x \leq \pi$, $0 \leq t \leq T$ the general term of the series of u is bounded by

$$C\left[|f_n| + \frac{\max_{0 \leq t \leq T} |F_n(t)|}{\lambda_n}\right].$$

By the Weierstrass M-test, the series of u converges uniformly on the strip

$$\{(x, t) \mid 0 \leq x \leq \pi, \ 0 \leq t \leq T\}.$$

Thus, the solution u is continuous on this strip and, in particular, at $t = 0$ we have $u(x, 0) = f(x)$. Hence, u is a classical solution.

Remark 6.44 In the hyperbolic case, the ODE for the coefficient a_n is of second order and has the form

$$m(t)a_n'' + \lambda_n a_n = F_n(t).$$

Example 6.45 Let $m \in \mathbb{N}$ and let $\omega \in \mathbb{R}$, and assume first that $\omega^2 \neq m^2\pi^2$. Solve the following wave problem:

$$
\begin{aligned}
u_{tt} - u_{xx} &= \sin m\pi x \, \sin \omega t & 0 < x < 1, \, t > 0, \\
u(0, t) = u(1, t) &= 0 & t \geq 0, \\
u(x, 0) &= 0 & 0 \leq x \leq 1, \\
u_t(x, 0) &= 0 & 0 \leq x \leq 1.
\end{aligned}
\tag{6.83}
$$

The related Sturm–Liouville problem is of the form

$$
\begin{aligned}
v'' + \lambda \, v &= 0, \\
v(0) = v(1) &= 0.
\end{aligned}
\tag{6.84}
$$

The eigenvalues are given by $\lambda_n = n^2\pi^2$, and the corresponding eigenfunctions are $v_n(x) = \sin n\pi x$, where $n = 1, 2, \ldots$. Therefore, the eigenfunction expansion of the solution u of (6.83) is

$$
u(x, t) = \sum_{n=1}^{\infty} T_n(t) \sin n\pi x.
\tag{6.85}
$$

In order to compute the coefficients $T_n(t)$, we (formally) substitute the series (6.85) into (6.83) and differentiate term by term. We find

$$
\sum_{n=1}^{\infty} (T_n'' + n^2\pi^2 \, T_n) \sin n\pi x = \sin m\pi x \, \sin \omega t.
\tag{6.86}
$$

Thus, for $n = m$ we need to solve the nonhomogeneous equation

$$
T_m'' + m^2\pi^2 T_m = \sin \omega t.
\tag{6.87}
$$

The corresponding initial conditions are zero. Therefore, the solution of this initial value problem is given by

$$
T_m(t) = \frac{1}{\omega^2 - m^2\pi^2} \left(\frac{\omega}{m\pi} \sin m\pi t - \sin \omega t \right).
\tag{6.88}
$$

For $n \neq m$ the corresponding ODE is

$$
T_n'' + n^2\pi^2 T_n = 0,
\tag{6.89}
$$

and again the initial conditions are zero. We conclude that $T_n(t) = 0$ for $n \neq m$, and

$$
u(x, t) = \frac{1}{\omega^2 - m^2\pi^2} \left(\frac{\omega}{m\pi} \sin m\pi t - \sin \omega t \right) \sin m\pi x.
\tag{6.90}
$$

As can be easily verified, u is a classical solution.

Assume now that $\omega^2 = m^2\pi^2$. Comparing the new problem with the previous one, we see that in solving the new problem the only difference occurs in (6.87). A simple way to derive the solution for $\omega^2 = m^2\pi^2$ from (6.90) is by letting $\omega \to m\pi$. Using L'Hospital's rule, we obtain

$$u(x, t) = \frac{1}{m\pi}\left(\frac{\sin m\pi t}{m\pi} - t\cos m\pi t\right)\sin m\pi x. \tag{6.91}$$

Let us discuss the important result that we have just obtained. Recall that the natural frequencies of the free string (without forcing) are $n\pi$ for $n = 1, 2, \ldots$. If the forcing frequency is not equal to one of the natural frequencies, the vibration of the string is a superposition of vibrations in the natural frequencies and in the forcing frequency, and the amplitude of the vibration is bounded. The energy provided by the external force to the string is divided between these two types of motion.

On the other hand, when the forcing frequency is equal to one of the natural frequencies, the amplitude of the vibrating string grows linearly in t and it is unbounded as $t \to \infty$. Of course, at some point the string will be ripped apart. The energy that is given to the string by the external force concentrates around one natural frequency and causes its amplitude to grow. This phenomenon is called *resonance*. It can partly explain certain cases where structures such as bridges and buildings collapse (see also Example 9.27). Note that the resonance phenomenon does not occur in the heat equation (see Exercise 5.7).

6.6 Nonhomogeneous boundary conditions

We now consider a general, one-dimensional, nonhomogeneous, parabolic initial boundary problem with nonhomogeneous boundary conditions (the hyperbolic case can be treated similarly). Let $u(x, t)$ be a solution of the problem

$$
\begin{aligned}
r(x)m(t)u_t - [(p(x)u_x)_x + q(x)u] &= F(x, t) \; a < x < b, \; t > 0, \\
B_a[u] = \alpha u(a, t) + \beta u_x(a, t) &= a(t) & t \geq 0, \\
B_b[u] = \gamma u(b, t) + \delta u_x(b, t) &= b(t) & t \geq 0, \\
u(x, 0) &= f(x) & a \leq x \leq b.
\end{aligned}
\tag{6.92}
$$

We already know how to use the eigenfunction expansion method to solve for homogeneous boundary conditions. We describe a simple technique for reducing the nonhomogeneous boundary conditions to the homogeneous case.

First we look for an auxiliary simple smooth function $w(x, t)$ satisfying (only) the given nonhomogeneous boundary conditions. In fact, we can always find such

Table 6.1.

Boundary condition	$w(x, t)$
Dirichlet: $u(0, t) = a(t)$, $u(L, t) = b(t)$	$w(x, t) = a(t) + \dfrac{x}{L}[b(t) - a(t)]$
Neumann: $u_x(0, t) = a(t)$, $u_x(L, t) = b(t)$	$w(x, t) = xa(t) + \dfrac{x^2}{2L}[b(t) - a(t)]$
Mixed: $u(0, t) = a(t)$, $u_x(L, t) = b(t)$	$w(x, t) = a(t) + xb(t)$
Mixed: $u_x(0, t) = a(t)$, $u(L, t) = b(t)$	$w(x, t) = (x - L)a(t) + b(t)$

a function that has the form

$$w(x, t) = (A_1 + B_1 x + C_1 x^2)a(t) + (A_2 + B_2 x + C_2 x^2)b(t). \tag{6.93}$$

Clearly, the function $v(x, t) = u(x, t) - w(x, t)$ should satisfy the homogeneous boundary conditions $B_a[v] = B_b[v] = 0$.

In the second step we check what are the PDE and the initial condition that v should satisfy in order for u to be a solution of the problem. By the superposition principle, it follows that v should be a solution of the following initial boundary problem

$$\begin{aligned}
r(x)m(t)v_t - ((p(x)v_x)_x + q(x)v) &= \tilde{F}(x, t) & a < x < b, \ t > 0, \\
B_a[v] = \alpha v(a, t) + \beta v_x(a, t) &= 0 & t \geq 0, \\
B_b[v] = \gamma v(b, t) + \delta v_x(b, t) &= 0 & t \geq 0, \\
v(x, 0) &= \tilde{f}(x) & a \leq x \leq b,
\end{aligned}$$

where

$$\tilde{F}(x, t) = F(x, t) - r(x)m(t)w_t + [(p(x)w_x)_x + q(x)w], \qquad \tilde{f}(x) = f(x) - w(x, 0).$$

Since this is exactly the kind of problem that was solved in the preceding section, we can proceed just as was explained there. To assist the reader in solving nonhomogeneous equations we present Table 6.1 where we list appropriate auxiliary functions w for the various boundary value problems.

We conclude this section with a final example in which we solve a nonhomogeneous heat problem.

Example 6.46 Consider the problem

$$\begin{aligned}
u_t - u_{xx} &= e^{-t} \sin 3x & 0 < x < \pi, \ t > 0, \\
u(0, t) = 0, \ u(\pi, t) &= 1 & t \geq 0, \\
u(x, 0) &= f(x) & 0 \leq x \leq \pi.
\end{aligned}$$

We shall solve the problem by the method of separation of variables. We shall also show that, under some regularity assumptions on f, the solution u is classical.

Recall that the eigenvalues and the corresponding eigenfunctions of the related Sturm–Liouville problem are of the form: $\{\lambda_n = n^2, \ v_n(x) = \sqrt{2/\pi}\sin nx\}_{n=1}^{\infty}$. In the first step, we reduce the problem to one with homogeneous boundary conditions. Using Table 6.1 we select the auxiliary function $w(x, t) = x/\pi$ that satisfies the given boundary conditions. Setting $v(x, t) = u(x, t) - w(x, t)$, then $v(x, t)$ is a solution of the problem:

$$v_t - v_{xx} = e^{-t}\sin 3x \qquad\qquad 0 < x < \pi, \ t > 0,$$
$$v(0, t) = 0, \quad v(\pi, t) = 0 \qquad t \geq 0,$$
$$v(x, 0) = f(x) - x/\pi \qquad\qquad 0 \leq x \leq \pi.$$

We assume that v is a classical solution that is a smooth function for $t > 0$. In particular, for a fixed $t > 0$ the eigenfunction expansion of $v(x, t)$ with respect the eigenfunctions of the related Sturm–Liouville problem converges uniformly and is of the form

$$v(x, t) = \sum_{n=1}^{\infty} a_n(t)\sin nx,$$

where

$$a_n(t) = \frac{2}{\pi}\int_0^{\pi} v(x, t)\sin nx \, dx.$$

Since by our assumption v is smooth, we can differentiate the function a_n with respect to t and then substitute the expansions for the derivatives into the PDE. We obtain

$$a_n'(t) = \frac{2}{\pi}\int_0^{\pi} (v_{xx}(x, t) + e^{-t}\sin 3x)\sin nx \, dx.$$

By Green's identity, with the functions $\sin nx$ and $v(x, t)$, and the operator $L[u] = \partial^2 u/\partial x^2$ we have

$$a_n'(t) = \frac{2}{\pi}\int_0^{\pi} (-n^2 v(x, t) + e^{-t}\sin 3x)\sin nx \, dx$$
$$= -n^2 a_n(t) + \frac{2e^{-t}}{\pi}\int_0^{\pi} \sin 3x \sin nx \, dx.$$

Consequently, a_n is a solution of the ODE

$$a_n' + n^2 a_n = \begin{cases} 0 & n \neq 3, \\ e^{-t} & n = 3. \end{cases}$$

The solution of this ODE is given by:

$$a_n(t) = \begin{cases} a_n(0)e^{-n^2 t} & n \neq 3, \\ 1/8e^{-t} + [a_3(0) - 1/8]e^{-9t} & n = 3. \end{cases}$$

The continuity of v at $t = 0$ and the initial condition f imply that

$$a_n(0) = \frac{2}{\pi} \int_0^\pi [f(x) - x/\pi] \sin nx \, dx,$$

and the proposed solution is

$$u(x, t) = \frac{x}{\pi} + \frac{1}{8}(e^{-t} - e^{-9t}) \sin 3x + \sum_{n=1}^{\infty} a_n(0) \sin nx \, e^{-n^2 t}. \tag{6.94}$$

It remains to show that u is indeed a classical solution. For this purpose, we assume further that $f(x) \in C^2([0, \pi])$ and satisfies the compatibility condition $f(0) = 0$, $f(\pi) = 1$. Under these assumptions, it follows from the convergence theorems that the eigenfunction expansion of the function of $f(x) - x/\pi$ converges uniformly to $f(x) - x/\pi$. Moreover, for the orthogonal system $\{\sin nx\}$ it is known [13] from classical Fourier analysis that, under the above conditions, the series $\sum |a_n(0)|$ converges.

We first prove that $u(x, t)$ satisfies the nonhomogeneous heat equation

$$u_t - u_{xx} = e^{-t} \sin 3x \qquad 0 < x < \pi, \ t > 0.$$

Since f is bounded on $[0, \pi]$, the Fourier coefficients $a_n(0)$ are bounded. For $t > \varepsilon > 0$, we formally differentiate the general term of the series of u twice with respect to x or once with respect to t. The obtained terms are bounded by $Cn^2 e^{-n^2 \varepsilon}$. Consequently, by the Weierstrass M-test, the corresponding series converges uniformly on the strip

$$\{(x, t) \, | \, 0 \leq x \leq \pi, \ \varepsilon \leq t\}.$$

Therefore, the term-by-term differentiation of the series of u is justified, and by our construction, u is a solution of the PDE.

For $0 \leq x \leq \pi$, $t \geq 0$, the general term of u is bounded by $|a_n(0)|$. Hence, by the Weierstrass M-test, the series of u converges uniformly on the strip

$$\{(x, t) \, | \, 0 \leq x \leq \pi, \ t \geq 0\}.$$

It follows that the series representing u is continuous there, and by substituting $x = 0, \pi$, we see that u satisfies the boundary conditions.

Similarly, substituting $t = 0$ implies that

$$u(x, 0) - \frac{x}{\pi} = \sum_{n=0}^{\infty} a_n(0) \sin nx,$$

which is the eigenfunction expansion of $(f(x) - x/\pi)$. By Proposition 6.30, the expansion converges uniformly to $(f(x) - x/\pi)$. In particular, $u(x, 0) = f(x)$. Thus, u is a classical solution.

In the special case where $f(x) = (x/\pi)^2$, then

$$a_n(0) = \frac{2}{\pi} \int_0^{\pi} [(x/\pi)^2 - x/\pi] \sin nx \, dx = \begin{cases} \dfrac{-8}{\pi^3(2k+1)^3} & n = 2k+1, \\ 0 & n = 2k. \end{cases}$$

Substituting $a_n(0)$ into (6.94) implies the solution

$$u(x, t) = \frac{x}{\pi} + \frac{1}{8}(e^{-t} - e^{-9t}) \sin 3x - \frac{8}{\pi^3} \sum_{k=0}^{\infty} \frac{\sin(2k+1)x}{(2k+1)^3} e^{-(2k+1)^2 t}$$

which is indeed a classical solution.

6.7 Exercises

6.1 Consider the following Sturm–Liouville problem

$$u'' + \lambda u = 0 \qquad 0 < x < 1,$$
$$u(0) - u'(0) = 0, \qquad u(1) + u'(1) = 0.$$

(a) Show that all the eigenvalues are positive.
(b) Solve the problem.
(c) Obtain an asymptotic estimate for large eigenvalues.

6.2 (a) Solve the Sturm–Liouville problem

$$(xu')' + \frac{\lambda}{x} u = 0 \qquad 1 < x < e,$$
$$u(1) = u'(e) = 0.$$

(b) Show directly that the sequence of eigenfunctions is orthogonal with respect the related inner product.

6.3 (a) Consider the Sturm–Liouville problem

$$(x^2 v')' + \lambda v = 0 \quad 1 < x < b, \qquad v(1) = v(b) = 0, \qquad (b > 1).$$

Find the eigenvalues and eigenfunctions of the problem.
Hint Show that the function

$$v(x) = x^{-1/2} \sin(\alpha \ln x)$$

is a solution of the ODE and satisfies the boundary condition $v(1) = 0$.

(b) Write a formal solution of the following heat problem

$$u_t = (x^2 u_x)_x \qquad 1 < x < b, \ t > 0, \tag{6.95}$$
$$u(1, t) = u(b, t) = 0 \qquad t \geq 0, \tag{6.96}$$
$$u(x, 0) = f(x) \qquad 1 \leq x \leq b. \tag{6.97}$$

6.4 Use the Rayleigh quotient to find a good approximation for the principal eigenvalue of the Sturm–Liouville problem

$$u'' + (\lambda - x^2)u = 0 \qquad 0 < x < 1,$$
$$u'(0) = u(1) = 0.$$

6.5 (a) Solve the Sturm–Liouville problem

$$((1 + x)^2 u')' + \lambda u = 0 \qquad 0 < x < 1,$$
$$u(0) = u(1) = 0.$$

(b) Show directly that the sequence of eigenfunctions is orthogonal with respect the related inner product.

6.6 Prove that all the eigenvalues of the following Sturm–Liouville problem are positive.

$$u'' + (\lambda - x^2)u = 0 \qquad 0 < x < 1,$$
$$u'(0) = u'(1) = 0.$$

6.7 (a) Solve the Sturm–Liouville problem

$$x^2 u'' + 2xu' + \lambda u = 0 \qquad 1 < x < e,$$
$$u(1) = u(e) = 0.$$

(b) Show directly that the sequence of eigenfunctions is orthogonal with respect the related inner product.

6.8 Prove that all the eigenfunctions of the following Sturm–Liouville problem are positive.

$$u'' + (\lambda - x^2)u = 0 \qquad 0 < x < \infty,$$
$$u'(0) = \lim_{x \to \infty} u(x) = 0.$$

6.9 Consider the eigenvalue problem

$$u'' + \lambda u = 0 \qquad -1 < x < 1, \tag{6.98}$$
$$u(1) + u(-1) = 0, \qquad u'(1) + u'(-1) = 0. \tag{6.99}$$

(a) Prove that for $u, v \in C^2([-1, 1])$ that satisfy the boundary conditions (6.99) we have

$$\int_{-1}^{1} [u''(x)v(x) - v''(x)u(x)] \, dx = 0.$$

(b) Show that all the eigenvalues are real.

(c) Find the eigenvalues and eigenfunctions of the problem.

(d) Determine the multiplicity of the eigenvalues.

(e) Explain if and how your answer for part (d) complies with the Sturm–Liouville theory.

6.10 Show that for $n \geq 0$, the eigenfunction of the nth eigenvalue of the Sturm–Liouville problem (6.78) has exactly n roots in $(0, 1)$.

6.11 Solve the problem

$$u_t - u_{xx} + u = 2t + 15 \cos 2x \qquad 0 < x < \pi/2,\ t > 0,$$
$$u_x(0, t) = u_x(\pi/2, t) = 0 \qquad t \geq 0,$$
$$u(x, 0) = 1 + \sum_{n=1}^{10} 3n \cos 2nx \qquad 0 \leq x \leq \pi/2.$$

6.12 The hyperbolic equation $u_{tt} + u_t - u_{xx} = 0$ describes wave propagation along telegraph lines. Solve the *telegraph equation* on $0 < x < 2, t > 0$ with the initial boundary conditions

$$u(0, t) = u(2, t) = 0 \qquad t \geq 0,$$
$$u(x, 0) = 0, \quad u_t(x, 0) = x \qquad 0 \leq x \leq 2.$$

6.13 Solve the problem

$$u_t - u_{xx} = \frac{x(1 + \pi t)}{\pi} \qquad 0 < x < \pi,\ t > 0,$$
$$u(0, t) = 2, \quad u(\pi, t) = t \qquad t \geq 0,$$
$$u(x, 0) = 2\left(1 - \frac{x^2}{\pi^2}\right) \qquad 0 \leq x \leq \pi.$$

6.14 (a) Solve the problem

$$u_t = u_{xx} - 4u \qquad 0 < x < \pi,\ t > 0,$$
$$u_x(0, t) = u(\pi, t) = 0 \qquad t \geq 0,$$
$$u(x, 0) = f(x) \qquad 0 \leq x \leq \pi,$$

for $f(x) = x^2 - \pi^2$.

(b) Solve the same problem for $f(x) = x - \cos x$.

(c) Are the solutions you found in (a) and (b) classical?

6.15 (a) Solve the problem

$$u_t - u_{xx} = 2t + (9t + 31)\sin\frac{3x}{2} \qquad 0 < x < \pi,\ t > 0,$$
$$u(0, t) = t^2, \quad u_x(\pi, t) = 1 \qquad t \geq 0,$$
$$u(x, 0) = x + 3\pi \qquad 0 \leq x \leq \pi.$$

(b) Is the solution classical?

6.16 (a) Solve the following periodic problem:

$$u_t - u_{xx} = 0 \qquad -\pi < x < \pi, \, t > 0,$$

$$u(-\pi, t) = u(\pi, t), \quad u_x(-\pi, t) = u_x(\pi, t) \qquad t \geq 0,$$

$$u(x, 0) = \begin{cases} 1 & -\pi \leq x \leq 0, \\ 0 & 0 \leq x \leq \pi. \end{cases}$$

(b) Is the solution classical?

6.17 Solve the problem

$$u_t - u_{xx} = 1 + x \cos t \qquad 0 < x < 1, \, t > 0,$$

$$u_x(0, t) = u_x(1, t) = \sin t \qquad t \geq 0,$$

$$u(x, 0) = 1 + \cos(2\pi x) \qquad 0 \leq x \leq 1.$$

6.18 (a) Solve the problem

$$u_{tt} + u_t - u_{xx} = 0 \qquad 0 < x < 2, \, t > 0,$$

$$u(0, t) = u(2, t) = 0 \qquad t \geq 0,$$

$$u(x, 0) = 0, \quad u_t(x, 0) = x \qquad 0 \leq x \leq 2.$$

(b) Is the solution classical?

6.19 Let $h > 0$. Solve the problem

$$u_t - u_{xx} + hu = 0 \qquad 0 < x < \pi, \, t > 0,$$

$$u(0, t) = 0, \, u(\pi, t) = 1 \qquad t \geq 0,$$

$$u(x, 0) = 0 \qquad 0 \leq x \leq \pi.$$

6.20 (a) Solve the problem

$$u_{tt} - 4u_{xx} = (1 - x) \cos t \qquad 0 < x < \pi, \, t > 0,$$

$$u_x(0, t) = \cos t - 1, \quad u_x(\pi, t) = \cos t \qquad t \geq 0,$$

$$u(x, 0) = \frac{x^2}{2\pi} \qquad 0 \leq x \leq \pi,$$

$$u_t(x, 0) = \cos 3x \qquad 0 \leq x \leq \pi.$$

(b) Is the solution classical?

6.21 Solve the nonhomogeneous heat problem

$$u_t - u_{xx} = t \cos(2001x) \qquad 0 < x < \pi, \, t > 0,$$

$$u_x(0, t) = u_x(\pi, t) = 0 \qquad t \geq 0,$$

$$u(x, 0) = \pi \cos 2x \qquad 0 \leq x \leq \pi.$$

6.22 Solve the nonhomogeneous heat problem

$$u_t = 13u_{xx} \qquad 0 < x < 1,\ t > 0,$$
$$u_x(0,t) = 0,\ u_x(1,t) = 1 \qquad t \geq 0,$$
$$u(x,0) = \frac{1}{2}x^2 + x \qquad 0 \leq x \leq 1.$$

6.23 Consider the heat problem

$$u_t - u_{xx} = g(x,t) \qquad 0 < x < 1,\ t > 0,$$
$$u_x(0,t) = u_x(1,t) = 0 \qquad t \geq 0,$$
$$u(x,0) = f(x) \qquad 0 \leq x \leq 1.$$

(a) Solve the problem for $f(x) = 3\cos(42\pi x),\ g(x,t) = e^{3t}\cos(17\pi x)$.
(b) Find $\lim_{t\to\infty} u(x,t)$ for

$$g(x,t) = 0, \quad f(x) = \frac{1}{1+x^2}.$$

6.24 Solve the nonhomogeneous wave problem

$$u_{tt} - u_{xx} = \cos 2t \cos 3x \qquad 0 < x < \pi,\ t > 0,$$
$$u_x(0,t) = u_x(\pi,t) = 0 \qquad t \geq 0,$$
$$u(x,0) = \cos^2 x, \quad u_t(x,0) = 1 \qquad 0 \leq x \leq \pi.$$

6.25 Solve the heat problem

$$u_t = ku_{xx} + \alpha \cos \omega t \qquad 0 < x < L,\ t > 0,$$
$$u_x(L,t) = u_x(L,t) = 0 \qquad t \geq 0,$$
$$u(x,0) = x \qquad 0 \leq x \leq L.$$

6.26 Solve the wave problem

$$u_{tt} = c^2 u_{xx} \qquad 0 < x < 1,\ t > 0,$$
$$u(0,t) = 1,\ u(1,t) = 2\pi \qquad t \geq 0,$$
$$u(x,0) = x + \pi, \quad u_t(x,0) = 0 \qquad 0 \leq x \leq 1.$$

6.27 Solve the radial problem

$$\frac{\partial u}{\partial t} = \frac{1}{r^2}\frac{\partial}{\partial r}\left(r^2 \frac{\partial u}{\partial r}\right) \qquad 0 < r < a,\ t > 0,$$
$$u(a,t) = a, \quad |u(0,t)| < \infty \qquad t \geq 0,$$
$$u(r,0) = r \qquad 0 \leq r \leq a.$$

Hint Use the substitution $\rho(r) = r R(r)$ to solve the related Sturm–Liouville problem.

6.28 Show that for the initial boundary value problem (6.92) it is possible to find an auxiliary function w which satisfies the boundary conditions and has the form of (6.93).

7

Elliptic equations

7.1 Introduction

We mentioned in Chapter 1 the central role played by the Laplace operator in the theory of PDEs. In this chapter we shall concentrate on elliptic equations, and, in particular, on the main prototype for elliptic equations, which is the Laplace equation itself:

$$\Delta u = 0. \tag{7.1}$$

We start by reviewing a few basic properties of elliptic problems. We then introduce the maximum principle, and also formulate a similar principle for the heat equation. We prove the uniqueness and stability of solutions to the Laplace equation in two ways. One approach is based on the maximum principle, and the other approach uses the method of Green's identities. The simplest solution method for the Laplace equation is the method of separation of variables. Indeed, this method is only applicable in simple domains, such as rectangles, disks, rings, etc., but these domains are often encountered in applications. Moreover, explicit solutions in simple domains provide an insight into the solution's structure in more general domains. Towards the end of the chapter we shall introduce Poisson's kernel formula.

7.2 Basic properties of elliptic problems

We limit the discussion in this chapter to functions $u(x, y)$ in two independent variables, although most of the analysis can be readily generalized to higher dimensions (see Chapter 9). We further limit the discussion to the case where the equation contains only the principal part, and this part is in a canonical form. Nevertheless, we allow for a nonhomogeneous term in the equation. We denote by D a planar domain (i.e. a nonempty connected and open set in $\mathbb{R}^2$). The Laplace equation is given by

$$\Delta u := u_{xx} + u_{yy} = 0 \qquad (x, y) \in D. \tag{7.2}$$

A function u satisfying (7.2) is called a *harmonic function*.

The Laplace equation is a special case of a more general equation:

$$\Delta u = F(x, y), \tag{7.3}$$

where F is a given function. Equation (7.3) was used by the French mathematician Simeon Poisson (1781–1840) in his studies of diverse problems in mechanics, gravitation, electricity, and magnetism. Therefore it is called *Poisson's equation*. In order to obtain a heuristic understanding of the results to be derived below, it is useful to provide Poisson's equation with a simple physical interpretation. For this purpose we recall from the discussion in Chapter 1 that the solution of Poisson's equation represents the distribution of temperature u in a domain D at equilibrium. The nonhomogeneous term F describes (up to a change of sign) the rate of heat production in D. For the benefit of readers who are familiar with the theory of electromagnetism, we point out that u could also be interpreted as the electric potential in the presence of a charge density $-F$.

In order to obtain a unique temperature distribution, we must provide conditions for the temperature (or temperature flux) at the boundary ∂D. There are several basic boundary conditions (see the discussion in Chapter 1).

Definition 7.1 The problem defined by Poisson's equation and the Dirichlet boundary condition

$$u(x, y) = g(x, y) \qquad (x, y) \in \partial D, \tag{7.4}$$

for a given function g, is called the *Dirichlet problem*. In Figure 7.1 we depict the problem schematically.

Definition 7.2 The problem defined by Poisson's equation and the Neumann boundary condition

$$\partial_n u(x, y) = g(x, y) \qquad (x, y) \in \partial D, \tag{7.5}$$

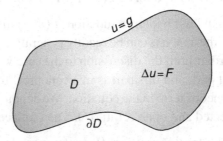

Figure 7.1 A schematic drawing for the Poisson equation with Dirichlet boundary conditions.

where g is a given function, $\hat{n}$ denotes the unit outward normal to ∂D, and ∂_n denotes a differentiation in the direction of $\hat{n}$ (i.e. $\partial_n = \hat{n} \cdot \nabla$), is called the *Neumann problem*.

Definition 7.3 The problem defined by Poisson's equation and the boundary condition of the third kind

$$u(x, y) + \alpha(x, y)\partial_n u(x, y) = g(x, y) \qquad (x, y) \in \partial D, \qquad (7.6)$$

where α and g are given functions, is called *a problem of the third kind* (it is also sometimes called the Robin problem).

The first question we have to address is whether there exists a solution to each one of the problems we just defined. This question is not at all easy. It has been considered by many great mathematicians since the middle of the nineteenth century. It was discovered that when the domain D is bounded and 'sufficiently smooth', then the Dirichlet problem, for example, does indeed have a solution. The precise definition of smoothness in this context and the general existence proof are beyond the scope of this book, and we refer the interested reader to [11]. It is interesting to point out that in applications one frequently encounters domains with corners (rectangles, for example). Near a corner the boundary is not differentiable; thus, we cannot always expect the solutions to be as smooth as we would like. In this chapter, we only consider classical solutions, i.e. the solutions are in the class $C^2(D)$. Some of the analysis we present requires further conditions on the behavior of the solutions near the boundary. For example, we sometimes have to limit ourselves to solutions in the class $C^1(\bar{D})$.

Consider now the Neumann problem. Since the temperature is in equilibrium, the heat flux through the boundary must be balanced by the temperature production inside the domain. This simple argument is the physical manifestation of the following statement.

Lemma 7.4 *A necessary condition for the existence of a solution to the Neumann problem is*

$$\int_{\partial D} g(x(s), y(s))ds = \int_D F(x, y)dxdy, \qquad (7.7)$$

where $(x(s), y(s))$ is a parameterization of ∂D.

Proof Let us first recall the vector identity $\Delta u = \vec{\nabla} \cdot \vec{\nabla} u$. Therefore we can write Poisson's equation as

$$\vec{\nabla} \cdot \vec{\nabla} u = F. \qquad (7.8)$$

Integrating both sides of the equation over D, and using Gauss' theorem, we obtain

$$\int_{\partial D} \vec{\nabla} u \cdot \hat{n} ds = \int_D F dx dy.$$

The lemma now follows from the definition of the directional derivative and from the boundary conditions.   □

For future reference it is useful to observe that for harmonic functions, i.e. solutions of the Laplace equation ($F = 0$), we have

$$\int_\Gamma \partial_n u ds = 0 \tag{7.9}$$

for any closed curve Γ that is fully contained in D.

Notice that we supplied just a single boundary condition for each one of the three problems we presented (Dirichlet, Neumann, third kind). Although we are dealing with second-order equations, the boundary conditions are quite different from the conditions we supplied in the hyperbolic case. There we provided two conditions (one on the solution and one on its derivative with respect to t) for each point on the line $t = 0$. The following example (due to Hadamard) demonstrates the difference between elliptic and hyperbolic equations on the upper half-plane. Consider Laplace's equation in the domain $-\infty < x < \infty$, $y > 0$, under the Cauchy conditions

$$u^n(x, 0) = 0, \quad u_y^n(x, 0) = \frac{\sin nx}{n} \qquad -\infty < x < \infty, \tag{7.10}$$

where n is a positive integer. It is easy to check that

$$u^n(x, y) = \frac{1}{n^2} \sin nx \sinh ny$$

is a harmonic function satisfying (7.10). Choosing n to be a very large number, the initial conditions describe an arbitrarily small perturbation of the trivial solution $u = 0$. On the other hand, the solution is not bounded at all in the half-plane $y > 0$. In fact, for any $y > 0$, the value of $\sup_{x \in \mathbb{R}} |u^n(x, y)|$ grows exponentially fast as $n \to \infty$. Thus the Cauchy problem for the Laplace equation is not stable and hence is not well posed with respect to the initial conditions (7.10).

Before developing a general theory, let us compute some special harmonic functions defined over the entire plane (except, maybe, for certain isolated points). We define a harmonic polynomial of degree n to be a harmonic function $P_n(x, y)$ of the form

$$P_n(x, y) = \sum_{0 \le i+j \le n} a_{i,j} x^i y^j.$$

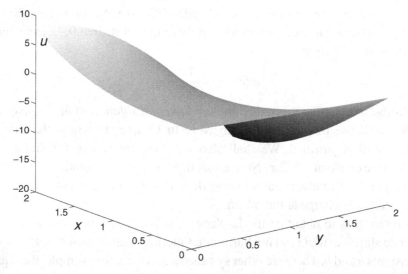

Figure 7.2 The surface of the harmonic polynomial $u(x, y) = x^3 - 3xy^2 - y$. Harmonic functions often have a saddle-like shape. This is a consequence of the maximum principle that we prove in Theorem 7.5.

For example, the functions $x - y, x^2 - y^2 + 2x, x^3 - 3xy^2 - y$ are harmonic polynomials of degree 1, 2 and 3 respectively. The graph of the harmonic polynomial $u(x, y) = x^3 - 3xy^2 - y$ is depicted in Figure 7.2. The subclass V_n of harmonic polynomials P_n^{H} of the form

$$P_n^{\mathrm{H}} = \sum_{i+j=n} a_{i,j} x^i y^j$$

is called the set of *homogeneous harmonic polynomials* of order n. In Exercise 7.9 we show that (somewhat surprisingly) for each $n > 0$ the dimension of the space V_n is exactly 2 (this result holds only in $\mathbb{R}^2$).

The most important solution of the Laplace equation over the plane is the solution that is symmetric about the origin (the radial solution). To find this solution it is convenient to use polar coordinates. We denote the polar variables by (r, θ), and the harmonic function by $w(r, \theta) = u(x(r, \theta), y(r, \theta))$. In Exercise 7.7 (a) we show that the Laplace equation takes the following form in polar coordinates:

$$\Delta w = w_{rr} + \frac{1}{r} w_r + \frac{1}{r^2} w_{\theta\theta} = 0. \tag{7.11}$$

Therefore the radial symmetric solution $w(r)$ satisfies

$$w'' + \frac{1}{r} w' = 0, \tag{7.12}$$

which is Euler (equidimensional) second-order ODE (see Section A.3). One solution is the constant function (a harmonic polynomial of degree 0, in fact), and the other solution is given by

$$w(r) = -\frac{1}{2\pi} \ln r. \tag{7.13}$$

The solution $w(r)$ in (7.13) is called the *fundamental solution* of the Laplace equation. We shall use this solution extensively in Chapter 8, where the title 'fundamental' will be justified. We shall also see there the reason for including the multiplicative constant $-1/2\pi$. Notice that the fundamental solution is not defined at the origin. The fundamental solution describes the electric potential due to a point-like electric charge at the origin.

It is interesting to note that the Laplace equation is symmetric with respect to coordinate shift: i.e. if $u(x, y)$ is a harmonic function, then so is $u(x - a, y - b)$ for any constants a and b. There are other symmetries as well; for example, the equation is symmetric with respect to rotations of the coordinate system, i.e. if $w(r, \theta)$ is harmonic, then $w(r, \theta + \gamma)$ is harmonic too for every constant γ. Another important symmetry concerns dilation of the coordinate system: if $u(x, y)$ is harmonic, then $u(x/\delta, y/\delta)$ is also harmonic for every positive constant δ.

7.3 The maximum principle

One of the central tools in the theory of (second-order) elliptic PDEs is the maximum principle. We first present a 'weak' form of this principle.

Theorem 7.5 (The weak maximum principle) *Let D be a bounded domain, and let $u(x, y) \in C^2(D) \cap C(\bar{D})$ be a harmonic function in D. Then the maximum of u in $\bar{D}$ is achieved on the boundary ∂D.*

Proof Consider a function $v(x, y) \in C^2(D) \cap C(\bar{D})$ satisfying $\Delta v > 0$ in D. We argue that v cannot have a local maximum point in D. To see why, recall from calculus that if $(x_0, y_0) \in D$ is a local maximum point of v, then $\Delta v \leq 0$, in contradiction to our assumption.

Since u is harmonic, the function $v(x, y) = u(x, y) + \varepsilon(x^2 + y^2)$ satisfies $\Delta v > 0$ for any $\varepsilon > 0$. Set $M = \max_{\partial D} u$, and $L = \max_{\partial D}(x^2 + y^2)$. From our argument about v it follows that $v \leq M + \varepsilon L$ in D. Since $u = v - \varepsilon(x^2 + y^2)$, it now follows that $u \leq M + \varepsilon L$ in D. Because ε can be made arbitrarily small, we obtain $u \leq M$ in D. $\square$

Remark 7.6 If u is harmonic in D, then $-u$ is harmonic there too. But for any set A and for any function u we have

$$\min_A u = -\max_A(-u).$$

Therefore the minimum of a harmonic function u is also obtained on the boundary ∂D.

The theorem we have just proved still does not exclude the possibility that the maximum (or minimum) of u is also attained at an internal point. We shall now prove a stronger result that asserts that if u is not constant, then the maximum (and minimum) cannot, in fact, be obtained at any interior point. For this purpose we need first to establish one of the marvelous properties of harmonic functions.

Theorem 7.7 (The mean value principle) *Let D be a planar domain, let u be a harmonic function there and let (x_0, y_0) be a point in D. Assume that B_R is a disk of radius R centered at (x_0, y_0), fully contained in D. For any $r > 0$ set $C_r = \partial B_r$. Then the value of u at (x_0, y_0) is the average of the values of u on the circle C_R:*

$$u(x_0, y_0) = \frac{1}{2\pi R} \oint_{C_R} u(x(s), y(s)) ds$$

$$= \frac{1}{2\pi} \int_0^{2\pi} u(x_0 + R\cos\theta, y_0 + R\sin\theta) d\theta. \qquad (7.14)$$

Proof Let $0 < r \le R$. We write $v(r, \theta) = u(x_0 + r\cos\theta, y_0 + r\sin\theta)$. We also define the integral of v with respect to θ:

$$V(r) = \frac{1}{2\pi r} \oint_{C_r} v ds = \frac{1}{2\pi} \int_0^{2\pi} v(r, \theta) d\theta.$$

Differentiating with respect to r we obtain

$$V_r(r) = \frac{1}{2\pi} \int_0^{2\pi} v_r(r, \theta) d\theta = \frac{1}{2\pi} \int_0^{2\pi} \frac{\partial}{\partial r} u(x_0 + r\cos\theta, y_0 + r\sin\theta) d\theta$$

$$= \frac{1}{2\pi r} \oint_{C_r} \partial_n u ds = 0,$$

where in the last equality we used (7.9). Hence $V(r)$ does not depend on r, and thus

$$u(x_0, y_0) = V(0) = \lim_{\rho \to 0} V(\rho) = V(r) = \frac{1}{2\pi r} \oint_{C_r} u(x(s), y(s)) ds$$

for all $0 < r \le R$. $\qquad \square$

Remark 7.8 It is interesting to note that the reverse statement is also true, i.e. a continuous function that satisfies the mean value property in some domain D is harmonic in D.

We prove next a slightly weaker result.

Theorem 7.9 *Let u be a function in $C^2(D)$ satisfying the mean value property at every point in D. Then u is harmonic in D.*

Proof Assume by contradiction that there is a point (x_0, y_0) in D where $\Delta u(x_0, y_0) \neq 0$. Without loss of generality assume $\Delta u(x_0, y_0) > 0$. Since $\Delta u(x, y)$ is a continuous function, then for a sufficiently small $R > 0$ there exists in D a disk B_R of radius R, centered at (x_0, y_0), such that $\Delta u > 0$ at each point in B_R. Denote the boundary of this disk by C_R. It follows that

$$
\begin{aligned}
0 < \frac{1}{2\pi} \int_{B_R} \Delta u \, \mathrm{d}x \mathrm{d}y &= \frac{1}{2\pi} \oint_{C_R} \partial_n u \, \mathrm{d}s \\
&= \frac{R}{2\pi} \int_0^{2\pi} \frac{\partial}{\partial R} u(x_0 + R\cos\theta, y_0 + R\sin\theta) \mathrm{d}\theta \\
&= \frac{R}{2\pi} \frac{\partial}{\partial R} \int_0^{2\pi} u(x_0 + R\cos\theta, y_0 + R\sin\theta) \mathrm{d}\theta \\
&= R \frac{\partial}{\partial R} [u(x_0, y_0)] = 0, \qquad (7.15)
\end{aligned}
$$

where in the fourth equality in (7.15) we used the assumption that u satisfies the mean value property. $\qquad\square$

As a corollary of the mean value theorem, we shall prove another maximum principle for harmonic functions.

Theorem 7.10 (The strong maximum principle) *Let u be a harmonic function in a domain D (here we also allow for unbounded D). If u attains it maximum (minimum) at an interior point of D, then u is constant.*

Proof Assume by contradiction that u obtains its maximum at some interior point q_0. Let $q \neq q_0$ be an arbitrary point in D. Denote by l a smooth orbit in D connecting q_0 and q (see Figure 7.3). In addition, denote by d_l the distance between l and ∂D.

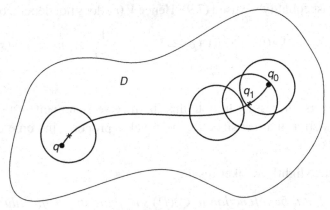

Figure 7.3 A construction for the proof of the strong maximum principle.

Consider a disk B_0 of radius $d_l/2$ around q_0. From the definition of d_l and from the mean value theorem, we infer that u is constant in B_0 (since the average of a set cannot be greater than all the objects of the set). Select now a point q_1 in $l \cap B_0$, and denote by B_1 the disk of radius $d_l/2$ centered at q_1. From our construction it follows that u also reaches its maximal value at q_1. Thus we obtain that u is constant also in B_1.

We continue in this way until we reach a disk that includes the point q. We conclude $u(q) = u(q_0)$, and since q is arbitrary, it follows that u is constant in D. Notice that we may choose the points $q_0, q_1, \ldots$, such that the process involves a finite number of disks $B_0, B_1, \ldots, B_{n_l}$ because the length of l is finite, and because all the disks have the same radius. □

Remark 7.11 The strong maximum theorem indeed guarantees that nonconstant harmonic functions cannot obtain their maximum or minimum in D. Notice that in unbounded domains the maximum (minimum) of u is not necessarily obtained in $\bar{D}$. For example, the function $\log(x^2 + y^2)$ is harmonic and positive outside the unit disk, and it vanishes on the domain's boundary. We also point out that the first proof of the maximum principle can be readily generalized to a large class of elliptic problems, while the mean value principle holds only for harmonic functions.

7.4 Applications of the maximum principle

We shall illustrate the importance of the maximum principle by using it to prove the uniqueness and stability of the solution to the Dirichlet problem.

Theorem 7.12 *Consider the Dirichlet problem in a bounded domain:*

$$\Delta u = f(x, y) \qquad (x, y) \in D,$$
$$u(x, y) = g(x, y) \qquad (x, y) \in \partial D.$$

The problem has at most one solution in $C^2(D) \cap C(\bar{D})$.

Proof Assume by contradiction that there exist two solutions u_1 and u_2. Denote their difference by $v = u_1 - u_2$. The problem's linearity implies that v is harmonic in D, and that it vanishes on ∂D. The weak maximum principle implies, then, $0 \leq v \leq 0$. Thus $v \equiv 0$. □

We note that the boundedness of D is essential. Consider, for instance, the following Dirichlet problem:

$$\Delta u = 0 \qquad x^2 + y^2 > 4, \tag{7.16}$$
$$u(x, y) = 1 \qquad x^2 + y^2 = 4. \tag{7.17}$$

It is easy to verify that the functions $u_1 \equiv 1$ and $u_2(x, y) = (\ln \sqrt{x^2 + y^2})/\ln 2$ both solve the problem.

Theorem 7.13 *Let D be a bounded domain, and let u_1 and u_2 be functions in $C^2(D) \cap C(\bar{D})$ that are solutions of the Poisson equation $\Delta u = f$ with the Dirichlet conditions g_1 and g_2, respectively. Set $M_g = \max_{\partial D} |g_1(x, y) - g_2(x, y)|$. Then*

$$\max_D |u_1(x, y) - u_2(x, y)| \leq M_g.$$

Proof Define $v = u_1 - u_2$. The construction implies that v is harmonic in D satisfying, $v = g_1 - g_2$ on ∂D. Therefore the maximum (and minimum) principle implies

$$\min_{\partial D}(g_1 - g_2) \leq v(x, y) \leq \max_{\partial D}(g_1 - g_2) \qquad \forall (x, y) \in D,$$

and the theorem follows.   □

7.5 Green's identities

We now develop another important tool for the analysis of elliptic problems – Green's identities. We shall use this tool to provide an alternative uniqueness proof for the Dirichlet problem, and, in addition, we shall prove the uniqueness of solutions to the Neumann problem and to problems of the third kind. The Green's identities method is similar to the energy method we used in Chapter 5, and to Green's formula, which we introduced in Chapter 6.

Our starting point is Gauss' (the divergence) theorem:

$$\int_D \vec{\nabla} \cdot \vec{\psi}(x, y) \, dxdy = \int_{\partial D} \vec{\psi}(x(s), y(s)) \cdot \hat{n} ds.$$

This theorem holds for any vector field $\vec{\psi} \in C^1(D) \cap C(\bar{D})$ and any bounded piecewise smooth domain D. Let u and v be two arbitrary functions in $C^2(D) \cap C^1(\bar{D})$. We consider several options for ψ in Gauss' theorem. Selecting

$$\vec{\psi} = \vec{\nabla} u,$$

we obtain (as we verified earlier)

$$\int_D \Delta u \, dxdy = \int_{\partial D} \partial_n u \, ds. \tag{7.18}$$

The selection $\vec{\psi} = v\vec{\nabla} u - u\vec{\nabla} v$ leads to

$$\int_D (v\Delta u - u\Delta v) \, dxdy = \int_{\partial D} (v\partial_n u - u\partial_n v) \, ds. \tag{7.19}$$

A third Green's identity

$$\int_D \vec{\nabla}u \cdot \vec{\nabla}v \, dxdy = \int_{\partial D} v \partial_n u ds - \int_D v \Delta u \, dxdy, \qquad (7.20)$$

is given as an exercise (see Exercise 7.1).

We applied the first Green's identity (7.18) to prove the mean value principle. We next apply the third Green's identity (7.20) to establish the general uniqueness theorem for Poisson's equation.

Theorem 7.14 *Let D be a smooth domain.*

(a) *The Dirichlet problem has at most one solution.*
(b) *If $\alpha \geq 0$, then the problem of the third kind has at most one solution.*
(c) *If u solves the Neumann problem, then any other solution is of the form $v = u + c$, where $c \in \mathbb{R}$.*

Proof We start with part (b) (part (a) is a special case of part (b)). Suppose u_1 and u_2 are two solutions of the problem of the third kind. Set $v = u_1 - u_2$. It is easy to see that v is a harmonic function in D, satisfying on ∂D the boundary condition

$$v + \alpha \partial_n v = 0.$$

Substituting $v = u$ in the third Green's identity (7.20), we obtain

$$\int_D |\vec{\nabla}v|^2 dxdy = - \int_{\partial D} \alpha \, (\partial_n v)^2 \, ds. \qquad (7.21)$$

Since the left hand side of (7.21) is nonnegative, and the right hand side is nonpositive, it follows that both sides must vanish. Hence $\nabla v = 0$ in D and $\alpha \partial_n v = -v = 0$ on ∂D. Therefore v is constant in D and it vanishes on ∂D. Thus $v \equiv 0$, and $u_1 \equiv u_2$.

The proof of part (c) is similar. We first notice that one cannot expect uniqueness in the sense of parts (a) and (b), since if u is a solution to the Neumann problem, then $u + c$ is a solution too for any constant c. Indeed, we now obtain from the identity (7.20)

$$\int_D |\vec{\nabla}v|^2 \, dxdy = 0,$$

implying that v is constant. On the other hand, since we have no constraint on the value of v on ∂D, we cannot determine the constant. We thus obtain

$$u_1 - u_2 = \text{constant}.$$

$\square$

7.6 The maximum principle for the heat equation

The maximum principle also holds for parabolic equations. Consider the heat equation for a function $u(x, y, z, t)$ in a three-dimensional bounded domain D:

$$u_t = k\Delta u \qquad (x, y, z) \in D \qquad t > 0, \tag{7.22}$$

where here we write $\Delta u = u_{xx} + u_{yy} + u_{zz}$. To formulate the maximum principle we define the domain

$$Q_T = \{(x, y, z, t) \mid (x, y, z) \in D, \ 0 < t \le T\}.$$

Notice that the time interval $(0, T)$ is arbitrary. It is convenient at this stage to define the *parabolic boundary* of Q_T:

$$\partial_P Q_T = \{D \times \{0\}\} \cup \{\partial D \times [0, T]\},$$

that is the boundary of Q_T, save for the top cover $D \times \{T\}$. We also denote by C_H the class of functions that are twice differentiable in Q_T with respect to (x, y, z), once differentiable with respect to t, and continuous in $\bar{Q}_T$. We can now state the (weak) maximum principle for the heat equation.

Theorem 7.15 *Let $u \in C_H$ be a solution to the heat equation (7.22) in Q_T. Then u achieves its maximum (minimum) on $\partial_P Q_T$.*

Proof We prove the statement with respect to the maximum of u. The proof with respect to the minimum of u follows at once, since if u satisfies the heat equation, so does $-u$. It is convenient to start with the following proposition.

Proposition 7.16 *Let v be a function in C_H satisfying $v_t - k\Delta v < 0$ in Q_T. Then v has no local maximum in Q_T. Moreover, v achieves its maximum in $\partial_P Q_T$.*

Proof of the proposition If v has a local maximum at some $q \in Q_T$, then $v_t(q) = 0$, implying $\Delta v(q) > 0$, which contradicts the assumption. Since v is continuous in the closed domain $\bar{Q}_T$, its maximum is achieved somewhere on the boundary $\partial(Q_T)$. If the maximum is achieved at a point $q = (x_0, y_0, z_0, T)$ on the top cover $D \times \{T\}$, then we must have $v_t(q) \ge 0$, and thus $\Delta v(q) > 0$. Again this contradicts the assumption on v, since (x_0, y_0, z_0) is a local maximum in D.

Returning to the maximum principle, we define (for $\varepsilon > 0$)

$$v(x, y, z, t) = u(x, y, z, t) - \varepsilon t.$$

Obviously

$$\max_{\partial_P Q_T} v \le M := \max_{\partial_P Q_T} u. \tag{7.23}$$

Since u satisfies the heat equation, it follows that $v_t - \Delta v < 0$ in Q_T. Proposition 7.14 and (7.23) imply that $v \le M$, hence for all points in Q_T we have $u \le M + \varepsilon T$. Because ε can be made arbitrarily small, we obtain $u \le M$.   □

As a direct consequence of the maximum principle we prove the following theorem, which guarantees the uniqueness and stability of the solution to the Dirichlet problem for the heat equation.

Theorem 7.17 *Let u_1 and u_2 be two solutions of the heat equation*

$$u_t - k\Delta u = F(x, t)   (x, y, z) \in D   0 < t < T,   (7.24)$$

with initial conditions $u_i(x, y, z, 0) = f_i(x, y, z)$, and boundary conditions

$$u_i(x, y, z, t) = h_i(x, y, z, t)   (x, y, z) \in \partial D   0 < t < T,$$

respectively. Set

$$\delta = \max_D |f_1 - f_2| + \max_{\partial D \times \{t > 0\}} |h_1 - h_2|.$$

Then

$$|u_1 - u_2| \le \delta   (x, y, z, t) \in \bar{Q}_T.$$

Proof Writing $w = u_1 - u_2$, the proof is the same as for the corresponding theorem for Poisson's equation. The special case $f_1 = f_2$, $h_1 = h_2$ implies at once the uniqueness part of the theorem.   □

Corollary 7.18 *Let*

$$u(x, t) = \sum_{n=1}^{\infty} B_n \sin \frac{n\pi x}{L} e^{-k(\frac{n\pi}{L})^2 t}   (7.25)$$

be the formal solution of the heat problem

$$u_t - k u_{xx} = 0   0 < x < L, \, t > 0,   (7.26)$$
$$u(0, t) = u(L, t) = 0   t \ge 0,   (7.27)$$
$$u(x, 0) = f(x)   0 \le x \le L.   (7.28)$$

If the series

$$f(x) - \sum_{n=1}^{\infty} B_n \sin \frac{n\pi x}{L}$$

converges uniformly on $[0, L]$, then the series (7.25) converges uniformly on $[0, L] \times [0, T]$, and u is a classical solution.

Proof Let $\varepsilon > 0$. By the Cauchy criterion for uniform convergence there exists N_ε such that for all $N_\varepsilon \leq k \leq l$ we have

$$\left| \sum_{n=k}^{l} B_n \sin \frac{n\pi x}{L} \right| < \varepsilon \qquad \forall x \in [0, L].$$

Note that

$$\sum_{n=k}^{l} B_n \sin \frac{n\pi x}{L} e^{-k(\frac{n\pi}{L})^2 t}$$

is a classical solution of the heat equation; hence by the maximum principle

$$\left| \sum_{n=k}^{l} B_n \sin \frac{n\pi x}{L} e^{-k(\frac{n\pi}{L})^2 t} \right| < \varepsilon \qquad \forall (x, t) \in [0, L] \times [0, T].$$

Invoking again the Cauchy criterion for uniform convergence, we infer that the series of (7.25) converges uniformly on $[0, L] \times [0, T]$ to the continuous function u. In particular, u satisfies the initial and boundary conditions. Since u satisfies the heat equation on $[0, L] \times (0, T]$, it follows that u is a classical solution. $\qquad \square$

We saw in Chapters 5 and 6 that the heat equation 'smoothes out' the initial conditions in the sense even if the initial data are not smooth, the solution is in class C^∞ for all $t > 0$. We examine now another aspect of the smoothing property of the heat equation. For simplicity let us return to the case of a single spatial variable x, so that our domain consists of an interval: $D = (a, b)$.

Proposition 7.19 *Let $u(x, t)$ be the solution of*

$$u_t = k u_{xx} \qquad a < x < b \qquad t > 0 \tag{7.29}$$

with Neumann's initial boundary value problem

$$u(x, 0) = f(x) \quad x \in (a, b), \qquad u_x(a, t) = u_x(b, t) = 0 \quad 0 < t < T.$$

Assume $f \in C^1([a, b])$ satisfying $f_x(a) = f_x(b) = 0$. Set $Q_T = (a, b) \times (0, T]$ for some $T > 0$. Then u satisfies

$$\max_{Q_T} |u_x| \leq \max_{(a,b)} |f'(x)|.$$

Proof We differentiate u with respect to x and define $w(x, t) = u_x(x, t)$. One can readily verify that w satisfies the heat equation with Dirichlet initial boundary conditions:

$$w(x, 0) = f'(x) \quad x \in (a, b), \qquad w(a, t) = w(b, t) = 0 \quad 0 < t < T.$$

Therefore Proposition 7.19 follows from the maximum principle for the heat equation. □

7.7 Separation of variables for elliptic problems

The method of separation of variables, introduced in Chapter 5 for the heat equation and for the wave equation, can in some cases also be applied to elliptic equations. Applying the method requires certain symmetries to hold, both for the equation and for the domain under study. We shall demonstrate how to use the method to solve the Laplace and Poisson equations in rectangles, in disks, and in circular sectors. Additional domains, such as an exterior of a disk or a ring will be dealt with in the exercises (see Exercises 7.11 and 7.20).

First, we prove the following general result.

Proposition 7.20 *Consider the Dirichlet problem*

$$\Delta u = 0 \qquad (x, y) \in D,$$
$$u(x, y) = g(x, y) \qquad (x, y) \in \partial D,$$

in a bounded domain D. Let

$$u(x, y) = \sum_{n=1}^{\infty} u_n(x, y) \tag{7.30}$$

be a formal solution of the problem, such that each $u_n(x, y)$ is a harmonic function in D and continuous in $\bar{D}$. If the series (7.30) converges uniformly on ∂D to g, then it converges uniformly on $\bar{D}$ and u is a classical solution of the problem.

Proof Let $\varepsilon > 0$. By the Cauchy criterion for uniform convergence, there exists N_ε such that for all $N_\varepsilon \leq k \leq l$ we have

$$\left| \sum_{n=k}^{l} u_n(x, y) \right| < \varepsilon$$

for all $(x, y) \in \partial D$. By the weak maximum principle

$$\left| \sum_{n=k}^{l} u_n(x, y) \right| < \varepsilon$$

for all $(x, y) \in \partial D$. Invoking again the Cauchy criterion for uniform convergence, we infer that the series of (7.30) converges uniformly on $\bar{D}$ to the continuous function u. In particular, u satisfies the boundary condition. Since each u_n satisfies the mean value property, the uniform convergence implies that u also satisfies the mean value property, and by Remark 7.8 u is harmonic in D. □

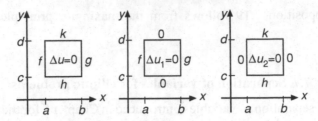

Figure 7.4 Separation of variables in rectangles.

7.7.1 Rectangles

Let u be the solution to the Dirichlet problem in a rectangular domain D (Figure 7.4):

$$\Delta u = 0 \qquad a < x < b, \ c < y < d, \tag{7.31}$$

with the boundary conditions

$$u(a, y) = f(y), \ u(b, y) = g(y), \ u(x, c) = h(x), \ u(x, d) = k(x). \tag{7.32}$$

We recall that the method of separation of variables is based on constructing an appropriate eigenvalue (Sturm–Liouville) problem. This, in turn, requires homogeneous boundary conditions. We thus split u into $u = u_1 + u_2$, where u_1 and u_2 are both harmonic in D, and where u_1 and u_2 satisfy the boundary conditions (see Figure 7.4)

$$u_1(a, y) = f(y), \ u_1(b, y) = g(y), \ u_1(x, c) = 0, \ u_1(x, d) = 0 \tag{7.33}$$

and

$$u_2(a, y) = 0, \ u_2(b, y) = 0, \ u_2(x, c) = h(x), \ u_2(x, d) = k(x). \tag{7.34}$$

We assume at this stage that the *compatibility condition*

$$f(c) = f(d) = g(c) = g(d) = h(a) = h(b) = k(a) = k(b) = 0 \tag{7.35}$$

holds. Since $u_1 + u_2$ satisfy (7.31) and the boundary conditions (7.32), the uniqueness theorem guarantees that $u = u_1 + u_2$.

The advantage of splitting the problem into two problems is that each one of the two new problems can be solved by separation of variables. Consider, for example, the problem for u_1. We shall seek a solution in the form of a sum of separated (nonzero) functions $U(x, y) = X(x)Y(y)$. Substituting such a solution into the Laplace equation (7.31), we obtain

$$X''(x) - \lambda X(x) = 0 \qquad a < x < b, \tag{7.36}$$

$$Y''(y) + \lambda Y(y) = 0 \qquad c < y < d. \tag{7.37}$$

The homogeneous boundary conditions imply that

$$Y(c) = Y(d) = 0. \tag{7.38}$$

Thus, we obtain a Sturm–Liouville problem for $Y(y)$. Solving (7.37)–(7.38), we derive a sequence of eigenvalues λ_n and eigenfunctions $Y_n(y)$. We can then substitute the sequence λ_n in (7.36) and obtain an associated sequence $X_n(x)$. The general solution u_1 is written formally as

$$u_1(x, y) = \sum_n X_n(x)Y_n(y).$$

The remaining boundary conditions for u_1 will be used to eliminate the two free parameters associated with X_n for each n, just as was done in Chapter 5. Instead of writing a general formula, we find it simpler to demonstrate the method via an example.

Example 7.21 Solve the Laplace equation in the rectangle $0 < x < b$, $0 < y < d$, subject to the Dirichlet boundary conditions

$$u(0, y) = f(y), \ u(b, y) = g(y), \ u(x, 0) = 0, \ u(x, d) = 0. \tag{7.39}$$

Recalling the notation u_1 and u_2 we introduced above, this problem gives rise to a Laplace equation with zero Dirichlet conditions for u_2. Therefore the uniqueness theorem implies $u_2 \equiv 0$. For u_1 we construct a solution consisting of an infinite combination of functions of the form $w(x, y) = X(x)Y(y)$. We thus obtain for $Y(y)$ the following Sturm–Liouville problem:

$$Y''(y) + \lambda Y(y) = 0 \quad 0 < y < d, \quad Y(0) = Y(d) = 0. \tag{7.40}$$

This problem was solved in Chapter 5. The eigenvalues and eigenfunctions are

$$\lambda_n = \left(\frac{n\pi}{d}\right)^2, \quad Y_n(y) = \sin\frac{n\pi}{d}y \quad n = 1, 2, \ldots. \tag{7.41}$$

The equation for the x-dependent factor is

$$X''(x) - \left(\frac{n\pi}{d}\right)^2 X(x) = 0 \quad 0 < x < b. \tag{7.42}$$

To facilitate the expansion of the boundary condition into a Fourier series, we select for (7.42) the fundamental system of solutions $\{\sinh[(n\pi/d)x], \sinh[(n\pi/d)(x - b)]\}$. Thus we write for u_1

$$u_1(x, y) = \sum_{n=1}^{\infty} \sin\frac{n\pi y}{d} \left[A_n \sinh\frac{n\pi}{d}x + B_n \sinh\frac{n\pi}{d}(x - b)\right]. \tag{7.43}$$

Substituting the expansion (7.43) into the nonhomogeneous boundary conditions of (7.39) we obtain:

$$g(y) = \sum_{n=1}^{\infty} A_n \sinh \frac{n\pi b}{d} \sin \frac{n\pi y}{d}, \quad f(y) = \sum_{n=1}^{\infty} B_n \sinh \frac{-n\pi b}{d} \sin \frac{n\pi y}{d}.$$

To evaluate the sequences $\{A_n\}, \{B_n\}$, expand $f(y)$ and $g(y)$ into generalized Fourier series:

$$f(y) = \sum_{n=1}^{\infty} \alpha_n \sin \frac{n\pi y}{d}, \quad \alpha_n = \frac{2}{d} \int_0^d f(y) \sin \frac{n\pi y}{d} dy,$$

$$g(y) = \sum_{n=1}^{\infty} \beta_n \sin \frac{n\pi y}{d}, \quad \beta_n = \frac{2}{d} \int_0^d g(y) \sin \frac{n\pi y}{d} dy.$$

This implies

$$A_n = \frac{\beta_n}{\sinh \dfrac{n\pi b}{d}}, \quad B_n = -\frac{\alpha_n}{\sinh \dfrac{n\pi b}{d}}.$$

We saw in Chapter 5 that the generalized Fourier series representing the solution to the heat equations converges exponentially fast for all $t > 0$. Moreover, the series for the derivatives of all orders converges too. On the other hand, the rate of convergence for the series representing the solution to the wave equation depends on the smoothness of the initial data, and singularities in the initial data are preserved by the solution. What is the convergence rate for the formal series representing the solution of the Laplace equation? Do we obtain a classical solution?

To answer these questions, let us consider the general term in the series (7.43). We assume that the functions $f(y)$ and $g(y)$ are piecewise differentiable, and that they satisfy the homogeneous Dirichlet conditions at the end points $y = 0, d$. Then the coefficients α_n and β_n satisfy $|\alpha_n| < C_1, |\beta_n| < C_2$, where C_1 and C_2 are constants that do not depend on n (actually, one can establish a far stronger result on the decay of α_n and β_n to zero as $n \to \infty$, but we do not need this result here). Consider a specific term

$$\frac{\alpha_n}{\sinh \dfrac{n\pi b}{d}} \sinh \frac{n\pi}{d}(x - b) \sin \frac{n\pi y}{d}$$

in the series that represents u_1, where (x, y) is some interior point in D. This term is of the order of $O(e^{-\frac{n\pi}{d}x})$ for large values of n. The same argument implies

$$\frac{\beta_n}{\sinh \dfrac{n\pi b}{d}} \sinh \frac{n\pi}{d}x \sin \frac{n\pi y}{d} = O(e^{\frac{n\pi}{d}(x-b)}).$$

Thus all the terms in the series (7.43) decay exponentially fast as $n \to \infty$. Similarly the series of derivatives of all orders also converges exponentially fast, since the kth derivative introduces an algebraic factor n^k into the nth term in the series, but this factor is negligible (for large n) in comparison with the exponentially decaying term. We point out, though, that the rate of convergence slows down as we approach the domain's boundary.

Example 7.22 Solve the Laplace equation in the square $0 < x, y < \pi$ subject to the Dirichlet condition

$$u(x, 0) = 1984, \quad u(x, \pi) = u(0, y) = u(\pi, y) = 0.$$

The problem involves homogeneous boundary conditions on the two boundaries parallel to the y axis, and a single nonhomogeneous condition on the boundary $y = 0$. Therefore we write the formal solution in the form

$$u(x, y) = \sum_{n=1}^{\infty} A_n \sin nx \sinh n(y - \pi).$$

Substituting the boundary condition $u(x, 0) = 1984$ we obtain

$$-\sum_{1}^{\infty} A_n \sin nx \sinh n\pi = 1984.$$

The generalized Fourier series of the constant function $f(x) = 1984$ is $1984 = \sum_{n=1}^{\infty} \alpha_n \sin nx$, where the coefficients $\{\alpha_n\}$ are given by

$$\alpha_n = \frac{2 \times 1984}{\pi} \int_0^{\pi} \sin nx dx = \frac{3968}{n\pi}(\cos 0 - \cos n\pi).$$

We thus obtain

$$A_n = \begin{cases} -\dfrac{7936}{(2k-1)\pi \sinh(2k-1)\pi} & n = 2k - 1, \\ 0 & n = 2k. \end{cases} \tag{7.44}$$

The solution to the problem is given formally by

$$u(x, y) = 7936 \sum_{n=1}^{\infty} \frac{\sin(2n-1)x \sinh(2n-1)(\pi - y)}{(2n-1)\pi \sinh(2n-1)\pi}. \tag{7.45}$$

The observant reader might have noticed that in the last example we violated the compatibility condition (7.35). This is the condition that guarantees that each of the problems we solve by separation of variables has a continuous solution. Nevertheless, we obtained a formal solution, and following our discussion above, we know that the solution converges at every interior point in the square. Furthermore, the convergence of the series, and all its derivatives at every interior point is exponentially fast. Why should we be bothered, then, by the violation of the compatibility condition?

The answer lies in the Gibbs phenomenon that we mentioned in Chapter 6. If we compute the solution near the problematic points $((0, 0), (1, 0)$ in the last example) by summing up finitely many terms in the series (7.45), we observe high frequency oscillations. The difficulty we describe here is relevant not only to analytical solutions, but also to numerical solutions. We emphasize that even if the boundary condition to the original problem leads to a continuous solution, the process of breaking the problem into several subproblems for the purpose of separating variables might introduce discontinuities into the subproblems!

We therefore present a method for transforming a Dirichlet problem with *continuous* boundary data that does not satisfy the compatibility condition (7.35) into another Dirichlet problem (with continuous boundary data) that does satisfy (7.35). Denote the harmonic function we seek by $u(x, y)$, and the Dirichlet boundary condition on the rectangle's boundary by g. We write u as a combination: $u(x, y) = v(x, y) - P_2(x, y)$, where P_2 is a second-order appropriate harmonic polynomial (that we still have to find), while v is a harmonic function. We construct the harmonic polynomial in such a way that v satisfies the compatibility condition (7.35) (i.e. v vanishes at the square's vertices). Denote the restriction of v and P_2 to the square's boundary by g_1 and g_2, respectively. We select P_2 so that the incompatibility of g at the square's vertices is included in g_2. Thus we obtain for v a compatible Dirichlet problem. To construct P_2 as just described, we write the general form of a second-order harmonic polynomial:

$$P_2(x, y) = a_1(x^2 - y^2) + a_2 xy + a_3 x + a_4 y + a_5. \tag{7.46}$$

For simplicity, and without loss of generality, consider the square $0 < x < 1$, $0 < y < 1$. Requiring g_1 to vanish at all the vertices leads to four equations for the five unknown coefficients of P_2:

$$\begin{aligned}
g(0, 0) + a_5 &= 0, \\
g(1, 0) + a_1 + a_3 + a_5 &= 0, \\
g(0, 1) - a_1 + a_4 + a_5 &= 0, \\
g(1, 1) + a_2 + a_3 + a_4 + a_5 &= 0.
\end{aligned} \tag{7.47}$$

We choose arbitrarily $a_1 = 0$ and obtain the solution:

$$\begin{aligned}
a_1 &= 0, \\
a_2 &= -g(1, 1) - g(0, 0) + g(1, 0) + g(0, 1), \\
a_3 &= g(0, 0) - g(1, 0), \\
a_4 &= g(0, 0) - g(0, 1), \\
a_5 &= -g(0, 0).
\end{aligned} \tag{7.48}$$

Having thrown all the incompatibilities into the (easy to compute) harmonic polynomial, it remains to find a harmonic function v that satisfies the compatible boundary conditions $g_1 = g + g_2$.

Example 7.23 Let $u(x, y)$ be the harmonic function in the unit square satisfying the Dirichlet conditions

$$u(x, 0) = 1 + \sin \pi x, \quad u(x, 1) = 2, \quad u(0, y) = u(1, y) = 1 + y.$$

Represent u as a sum of a harmonic polynomial, and a harmonic function $v(x, y)$ that satisfies the compatibility condition (7.35).

We compute the appropriate harmonic polynomial. Solving the algebraic system (7.48) we get

$$a_1 = a_2 = a_3 = 0, \quad a_4 = -1, \quad a_5 = -1.$$

Hence the harmonic polynomial is $P_2(x, y) = -1 - y$. Define now

$$v(x, y) = u(x, y) - (1 + y).$$

Our construction implies that v is the harmonic function satisfying the Dirichlet data

$$v(x, 0) = \sin \pi x, \quad v(x, 1) = v(0, y) = v(1, y) = 0.$$

Indeed, the compatibility condition holds for v. Finally, we obtain that $v(x, y) = \sin \pi x \sinh(\pi - y)/\sinh \pi$, and therefore

$$u(x, y) = \sin \pi x \frac{\sinh(\pi - y)}{\sinh \pi} + 1 + y.$$

We end this section by solving a Neumann problem in a square.

Example 7.24 Find a harmonic function $u(x, y)$ in the square $0 < x, y < \pi$ satisfying the Neumann boundary conditions

$$u_y(x, \pi) = x - \pi/2, \quad u_x(0, y) = u_x(\pi, y) = u_y(x, 0) = 0. \tag{7.49}$$

The first step in solving a Neumann problem is to verify that the necessary condition for existence holds. In the current case, the integral $\int_{\partial D} \partial_n u \, ds$ is equal to $\int_0^\pi (x - \pi/2) dx$, which indeed vanishes.

The nature of the boundary conditions implies separated solutions of the form $U_n(x, y) = \cos nx \cosh ny$, where $n = 0, 1, 2 \ldots$. Thus the formal solution is

$$u(x, y) = A_0 + \sum_{n=1}^{\infty} A_n \cos nx \cosh ny. \tag{7.50}$$

The function u represented by (7.50) formally satisfies the equation and all the homogeneous boundary conditions. Substituting this solution into the nonhomogeneous boundary conditions on the edge $y = \pi$ leads to

$$\sum_{n=1}^{\infty} n A_n \sinh n\pi \cos nx = x - \pi/2 \qquad 0 < x < \pi.$$

We therefore expand $x - \pi/2$ into the following generalized Fourier series:

$$x - \pi/2 = \sum_{n=1}^{\infty} \beta_n \cos nx,$$

$$\beta_n = \frac{2}{\pi} \int_0^{\pi} (x - \pi/2) \cos nx\, dx = \begin{cases} -4/\pi n^2 & n = 1, 3, 5, \ldots \\ 0 & n = 0, 2, 4, \ldots \end{cases}.$$

Thus

$$u(x, y) = A_0 - \frac{4}{\pi} \sum_{n=1}^{\infty} \frac{\cos(2n-1)x \cosh(2n-1)y}{(2n-1)^3 \sinh(2n-1)\pi}. \qquad (7.51)$$

The graph of u is depicted in Figure 7.5. Notice that the additive constant A_0 is not determined by the problem's conditions. This was expected due to the nonuniqueness of the Neumann problem as was discussed in Section 7.5.

Remark 7.25 When we considered the Dirichlet problem earlier, we sometimes had to divide the problem into two subproblems, each of which involved homogeneous Dirichlet conditions on two opposite edges of the rectangle. A similar division is sometimes needed for the Neumann problem. Here, however, a

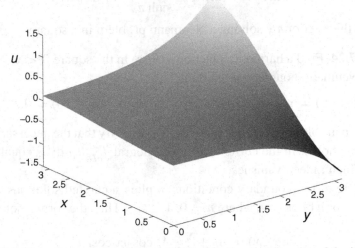

Figure 7.5 The graph of $u(x, y)$ from (7.51). Observe that in spite of the intricate form of the Fourier series, the actual shape of the surface is very smooth. We also see that u achieves its maximum and minimum on the boundary.

fundamental difficulty might arise: while the original problem presumably satisfies the necessary existence condition (otherwise the problem is not solvable at all!), it is not guaranteed that each of the *subproblems* will satisfy this condition! To demonstrate the difficulty and to propose a remedy for it, we look at it in some detail.

Consider the Neumann problem for the Laplace equation:

$$\Delta u = 0 \quad x \in \Omega, \tag{7.52}$$

$$\partial_n u = g \quad x \in \partial\Omega. \tag{7.53}$$

We assume, of course, that the condition $\int_{\partial\Omega} g\,ds = 0$ holds. Split the boundary of Ω into two parts $\partial\Omega = \partial_1\Omega \cup \partial_2\Omega$. Define $u = u_1 + u_2$, where u_1, u_2 are both harmonic in Ω and satisfy the boundary conditions

$$\partial_n u_1 = \begin{cases} g & x \in \partial_1\Omega, \\ 0 & x \in \partial_2\Omega. \end{cases} \qquad \partial_n u_2 = \begin{cases} 0 & x \in \partial_1\Omega, \\ g & x \in \partial_2\Omega. \end{cases}$$

The difficulty is that now the existence condition may not hold separately for u_1 and u_2. We overcome this by the same method we used earlier to take care of the Gibbs phenomenon. We add to (and subtract from) the solution a harmonic polynomial. We use a harmonic polynomial $P(x, y)$ that satisfies $\int_{\partial_1\Omega} \partial_n P(x, y)\,ds \neq 0$. Assume, for example, that the harmonic polynomial $x^2 - y^2$ satisfies this condition. We then search for harmonic functions v_1 and v_2 that satisfy the following Neumann conditions:

$$\partial_n v_1 = \begin{cases} g + a\partial_n(x^2 - y^2) & x \in \partial_1\Omega, \\ 0 & x \in \partial_2\Omega, \end{cases} \qquad \partial_n v_2 = \begin{cases} 0 & x \in \partial_1\Omega, \\ g + a\partial_n(x^2 - y^2) & x \in \partial_2\Omega. \end{cases}$$

We choose the parameter a such that the solvability condition holds for v_1. Since the original problem is assumed to be solvable and also the harmonic polynomial, by its very existence, satisfies the compatibility condition, it follows that v_2 must satisfy that condition too. Finally, we observe that $u = v_1 + v_2 - a(x^2 - y^2)$.

7.7.2 Circular domains

Another important domain where the Laplace equation can be solved by separation of variables is a disk. Let B_a be a disk of radius a around the origin. We want to find the function $u(x, y)$ that solves the Dirichlet problem

$$\Delta u = 0 \quad (x, y) \in B_a, \tag{7.54}$$

$$u(x, y) = g(x, y) \quad (x, y) \in \partial B_a. \tag{7.55}$$

It is convenient to solve the equation in polar coordinates in order to use the symmetry of the domain. We thus denote the polar coordinates by (r, θ), and the unknown function is written as $w(r, \theta) = u(x(r, \theta), y(r, \theta))$. We mentioned in Section 7.2 that w satisfies the equation

$$\Delta w = w_{rr} + \frac{1}{r} w_r + \frac{1}{r^2} w_{\theta\theta} = 0. \tag{7.56}$$

Consequently, we have to solve (7.56) in the domain

$$B_a = \{(r, \theta) \mid 0 < r < a, \ 0 \le \theta \le 2\pi\}.$$

The PDE is subject to the boundary condition

$$w(a, \theta) = h(\theta) = g(x(a, \theta), y(a, \theta)), \tag{7.57}$$

and to the additional obvious requirement that $\lim_{r \to 0} w(r, \theta)$ exists and is finite (the origin needs special attention, since it is a singular point in polar coordinates). We seek a solution of the form $w(r, \theta) = R(r)\Theta(\theta)$. Substituting this function into (7.56), and using the usual arguments from Chapter 5, we obtain a pair of equations for R and Θ:

$$r^2 R''(r) + r R'(r) - \lambda R(r) = 0 \quad 0 < r < a, \tag{7.58}$$
$$\Theta''(\theta) + \lambda \Theta(\theta) = 0. \tag{7.59}$$

The equation for Θ holds at the interval $(0, 2\pi)$. In order that the solution $w(r, \theta)$ be of class C^2, we need to impose two periodicity conditions:

$$\Theta(0) = \Theta(2\pi), \quad \Theta'(0) = \Theta'(2\pi). \tag{7.60}$$

Notice that (7.59) and (7.60) together also imply the periodicity of the second derivative with respect to θ.

The general solution to the Sturm–Liouville problem (7.59)–(7.60) is given (see Chapter 6) by the sequence

$$\Theta_n(\theta) = A_n \cos n\theta + B_n \sin n\theta, \quad \lambda_n = n^2, \ 0, 1, 2, \dots. \tag{7.61}$$

Substituting the eigenvalues λ_n into (7.58) yields a second-order Euler (equidimensional) ODE for R (see Subsection A.3):

$$r^2 R_n'' + r R_n' - n^2 R_n = 0. \tag{7.62}$$

The solutions of these equations are given (except for $n = 0$) by appropriate powers of the independent variable r:

$$R_n(r) = C_n r^n + D_n r^{-n}, \quad n = 1, 2, \dots. \tag{7.63}$$

In the special case $n = 0$ we obtain

$$R_0(r) = C_0 + D_0 \ln r. \tag{7.64}$$

Observe that the functions r^{-n}, $n = 1, 2, \ldots$ and the function $\ln r$ are singular at the origin $(r = 0)$. Since we only consider smooth solutions, we impose the condition

$$D_n = 0 \qquad n = 0, 1, 2, \ldots.$$

We still have to satisfy the boundary condition (7.57). For this purpose we form the superposition

$$w(r, \theta) = \frac{\alpha_0}{2} + \sum_{n=1}^{\infty} r^n (\alpha_n \cos n\theta + \beta_n \sin n\theta). \tag{7.65}$$

Formally differentiating this series term-by-term, we verify that (7.65) is indeed harmonic. Imposing the boundary condition (7.57), and using the classical Fourier formula (Chapter 6), we obtain

$$\alpha_0 = \frac{1}{\pi} \int_0^{2\pi} h(\varphi) d\varphi,$$

$$\alpha_n = \frac{1}{\pi a^n} \int_0^{2\pi} h(\varphi) \cos n\varphi d\varphi, \quad \beta_n = \frac{1}{\pi a^n} \int_0^{2\pi} h(\varphi) \sin n\varphi d\varphi \; n \geq 1. \tag{7.66}$$

Example 7.26 Solve the Laplace equation in the unit disk subject to the boundary conditions $w(r, \theta) = y^2$ on $r = 1$.

Observe that on the boundary $y^2 = \sin^2 \theta$. All we have to do is to compute the classical Fourier expansion of the function $\sin^2 \theta$. This expansion is readily performed in light of the identity $\sin^2 \theta = \frac{1}{2}(1 - \cos 2\theta)$. Thus the Fourier series is finite, and the required harmonic function is $w(r, \theta) = \frac{1}{2}(1 - r^2 \cos 2\theta)$, or, upon returning to Cartesian coordinates, $u(x, y) = \frac{1}{2}(1 - x^2 + y^2)$.

Before proceeding to other examples, it is worthwhile to examine the convergence properties of the formal Fourier series we constructed in (7.65). We write $M = (1/\pi) \int_0^{2\pi} |h(\theta)| d\theta$. The Fourier formulas (7.66) imply the inequalities

$$|\alpha_n|, \; |\beta_n| \leq M a^{-n}.$$

Hence the nth term in the Fourier series (7.65) is bounded by $2M(r/a)^n$; thus the series converges for all $r < a$, and even converges uniformly in any disk of radius $\tilde{a} < a$. In fact, we can use the same argument to show also that the series of derivatives of any order converges uniformly in any disk of radius $\tilde{a} < a$ to the appropriate derivative of the solution. Moreover, if $h(\theta)$ is a periodic piecewise differentiable and continuous function, then by Proposition 6.30, its Fourier expansion converges uniformly. It follows from Proposition 7.20 that w is a classical solution. Observe, however, that the rate of convergence deteriorates when we approach the boundary.

The method of separation of variables can be used for other domains with a symmetric polar shape. For example, in Exercise 7.20 we solve the Dirichlet problem in the domain bounded by concentric circles. Interestingly, one can also separate variables in the domain bounded by two nonconcentric circles, but this requires the introduction of a special coordinate system, called *bipolar*, and the computations are somewhat more involved. In Exercise 7.11 the Dirichlet problem in the exterior of a disk is solved. We now demonstrate the solution of the Dirichlet problem in a circular sector.

Example 7.27 Find the harmonic function $w(r, \theta)$ in the sector

$$D_\gamma = \{(r, \theta)|\ 0 < r < a,\ 0 < \theta < \gamma\}$$

that satisfies on the sector's boundary the Dirichlet condition

$$w(a, \theta) = g(\theta)\ 0 \le \theta \le \gamma, \quad w(r, 0) = w(r, \gamma) = 0\ 0 \le r \le a. \tag{7.67}$$

The process of obtaining separated solutions and an appropriate eigenvalue problem is similar to the previous case of the Laplace equation in the entire disk. Namely, we again seek solutions of the form $R(r)\Theta(\theta)$, where the Sturm–Liouville equation is again (7.59) for Θ, and the equation for the radial component is again (7.58). The difference is in the boundary condition for the Θ equation. Unlike the periodic boundary conditions that we encountered for the problem in the full disk, we now have Dirichlet boundary conditions $\Theta(0) = \Theta(\gamma) = 0$. Therefore the sequences of eigenfunctions and eigenvalues are now given by

$$\Theta_n(\theta) = A_n \sin \frac{n\pi}{\gamma}\theta, \quad \lambda_n = \left(\frac{n\pi}{\gamma}\right)^2 \quad n = 1, 2, \ldots.$$

Substituting the eigenvalues λ_n into (7.58), and keeping only the solutions that are bounded in the origin, we obtain

$$w_n(r, \theta) = \sin \frac{n\pi\theta}{\gamma} r^{n\pi/\gamma}.$$

Hence, the formal solution is given by the series

$$w(r, \theta) = \sum_{n=1}^{\infty} \alpha_n \sin \frac{n\pi\theta}{\gamma} r^{n\pi/\gamma}. \tag{7.68}$$

On $r = a,\ 0 < \theta < \gamma$ we have

$$g(\theta) = \sum_{n=1}^{\infty} \alpha_n a^{n\pi/\gamma} \sin \frac{n\pi\theta}{\gamma},$$

therefore,

$$\alpha_n = \frac{2a^{-n\pi/\gamma}}{\gamma} \int_0^\gamma g(\varphi) \sin \frac{n\pi\varphi}{\gamma} \mathrm{d}\varphi.$$

Remark 7.28 Consider the special case in which $\gamma = 2\pi$, and write explicitly the solution (7.68):

$$w(r, \theta) = \sum_{n=1}^\infty \alpha_n \sin \frac{n\theta}{2} r^{n/2}. \tag{7.69}$$

Observe that even though the sector now consists of the entire disk, the solution (7.69) is completely different from the solution we found earlier for the Dirichlet problem in the disk. The reason for the difference is that the boundary condition (7.67) on the sector's boundary is fundamentally different from the periodic boundary condition. The condition (7.67) singles out a specific curve in the disk, and, thus, breaks the disk's symmetry.

Remark 7.29 We observe that, in general, some derivatives (in fact, most of them) of the solution (7.68) are singular at the origin (the sector's vertex). We cannot require the solution to be as smooth there as we wish. This singularity has important physical significance. The Laplace equation in a sector is used to model cracks in the theory of elasticity. The singularity we observe indicates a concentration of large stresses at the vertex.

We end this section by demonstrating the separation of variables method for the Poisson equation in a disk with Dirichlet boundary conditions. The problem is thus to find a function $w(r, \theta)$, satisfying

$$\Delta w = F(r, \theta) \qquad 0 < r < a, \ 0 \le \theta \le 2\pi,$$

together with the boundary condition $w(a, \theta) = g(\theta)$. In light of the general technique we developed in Chapter 6 to solve nonhomogeneous equations, we seek a solution in the form

$$w(r, \theta) = \frac{f_0(r)}{2} + \sum_{n=1}^\infty [f_n(r) \cos n\theta + g_n(r) \sin n\theta].$$

Similarly we expand F into a Fourier series

$$F(r, \theta) = \frac{\delta_0(r)}{2} + \sum_{n=1}^\infty [\delta_n(r) \cos n\theta + \epsilon_n(r) \sin n\theta].$$

Substituting these two Fourier series into the Poisson equation, and comparing the associated coefficients, we find

$$f_n'' + \frac{1}{r} f_n' - \frac{n^2}{r^2} f_n = \delta_n(r) \quad n = 0, 1, \dots, \tag{7.70}$$

$$g_n'' + \frac{1}{r} g_n' - \frac{n^2}{r^2} g_n = \epsilon_n(r) \quad n = 1, 2, \dots. \tag{7.71}$$

The general solutions of these equations can be written as

$$f_n(r) = A_n r^n + \tilde{f}_n(r), \quad g_n(r) = B_n r^n + \tilde{g}_n(r),$$

where $\tilde{f}_n$ and $\tilde{g}_n$ are particular solutions of the appropriate nonhomogeneous equations. In fact, it is shown in Exercise 7.18 that the solutions of (7.70)–(7.71), that satisfy the homogeneous boundary conditions $\tilde{f}_n(a) = \tilde{g}_n(a) = 0$, and that are bounded at the origin, can be written as

$$\tilde{f}_n(r) = \int_0^r K_1^{(n)}(r, a, \rho) \delta_n(\rho) \rho \, d\rho + \int_r^a K_2^{(n)}(r, a, \rho) \delta_n(\rho) \rho \, d\rho, \tag{7.72}$$

$$\tilde{g}_n(r) = \int_0^r K_1^{(n)}(r, a, \rho) \epsilon_n(\rho) \rho \, d\rho + \int_r^a K_2^{(n)}(r, a, \rho) \epsilon_n(\rho) \rho \, d\rho, \tag{7.73}$$

where

$$K_1^{(0)} = \log \frac{r}{a}, \quad K_2^{(0)} = \log \frac{\rho}{a}, \tag{7.74}$$

$$K_1^{(n)} = \frac{1}{2n} \left[\left(\frac{r}{a} \right)^n - \left(\frac{a}{r} \right)^n \right] \left(\frac{a}{a} \right)^n, \quad K_2^{(n)} = \frac{1}{2n} \left[\left(\frac{\rho}{a} \right)^n - \left(\frac{a}{\rho} \right)^n \right] \left(\frac{r}{a} \right)^n \quad n \geq 1. \tag{7.75}$$

We thus constructed a solution of the form

$$w(r, \theta) = \frac{A_0 + \tilde{f}_0(r)}{2} + \sum_{n=1}^{\infty} \left\{ [A_n r^n + \tilde{f}_n(r)] \cos n\theta + [B_n r^n + \tilde{g}_n(r)] \sin n\theta \right\}. \tag{7.76}$$

To find the coefficients $\{A_n, B_n\}$ we substitute this solution into the boundary conditions

$$w(a, \theta) = \frac{A_0 + \tilde{f}_0(a)}{2} + \sum_{n=1}^{\infty} \left\{ [A_n a^n + \tilde{f}_n(a)] \cos n\theta + [B_n a^n + \tilde{g}_n(a)] \sin n\theta \right\}$$

$$= \frac{\alpha_0}{2} + \sum_{n=1}^{\infty} (\alpha_n \cos n\theta + \beta_n \sin n\theta) = g(\theta).$$

The required coefficients can be written as

$$A_n = \frac{\alpha_n - 2\tilde{f}_n(a)}{a^n} = \frac{\alpha_n}{a^n} \quad n = 0, 1, \dots, \quad B_n = \frac{\beta_n - \tilde{g}_n(a)}{a^n} = \frac{\beta_n}{a^n} \quad n = 1, 2, \dots.$$

Example 7.30 Solve the Poisson equation

$$\Delta w = 8r \cos \theta \qquad 0 \le r < 1, \ 0 \le \theta \le 2\pi,$$

subject to the boundary conditions $w(1, \theta) = \cos^2 \theta$.

One can verify that

$$\tilde{f}_n(r) = \begin{cases} r^3 & n = 1, \\ 0 & n \ne 1, \end{cases} \tag{7.77}$$

and $\tilde{g}_n(r) = 0$ for every n, are particular solutions to the nonhomogeneous equations (7.70)–(7.71). Therefore the general solution can be written as

$$w(r, \theta) = \frac{A_0}{2} + (A_1 r + r^3) \cos \theta + B_1 r \sin \theta + \sum_{n=2}^{\infty} (A_n r^n \cos n\theta + B_n r^n \sin n\theta).$$

We use the identity $\cos^2 \theta = \frac{1}{2}(1 + \cos 2\theta)$ to obtain the expansion of the boundary condition into a Fourier series. Therefore,

$$A_0 = 1, \ A_1 = -1, \ A_2 = \frac{1}{2}, \ A_n = 0 \quad \forall n \ne 0, 1, 2, \quad B_n = 0 \quad \forall n = 1, 2, 3 \ldots,$$

and the solution is

$$w(r, \theta) = \frac{1}{2} + (r^3 - 1) \cos \theta + \frac{r^2}{2} \cos 2\theta.$$

7.8 Poisson's formula

One of the important tools in the theory of PDEs is the *integral representation* of solutions. An integral representation is a formula for the solution of a problem in terms of an integral depending on a *kernel function*. We need to compute the kernel function just once for a given equation, a given domain, and a given type of boundary condition. We demonstrate now an integral representation for the Laplace equation in a disk of radius a with Dirichlet boundary conditions (7.54)–(7.55). We start by rewriting the solution as a Fourier series (see (7.65)), using (7.66):

$$w(r, \theta) = \frac{1}{2\pi} \int_0^{2\pi} h(\varphi) d\varphi$$

$$+ \frac{1}{\pi} \sum_{n=1}^{\infty} \left(\frac{r}{a}\right)^n \int_0^{2\pi} h(\varphi)(\cos n\varphi \cos n\theta + \sin n\varphi \sin n\theta) d\varphi. \tag{7.78}$$

Consider $r < \tilde{a} < a$. Since the series converges uniformly there, we can interchange the order of summation and integration, and obtain

$$w(r, \theta) = \frac{1}{\pi} \int_0^{2\pi} h(\varphi) \left[\frac{1}{2} + \sum_{n=1}^{\infty} \left(\frac{r}{a}\right)^n \cos n(\theta - \varphi) \right] d\varphi. \tag{7.79}$$

The summation of the infinite series

$$\frac{1}{2} + \sum_{n=1}^{\infty} \left(\frac{r}{a}\right)^n \cos n(\theta - \varphi)$$

requires a little side calculation. Define for this purpose $z = \rho e^{i\alpha}$ and evaluate (for $\rho < 1$) the geometric sum

$$\frac{1}{2} + \sum_{1}^{\infty} z^n = \frac{1}{2} + \frac{z}{1 - z} = \frac{1 - \rho^2 + 2i\rho \sin \alpha}{2(1 - 2\rho \cos \alpha + \rho^2)}.$$

Since $z^n = \rho^n(\cos n\alpha + i \sin n\alpha)$, we conclude upon separating the real and imaginary parts that

$$\frac{1}{2} + \sum_{1}^{\infty} \rho^n \cos n\alpha = \frac{1 - \rho^2}{2(1 - 2\rho \cos \alpha + \rho^2)}. \tag{7.80}$$

Returning to (7.79) using $\rho = r/a$, $\alpha = \theta - \varphi$, we obtain the *Poisson formula*

$$w(r, \theta) = \frac{1}{2\pi} \int_0^{2\pi} K(r, \theta; a, \varphi) h(\varphi) d\varphi, \tag{7.81}$$

where the kernel K, given by

$$K(r, \theta; a, \varphi) = \frac{a^2 - r^2}{a^2 - 2ar \cos(\theta - \varphi) + r^2}, \tag{7.82}$$

is called *Poisson's kernel*. This is a very useful formula. The kernel describes a universal solution for the Laplace equation in a disk. All we have to do (at least in theory), is to substitute the boundary condition into (7.81) and carry out the integration. Moreover, the formula is valid for any integrable function h. It turns out that one can derive similar representations not just in disks, but also in arbitrary smooth domains. We shall elaborate on this issue in Chapter 8.

As another example of an integral representation for harmonic functions, we derive the Poisson formula for the Neumann problem in a disk. Let $w(r, \theta)$ be a harmonic function in the disk $r < a$, satisfying on the disk's boundary the Neumann condition $\partial w(a, \theta)/\partial r = g(\theta)$. We assume, of course, that the solvability condition $\int_0^{2\pi} g(\theta) d\theta = 0$ holds. Recall that the general form of a harmonic function in the disk is

$$w(r, \theta) = \frac{\alpha_0}{2} + \sum_{n=1}^{\infty} r^n (\alpha_n \cos n\theta + \beta_n \sin n\theta).$$

Note that the coefficient α_0 is arbitrary, and cannot be retrieved from the boundary conditions (cf. the uniqueness theorem for the Neumann problem). To find the coefficients $\{\alpha_n, \beta_n\}$, substitute the solution into the boundary conditions and obtain

$$\alpha_n = \frac{1}{n\pi a^{n-1}} \int_0^{2\pi} g(\varphi) \cos n\varphi \, d\varphi, \quad \beta_n = \frac{1}{n\pi a^{n-1}} \int_0^{2\pi} g(\varphi) \sin n\varphi \, d\varphi \quad n = 1, 2, \ldots.$$

$$(7.83)$$

Hence

$$w(r, \theta) = \frac{\alpha_0}{2} + \frac{a}{\pi} \int_0^{2\pi} K_N(r, \theta; a, \varphi) g(\varphi) d\varphi, \qquad (7.84)$$

where

$$K(r, \theta; a, \varphi) = \sum_{n=1}^{\infty} \frac{1}{n} \left(\frac{r}{a}\right)^n \cos n(\theta - \varphi). \qquad (7.85)$$

Because of the $1/n$ factor we cannot use the summation formula (7.80) directly. Instead we perform another quick side calculation. Notice that by a process like the one leading to (7.80) one can derive

$$\sum_{n=1}^{\infty} \rho^{n-1} \cos n\alpha = \frac{\cos\alpha - \rho}{1 - 2\rho \cos\alpha + \rho^2}.$$

Therefore after an integration with respect to ρ we obtain the Poisson kernel for the Neumann problem in a disk:

$$K_N(r, \theta; a, \varphi) = \sum_{n=1}^{\infty} \frac{1}{n} \left(\frac{r}{a}\right)^n \cos n(\theta - \varphi)$$

$$= -\frac{1}{2} \ln \left[1 - 2\frac{r}{a} \cos(\theta - \varphi) + \left(\frac{r}{a}\right)^2 \right]. \qquad (7.86)$$

Remark 7.31 It is interesting to note that both Poisson's kernels that we have computed have a special dependency on the angular variable θ. In both cases we have

$$K(r, \theta; a, \varphi) = \tilde{K}(r, a, \theta - \varphi),$$

namely, the dependency is only through the difference between θ and φ. This property is a consequence of the symmetry of the Laplace equation and the circular domain with respect to rotations.

Remark 7.32 Poisson's formula provides, as a by-product, another proof for the mean value principle. In fact, the formula is valid with respect to any circle around any point in a given domain (provided that the circle is fully contained in the domain). Indeed, if we substitute $r = 0$ into (7.81), we obtain at once the mean value principle.

When we solved the Laplace equation in a rectangle or in a disk, we saw that the solution is in $C^\infty(D)$. That is, the solution is differentiable infinitely many times at any *interior* point. Let us prove this property for any domain.

Theorem 7.33 (Smoothness of harmonic functions) *Let $u(x, y)$ be a harmonic function in D. Then $u \in C^\infty(D)$.*

Proof Denote by p an interior point in D, and construct a coordinate system centered at p. Let B_a be a disk of radius a centered at p, fully contained in D. Write Poisson's formula for an arbitrary point (x, y) in B_a. We can differentiate under the integral sign arbitrarily many times with respect to r or with respect to θ, and thus establish the theorem. $\square$

7.9 Exercises

7.1 Prove Green's identity (7.20).

7.2 Prove uniqueness for the Dirichlet and Neumann problems for the *reduced Helmholtz equation*

$$\Delta u - ku = 0$$

in a bounded planar domain D, where k is a positive constant.

7.3 Find the solution $u(x, y)$ of the reduced Helmholtz equation $\Delta u - ku = 0$ (k is a positive parameter) in the square $0 < x, y < \pi$, where u satisfies the boundary condition

$$u(0, y) = 1, \ u(\pi, y) = u(x, 0) = u(x, \pi) = 0.$$

7.4 Solve the Laplace equation $\Delta u = 0$ in the square $0 < x, y < \pi$, subject to the boundary condition

$$u(x, 0) = u(x, \pi) = 1, \ u(0, y) = u(\pi, y) = 0.$$

7.5 Let $u(x, y)$ be a nonconstant harmonic function in the disk $x^2 + y^2 < R^2$. Define for each $0 < r < R$

$$M(r) = \max_{x^2+y^2=r^2} u(x, y).$$

Prove that $M(r)$ is a monotone increasing function in the interval $(0, R)$

7.6 Verify that the solution of the Dirichlet problem defined in Example 7.23 is classical.

7.7 (a) Compute the Laplace equation in a polar coordinate system.

(b) Find a function u, harmonic in the disk $x^2 + y^2 < 6$, and satisfying $u(x, y) = y + y^2$ on the disk's boundary. Write your answer in a Cartesian coordinate system.

7.8 (a) Solve the problem

$$
\begin{aligned}
\Delta u &= 0 & 0 < x < \pi, \ 0 < y < \pi, \\
u(x, 0) &= u(x, \pi) = 0 & 0 \le x \le \pi, \\
u(0, y) &= 0 & 0 \le y \le \pi, \\
u(\pi, y) &= \sin y & 0 \le y \le \pi.
\end{aligned}
$$

(b) Is there a point $(x, y) \in \{(x, y) \mid 0 < x < \pi, \ 0 < y < \pi\}$ such that $u(x, y) = 0$?

7.9 A harmonic function of the form

$$
P_n(x, y) = \sum_{i+j=n} a_{i,j} x^i y^j
$$

is called a *homogeneous harmonic polynomial* of degree n. Denote the space of homogeneous harmonic polynomials of degree n by V_n. What is the dimension of V_n?
Hint Use the polar form of the Laplace equation.

7.10 Consider the Laplace equation $\Delta u = 0$ in the domain $0 < x, y < \pi$ with the boundary condition

$$
u_y(x, \pi) = x^2 - a, \ u_x(0, y) = u_x(\pi, y) = u_y(x, 0) = 0.
$$

Find all the values of the parameter a for which the problem is solvable. Solve the problem for these values of a.

7.11 Solve the Laplace equation $\Delta u = 0$ in the domain $x^2 + y^2 > 4$, subject to the boundary condition $u(x, y) = y$ on $x^2 + y^2 = 4$, and the decay condition $\lim_{|x|+|y| \to \infty} u(x, y) = 0$.

7.12 Solve the problem

$$
\begin{aligned}
u_{xx} + u_{yy} &= 0 & 0 < x < 2\pi, \ -1 < y < 1, \\
u(x, -1) = 0, \quad u(x, 1) &= 1 + \sin 2x & 0 \le x \le 2\pi, \\
u_x(0, y) = u_x(2\pi, y) &= 0 & -1 < y < 1.
\end{aligned}
$$

7.13 Prove that every nonnegative harmonic function in the disk of radius a satisfies

$$
\frac{a - r}{a + r} u(0, 0) \le u(r, \theta) \le \frac{a + r}{a - r} u(0, 0).
$$

Remark This result is called the *Harnack inequality*.

7.14 Let D be the domain $D = \{(x, y) \mid x^2 + y^2 < 4\}$. Consider the Neumann problem

$$
\begin{aligned}
\Delta u &= 0 & (x, y) \in D, \\
\partial_n u &= \alpha x^2 + \beta y + \gamma & (x, y) \in \partial D,
\end{aligned}
$$

where α, β, and γ are real constants.

(a) Find the values of α, β, γ for which the problem is not solvable.

(b) Solve the problem for those values of α, β, γ for which a solution does exist.

7.15 Let $D = \{(x, y) \mid 0 < x < \pi, \ 0 < y < \pi\}$. Denote its boundary by ∂D.

(a) Assume $v_{xx} + v_{yy} + xv_x + yv_y > 0$ in D. Prove that v has no local maximum in D.

(b) Consider the problem

$$u_{xx} + u_{yy} + xu_x + yu_y = 0 \quad (x, y) \in D,$$
$$u(x, y) = f(x, y) \quad (x, y) \in \partial D,$$

where f is a given continuous function. Show that if u is a solution, then the maximum of u is achieved on the boundary ∂D.

Hint Use the auxiliary function $v_\varepsilon(x, y) = u(x, y) + \varepsilon x^2$.

(c) Show that the problem formulated in (b) has at most one solution.

7.16 Let $u(x, y)$ be a smooth solution for the Dirichlet problem

$$\Delta u + \vec{V} \cdot \nabla u = F \quad (x, y) \in D,$$
$$u(x, y) = g(x, y) \quad (x, y) \in \partial D,$$

where $F > 0$ in D, $g < 0$ on ∂D and $\vec{V}(x, y)$ is a smooth vector field in D. Show that $u(x, y) < 0$ in D.

7.17 (a) Solve the equation $u_t = 2u_{xx}$ in the domain $0 < x < \pi, t > 0$ under the initial boundary value conditions

$$u(0, t) = u(\pi, t) = 0, \quad u(x, 0) = f(x) = x(x^2 - \pi^2).$$

(b) Use the maximum principle to prove that the solution in (a) is a classical solution.

7.18 Prove that the formulas (7.72)–(7.75) describe solutions of (7.70)–(7.71) that are bounded at the origin and vanish at $r = a$.

7.19 Let $u(r, \theta)$ be a harmonic function in the disk

$$D = \{(r, \theta) \mid 0 \le r < R, -\pi < \theta \le \pi\},$$

such that u is continuous in the closed disk $\bar{D}$ and satisfies

$$u(R, \theta) = \begin{cases} \sin^2 2\theta & |\theta| \le \pi/2, \\ 0 & \pi/2 < |\theta| \le \pi. \end{cases}$$

(a) Evaluate $u(0, 0)$ without solving the PDE.

(b) Show that the inequality $0 < u(r, \theta) < 1$ holds at each point (r, θ) in the disk.

7.20 Find a function $u(r, \theta)$ harmonic in $\{2 < r < 4, \ 0 \le \theta \le 2\pi\}$, satisfying the boundary condition

$$u(2, \theta) = 0, \quad u(4, \theta) = \sin \theta.$$

7.21 Let $u(x, t)$ be a solution of the problem

$$u_t - u_{xx} = 0 \qquad Q_T = \{(x, t) \mid 0 < x < \pi, \ 0 < t \le T\},$$
$$u(0, t) = u(\pi, t) = 0 \qquad 0 \le t \le T,$$
$$u(x, 0) = \sin^2(x) \qquad 0 \le x \le \pi.$$

Use the maximum principle to prove that $0 \le u(x, t) \le e^{-t} \sin x$ in the rectangle Q_T.

7.22 Let $u(x, y)$ be the harmonic function in $D = \{(x, y) \mid x^2 + y^2 < 36\}$ which satisfies
on ∂D the Dirichlet boundary condition

$$u(x, y) = \begin{cases} x & x < 0 \\ 0 & \text{otherwise.} \end{cases}$$

(a) Prove that $u(x, y) < \min\{x, 0\}$ in D.
Hint Prove that $u(x, y) < x$ and that $u(x, y) < 0$ in D.
(b) Evaluate $u(0, 0)$ using the mean value principle.
(c) Using Poisson's formula evaluate $u(0, y)$ for $0 \le y < 6$.
(d) Using the separation of variables method, find the solution u in D.
(e) Is the solution u classical?

8

Green's functions and integral representations

8.1 Introduction

Integral representations play a central role in various fields of pure and applied mathematics, theoretical physics, and engineering. Many boundary value problems and initial boundary value problems can be solved using integral kernels. In such a case, we usually have an explicit formula for the solution as a (finite) sum of integrals involving integral kernels and the associated side conditions (which are the given data). The integral kernel depends on the differential operator, the type of given boundary condition, and the domain. It should be computed just once for any given type of problem. Hence, given an integral representation for a differential problem, we can find the solution for a specific choice of associated conditions by carrying out just a small number of integrations.

A typical example of an integral representation is the Poisson formula (7.81) which is an explicit integral representation for solutions of the Dirichlet problem for the Laplace equation in a disk. Note that the d'Alembert formula (4.17) for solving the (one-dimensional) Cauchy problem for the nonhomogeneous wave equation with zero initial conditions is also an integral representation.

In this chapter, we present some further examples of integral representations for the Laplace operator and for the heat equation. The integral kernel for the Laplace operator is called Green's function in honor of the great English mathematician George Green (1793–1841)[1].

Even for these equations, the corresponding integral kernels can be computed explicitly only in a few special cases. Nevertheless, we can obtain many qualitative properties of the kernels for a large number of important problems. These properties can teach us a great deal about the qualitative properties of the solutions of

---

[1] George Green is best known for his famous formula and for his discovery of the Green function. He was an autodidact with only four terms of elementary school education [1]. Green achieved some of his great results while working full time as a miller. At the age of 40 he left his mill and was accepted as an undergraduate student at Cambridge University. Green died before the importance of his results was recognized.

the problem (their structure, regularity, asymptotic behavior, positivity, stability, etc.).

The current chapter is devoted to two-dimensional elliptic and parabolic problems. In Chapter 9, we extend the discussion to integral representations for elliptic and parabolic problems in higher dimensions, and to the wave equation in the Euclidean plane and space.

8.2 Green's function for Dirichlet problem in the plane

Consider the Dirichlet problem for the Poisson equation

$$
\begin{aligned}
\Delta u = f \quad & D, \\
u = g \quad & \partial D,
\end{aligned}
\tag{8.1}
$$

where D is a bounded planar domain with a smooth boundary ∂D. The fundamental solution of the Laplace equation plays an important role in our discussion. Recall that this fundamental solution is defined by

$$
\Gamma(x, y) = -\frac{1}{2\pi} \ln r = -\frac{1}{4\pi} \ln(x^2 + y^2).
\tag{8.2}
$$

The fundamental solution is harmonic in the punctured plane, and it is a radially symmetric function with a singularity at the origin. Fix a point $(\xi, \eta) \in \mathbb{R}^2$. Note that if $u(x, y)$ is harmonic, then $u(x - \xi, y - \eta)$ is also harmonic for every fixed pair (ξ, η). We use the notation

$$
\Gamma(x, y; \xi, \eta) := \Gamma(x - \xi, y - \eta).
$$

We call $\Gamma(x, y; \xi, \eta)$ the *fundamental solution of the Laplace equation with a pole at* (ξ, η). The reader can check that

$$
\left| \frac{\partial \Gamma}{\partial x}(x, y; \xi, \eta) \right| + \left| \frac{\partial \Gamma}{\partial y}(x, y; \xi, \eta) \right| \leq \frac{C}{\sqrt{(x - \xi)^2 + (y - \eta)^2}},
\tag{8.3}
$$

$$
\left| \frac{\partial^2 \Gamma}{\partial x^2}(x, y; \xi, \eta) \right| + \left| \frac{\partial^2 \Gamma}{\partial x \partial y}(x, y; \xi, \eta) \right| + \left| \frac{\partial^2 \Gamma}{\partial y^2}(x, y; \xi, \eta) \right| \leq \frac{C}{(x - \xi)^2 + (y - \eta)^2}.
\tag{8.4}
$$

The function $\Gamma(x, y; \xi, \eta)$ is harmonic for any point (x, y) in the plane such that $(x, y) \neq (\xi, \eta)$. For $\varepsilon > 0$, set

$$
B_\varepsilon := \{(x, y) \in D \mid \sqrt{(x - \xi)^2 + (y - \eta)^2} < \varepsilon\}, \qquad D_\varepsilon := D \setminus B_\varepsilon.
$$

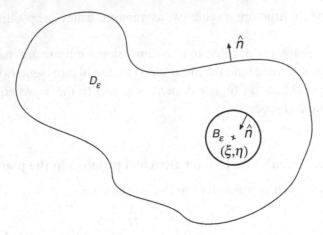

Figure 8.1 A drawing for the construction of Green's function.

Let $u \in C^2(\bar{D})$. We use the second Green identity (7.19) in the domain D_ε where the function $v(x, y) = \Gamma(x, y; \xi, \eta)$ is harmonic to obtain

$$\int_{D_\varepsilon} (\Gamma \Delta u - u \Delta \Gamma) dx dy = \int_{\partial D_\varepsilon} (\Gamma \partial_n u - u \partial_n \Gamma) ds.$$

Therefore,

$$\int_{D_\varepsilon} \Gamma \Delta u \, dx dy = \int_{\partial D} (\Gamma \partial_n u - u \partial_n \Gamma) ds + \int_{\partial B_\varepsilon} (\Gamma \partial_n u - u \partial_n \Gamma) ds.$$

Let ε tend to zero, recalling that the outward normal derivative (with respect to the domain D_ε) on the boundary of B_ε is the inner radial derivative pointing towards the pole (ξ, η) (see Figure 8.1).

Using estimates (8.3)–(8.4) we obtain

$$\left| \int_{\partial B_\varepsilon} \Gamma \partial_n u \, ds \right| \leq C\varepsilon |\ln \varepsilon| \to 0 \qquad \text{as } \varepsilon \to 0,$$

$$\int_{\partial B_\varepsilon} u \partial_n \Gamma \, ds = \frac{1}{2\pi\varepsilon} \int_{\partial B_\varepsilon} u \, ds \to u(\xi, \eta) \qquad \text{as } \varepsilon \to 0.$$

Therefore,

$$u(\xi, \eta) = \int_{\partial D} [\Gamma(x - \xi, y - \eta) \partial_n u - u \partial_n \Gamma(x - \xi, y - \eta)] ds$$

$$- \int_D \Gamma(x - \xi, y - \eta) \Delta u \, dx dy. \tag{8.5}$$

Formula (8.5) is called *Green's representation formula*, and the function

$$\Gamma[f](\xi, \eta) := -\int_D \Gamma(x - \xi, y - \eta) f(x, y) \, \mathrm{d}x \mathrm{d}y$$

is called the *Newtonian potential* of f.

The following corollary has already been proved using another approach.

Corollary 8.1 *If u is harmonic in a domain D, then u is infinitely differentiable in D.*

Proof By Green's representation formula

$$u(\xi, \eta) = \int_{\partial D} [\Gamma(x - \xi, y - \eta) \partial_n u - u \partial_n \Gamma(x - \xi, y - \eta)] \mathrm{d}s.$$

The integrand is an infinitely differentiable function of ξ and η inside D. Interchanging the order of integration and differentiation we obtain the claim. $\square$

Corollary 8.2 *Let $u \in C^2(\mathbb{R}^2)$ be a function that vanishes identically outside a disk (in other words, u has a compact support in $\mathbb{R}^2$). Then*

$$u(\xi, \eta) = -\int_{\mathbb{R}^2} \Gamma(x - \xi, y - \eta) \Delta u(x, y) \, \mathrm{d}x \mathrm{d}y. \qquad (8.6)$$

Let us discuss in some detail the nature of the function $\Delta \Gamma(x - \xi, y - \eta)$. It is clear that $\Delta \Gamma(x - \xi, y - \eta) = 0$ for all $(x, y) \neq (\xi, \eta)$. On the other hand, if we (formally) carry out integration by parts of (8.6) for u with a compact support, we obtain

$$u(\xi, \eta) = -\int_{\mathbb{R}^2} \Delta \Gamma(x - \xi, y - \eta) u(x, y) \, \mathrm{d}x \mathrm{d}y. \qquad (8.7)$$

Therefore, the "function"

$$\delta(x - \xi, y - \eta) := -\Delta \Gamma(x - \xi, y - \eta)$$

vanishes at all points $(x, y) \neq (\xi, \eta)$, but its integral against any smooth function u is not zero; rather it reproduces the value of u at the point (ξ, η). For the particular case $(\xi, \eta) = (0, 0)$, we write

$$-\Delta \Gamma(x, y) = \delta(x, y).$$

It is clear that δ is not a function in the classical sense. It is a mathematical object called a *distribution*. The distribution δ (which in the folklore is often termed the *delta function*) is called the *Dirac distribution*, and it is characterized by the following formal expression:

$$u(\xi, \eta) = \int_{\mathbb{R}^2} \delta(x - \xi, y - \eta) u(x, y) \, dx dy \tag{8.8}$$

for any smooth function u with a compact support in $\mathbb{R}^2$.

We may characterize the delta function as the limit of certain sequences of smooth functions with a compact support. For example, consider a smooth nonnegative function ρ on $\mathbb{R}^2$, vanishing outside the unit ball, and satisfying

$$\int_{\mathbb{R}^2} \rho(\vec{x}) \, d\vec{x} = 1.$$

Fix $\vec{y} \in \mathbb{R}^2$ and let $\varepsilon > 0$. Define the function

$$\rho_\varepsilon(\vec{x}) := \varepsilon^{-2} \rho \left(\frac{\vec{x} - \vec{y}}{\varepsilon} \right).$$

Note that ρ_ε is supported in a ball of radius ε around $\vec{y}$ and satisfies

$$\int_{\mathbb{R}^2} \rho_\varepsilon(\vec{x}) \, d\vec{x} = 1.$$

For any smooth function u with a compact support in $\mathbb{R}^2$ we have

$$\lim_{\varepsilon \to 0_+} \int_{\mathbb{R}^2} \rho_\varepsilon(\vec{x}) u(\vec{x}) \, d\vec{x} = u(\vec{y}) = \int_{\mathbb{R}^2} \delta(\vec{x} - \vec{y}) u(\vec{x}) \, d\vec{x}. \tag{8.9}$$

We say that ρ_ε *converges in the sense of distribution* to the delta function at $\vec{y}$ as $\varepsilon \to 0$, and ρ_ε is called an approximation of the delta function.

A standard example of such an approximation of the delta function is given by

$$\rho(\vec{x}) := \begin{cases} c \exp \left(\frac{1}{|\vec{x}|^2 - 1} \right) & |\vec{x}| \leq 1, \\ 0 & \text{otherwise,} \end{cases} \tag{8.10}$$

where c is a positive constant (see Exercise 8.7).

The reader may recall from linear algebra the notions of adjoint matrix and adjoint of a linear operator. We introduce now the definition of the adjoint operator of a given differential operator L. The relation to the algebraic notion will be explained later. We also give below the general definition of a fundamental solution.

Definition 8.3 Let

$$L[u] = \sum_{0 \leq i+j \leq m} a_{ij}(x, y) \partial_x^i \partial_y^j u$$

be a linear differential operator of order m with smooth coefficients a_{ij} that is defined on $\mathbb{R}^2$.

(a) The operator

$$L^*[v] = \sum_{0 \le i+j \le m} (-1)^{i+j} \partial_x^i \partial_y^j (a_{ij}(x, y)v)$$

is called the *formal adjoint operator* of L.

(b) Fix $(\xi, \eta) \in \mathbb{R}^2$. Suppose that a function $v(x, y; \xi, \eta)$ satisfies

$$L[v] = \delta(x - \xi, y - \eta)$$

in the sense of distributions, which means that

$$\phi(\xi, \eta) = \int_{\mathbb{R}^2} v(x, y; \xi, \eta) L^*[\phi(x, y)] \, dxdy$$

for any smooth function ϕ with a compact support. Then v is called the *fundamental solution of the equation* $L[u] = 0$ *with a pole at* (ξ, η).

Example 8.4 (a) The formal adjoint of the Laplace operator $L[u] = \Delta u$ is given by $L^*[u] = \Delta u = L[u]$. If $L^* = L$, the operator L is said to be *formally selfadjoint*.

(b) It can be checked that a Sturm–Liouville operator on an interval I, the wave operator $L[u] = u_{tt} - c^2 \Delta u$, and the biharmonic operator $L[u] = \Delta^2 u$ are also formally selfadjoint.

(c) The formal adjoint of the heat operator $L[u] = u_t - \Delta u$ is the *backward heat operator* $L^*[u] = -u_t - \Delta u$.

Remark 8.5 (a) Let $u, v \in C_0^\infty(\mathbb{R}^2)$, where $C_0^\infty(D)$ is the space of all smooth (infinitely differentiable) functions with a compact support in D. Integrating by parts, one can verify that

$$\int_{\mathbb{R}^2} L[u(x, y)] v(x, y) \, dxdy = \int_{\mathbb{R}^2} u(x, y) L^*[v(x, y)] \, dxdy.$$

This means that the operator L^* is the (algebraic) adjoint of the operator L on the space $C_0^\infty(\mathbb{R}^2)$ with respect to the inner product $\langle u, v \rangle = \int_{\mathbb{R}^2} u(x, y)v(x, y) \, dxdy$.

(b) The fundamental solution is not uniquely defined. If v is a fundamental solution, then $v + w$ is also fundamental solution for any w that solves the homogeneous equation $L[u] = 0$.

(c) If the operator L is a linear operator with constant coefficients, and if $v(x, y)$ is a fundamental solution with a pole at $(0, 0)$, then $v(x - \xi, y - \eta)$ is a fundamental solution with a pole at (ξ, η).

(d) We showed above that Γ is a fundamental solution of the Laplace operator $-\Delta$ on $\mathbb{R}^2$, and the significance of the factor $-1/2\pi$ in the definition of Γ is now clear.

Consider again the Dirichlet problem (8.1). Green's representation formula (8.5) enables us to compute the value of $u(\xi, \eta)$ for all $(\xi, \eta) \in D$ if we know Δu in D, and the values of u and $\partial_n u$ on the boundary of D. But for the Dirichlet problem for the Poisson equation the values of $\partial_n u$ are not given on ∂D. Therefore, in order to obtain an integral representation for the Dirichlet problem, we have to modify (8.5). Let $h(x, y; \xi, \eta)$ be a solution (that depends on the parameter $(\xi, \eta) \in D$) of the following Dirichlet problem:

$$\Delta h(x, y; \xi, \eta) = 0 \quad (x, y) \in D, \qquad h(x, y; \xi, \eta) = \Gamma(x, y; \xi, \eta) \quad (x, y) \in \partial D.$$
(8.11)

By the second Green's identity (7.19),

$$-\int_D h(x, y; \xi, \eta) \Delta u(x, y) \, dxdy$$

$$= \int_{\partial D} [u(x, y) \partial_n h(x, y; \xi, \eta) - h(x, y; \xi, \eta) \partial_n u(x, y)] \, ds$$

$$= \int_{\partial D} [u(x, y) \partial_n h(x, y; \xi, \eta) - \Gamma(x - \xi, y - \eta) \partial_n u(x, y)] \, ds. \quad (8.12)$$

We introduce now the following important definition.

Definition 8.6 The *Green function* of the domain D for the Laplace operator and the Dirichlet boundary condition is given by

$$G(x, y; \xi, \eta) := \Gamma(x, y; \xi, \eta) - h(x, y; \xi, \eta) \quad (x, y), (\xi, \eta) \in D, (x, y) \neq (\xi, \eta),$$
(8.13)

where h is the solution of (8.11)

It follows that the Green function satisfies

$$\Delta G(x, y; \xi, \eta) = -\delta(x - \xi, y - \eta) \quad (x, y) \in D,$$
$$G(x, y; \xi, \eta) = 0 \qquad\qquad (x, y) \in \partial D.$$
(8.14)

We now add (8.12) and (8.5) to obtain

$$u(\xi, \eta) = -\int_{\partial D} \partial_n G(x, y; \xi, \eta) u(x, y) \, ds - \int_D G(x, y; \xi, \eta) \Delta u(x, y) \, dxdy.$$
(8.15)

Substituting the given data into (8.15), we finally arrive at the following integral representation formula for solutions of the Dirichlet problem for the Poisson equation.

Theorem 8.7 *Let D be a smooth bounded domain, $f \in C(\bar{D})$, and $g \in C(\partial D)$. Let $u \in C^2(\bar{D})$ be a solution of the Dirichlet problem*

$$\Delta u = f \quad (x, y) \in D,$$
$$u = g \qquad (x, y) \in \partial D.$$
(8.16)

Then

$$u(\xi, \eta) = -\int_{\partial D} \partial_n G(x, y; \xi, \eta) g(x, y) \, ds - \int_D G(x, y; \xi, \eta) f(x, y) \, dxdy. \quad (8.17)$$

The representation formula (8.17) involves two integral kernels:

(1) The Green function $G(x, y; \xi, \eta)$, which is defined for all $(x, y), (\xi, \eta) \in D$, $(x, y) \neq (\xi, \eta)$.

(2) $K(x, y; \xi, \eta) := -\partial_n G(x, y; \xi, \eta)$, which is the inward normal derivative of the Green function on the boundary of the domain D. Therefore, the kernel K is defined for $(x, y) \in \partial D, (\xi, \eta) \in D$.

Definition 8.8 The function $K(x, y; \xi, \eta)$ is called the *Poisson kernel* of the Laplace operator and the Dirichlet problem on D.

Remark 8.9 The reader should show as an exercise that the Poisson kernel that was obtained in Section 7.8 (for the special case of a disk) is indeed the normal derivative of the corresponding Green function.

Theorem 8.7 enables us to solve the Dirichlet problem in a domain D provided that the Green function is known, and that it is a priori known that the solution is in $C^2(D) \cap C^1(\bar{D})$. This additional regularity is indeed ensured if f, g, and ∂D are sufficiently smooth.

The Green function can be computed explicitly only for a small number of domains. Some examples of such domains will be presented below and in the exercises. Nevertheless, the Green function is a very useful tool in the study of the Dirichlet problem, and therefore we present now its main properties.

The *uniqueness* of the Green function follows directly from the uniqueness of the function h, i.e. from the uniqueness of the solution of the Dirichlet problem for the Laplace equation in D (Theorem 7.12). On the other hand, the *existence* of the Green function for a domain D follows from the existence of a solution of the Dirichlet problem for the Laplace equation in the domain D. The study of the existence theorem for a smooth bounded domain D is outside the scope of this book, but the standard proof relies heavily on the existence of a solution for the special case of a disk. Recall that the existence theorem for the disk was proved independently in Section 7.8 using the Poisson formula. It follows that the existence of the Green function is not based on a circular argumentation.

Theorem 8.10 *The Green function for the Dirichlet problem is symmetric in the sense that*

$$G(x, y; \xi, \eta) = G(\xi, \eta; x, y)$$

for all $(x, y), (\xi, \eta) \in D$ *such that* $(x, y) \neq (\xi, \eta)$.

Proof Fix two points $(x, y), (\xi, \eta) \in D$ such that $(x, y) \neq (\xi, \eta)$, and let

$$v(\sigma, \tau) := G(\sigma, \tau; x, y), \qquad w(\sigma, \tau) := G(\sigma, \tau; \xi, \eta).$$

The functions v and w are harmonic in $D \setminus \{(x, y), (\xi, \eta)\}$ and vanish on ∂D. We again use the second Green identity (7.19) for the domain $\tilde{D}_\varepsilon$ which contains all the points in D such that their distances from the poles (x, y) and (ξ, η) are larger than ε. We have

$$\int_{\partial B((x,y);\varepsilon)} (w\partial_n v - v\partial_n w)ds(\sigma, \tau) = \int_{\partial B((\xi,\eta);\varepsilon)} (v\partial_n w - w\partial_n v)ds(\sigma, \tau). \quad (8.18)$$

Using the estimates (8.3)–(8.4) we infer that

$$\lim_{\varepsilon \to 0} \int_{\partial B((x,y);\varepsilon)} |v\partial_n w|ds(\sigma, \tau) = \lim_{\varepsilon \to 0} \int_{\partial B((\xi,\eta);\varepsilon)} |w\partial_n v|ds(\sigma, \tau) = 0; \quad (8.19)$$

but

$$\lim_{\varepsilon \to 0} \int_{\partial B((x,y);\varepsilon)} w\partial_n v \, ds(\sigma, \tau) = w(x, y), \quad \lim_{\varepsilon \to 0} \int_{\partial B((\xi,\eta);\varepsilon)} v\partial_n w \, ds(\sigma, \tau) = v(\xi, \eta). \quad (8.20)$$

Letting $\varepsilon \to 0$ in (8.18) and using (8.19) and (8.20), we obtain

$$G(x, y; \xi, \eta) = w(x, y) = v(\xi, \eta) = G(\xi, \eta; x, y).$$

$\square$

Theorem 8.11 *(a) Fix $(x, y) \in D$. The Green function $G(x, y; \xi, \eta)$, considered as a function of (ξ, η), is a positive harmonic function in the domain $D \setminus \{(x, y)\}$ which vanishes on ∂D.*
(b) Fix $(x, y) \in \partial D$. The Poisson kernel $K(x, y; \xi, \eta)$, considered as a function of (ξ, η), is a positive harmonic function in the domain D which vanishes on $\partial D \setminus \{(x, y)\}$.

Proof We only sketch the proof.

(a) The fact that G, as a function of (ξ, η), is harmonic and vanishes on the boundary follows directly from the symmetry of G. Since G is positive near the pole and vanishes on the boundary, the weak maximum principle implies that for $\varepsilon > 0$ sufficiently small, the function G is positive also in $D \setminus B_\varepsilon$, where B_ε is the open disk of radius ε with a center on the pole (x, y).
(b) Since G vanishes on the boundary and is positive on D, it follows that on the boundary its inward normal derivative (i.e. K) is nonnegative. The proof of the strict positivity of K will not be given here. The Poisson kernel is a derivative of a harmonic function. In other word, K is a limit of a family of harmonic functions which implies that it is harmonic. $\square$

Corollary 8.12 *Let D be a smooth bounded domain. Let f be a nonpositive continuous function in D, and let g be a nonnegative continuous function on ∂D, such*

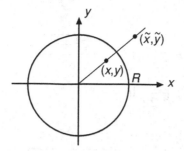

Figure 8.2 The inverse of a point with respect to the circle.

that at least one of these two functions is not identically zero. Then the solution u of the Dirichlet problem (8.1) is a positive function in D.

Proof The proof follows directly from Theorems 8.7 and 8.11. □

Proposition 8.13 *Let D_1, D_2 be (planar) smooth bounded domains such that $D_1 \subset D_2$. Let G_i be the Green function of the domain D_i, where $i = 1, 2$. Then*

$$0 \le G_1(x, y; \xi, \eta) \le G_2(x, y; \xi, \eta) \qquad (x, y), (\xi, \eta) \in D_1.$$

Proof Fix $(\xi, \eta) \in D_1$, and let B_ε be the open disk of radius ε centered at (ξ, η). Since

$$\lim_{(x,y) \to (\xi,\eta)} \frac{G_1(x, y; \xi, \eta)}{G_2(x, y; \xi, \eta)} = 1,$$

it follows that for any $\delta > 1$ there exists $\varepsilon > 0$ such that $0 \le G_1(x, y; \xi, \eta) \le \delta G_2(x, y; \xi, \eta)$ in a disk B_ε. Theorem 8.11 and the weak maximum principle in the domain $D_1 \setminus B_\varepsilon$ imply that $0 \le G_1(x, y; \xi, \eta) \le \delta G_2(x, y; \xi, \eta)$ on $D_1 \setminus B_\varepsilon$. Letting $\delta \to 1$, it follows that

$$0 \le G_1(x, y; \xi, \eta) \le G_2(x, y; \xi, \eta) \qquad (x, y), (\xi, \eta) \in D_1.$$

□

Example 8.14 Let B_R be the disk of radius R centered at the origin. We want to compute the Green function of B_R, and derive from it the Poisson kernel. We use the *reflection principle.*

Let $(x, y) \in B_R$. The point

$$(\tilde{x}, \tilde{y}) := \frac{R^2}{x^2 + y^2}(x, y)$$

is called the *inverse point* of (x, y) with respect to the circle ∂B_R (see Figure 8.2). It is convenient to define the ideal point ∞ as the inverse of the origin. Define

$$G_R(x, y; \xi, \eta) := \begin{cases} \Gamma(x - \xi, y - \eta) - \Gamma\left(\dfrac{\sqrt{\xi^2 + \eta^2}}{R}(x - \tilde{\xi}, y - \tilde{\eta})\right) & (\xi, \eta) \neq (0, 0), \\ \Gamma(x, y) + \dfrac{1}{2\pi} \ln R & (\xi, \eta) = (0, 0), \end{cases}$$

$$(8.21)$$

and set

$$r = \sqrt{(x - \xi)^2 + (y - \eta)^2}, \quad r^* = \sqrt{\left(x - \frac{R^2}{\rho^2}\xi\right)^2 + \left(y - \frac{R^2}{\rho^2}\eta\right)^2}, \quad \rho = \sqrt{\xi^2 + \eta^2}.$$

An elementary calculation implies that

$$G_R(x, y; \xi, \eta) = \begin{cases} -\dfrac{1}{2\pi} \ln \dfrac{Rr}{\rho r^*} & (\xi, \eta) \neq (0, 0), \\ -\dfrac{1}{2\pi} \ln \dfrac{r}{R} & (\xi, \eta) = (0, 0), \end{cases}$$

$$(8.22)$$

and that G_R satisfies all the properties of the Green function. Moreover, it can be checked that the radial derivative of G_R on the circle of radius R is the Poisson kernel, which was calculated using a completely different approach in Section 7.8 (see Exercise 8.1).

Example 8.15 Denote by $\mathbb{R}^2_+ := \{(x, y) \mid y > 0\}$ the open upper half-plane. Although this is an unbounded domain, it is possible to use the reflection principle again to obtain the corresponding Green function. Let $(x, y) \in \mathbb{R}^2_+$. The point

$$(\tilde{x}, \tilde{y}) := (x, -y)$$

is called the *inverse point* of (x, y) with respect to the real line. It can be readily verified that the function

$$G(x, y; \xi, \eta) := \Gamma(x - \xi, y - \eta) - \Gamma(x - \tilde{\xi}, y - \tilde{\eta})$$

$$= -\frac{1}{2\pi} \ln \frac{\sqrt{(x - \xi)^2 + (y - \eta)^2}}{\sqrt{(x - \xi)^2 + (y + \eta)^2}} \qquad (8.23)$$

satisfies all the properties of the Green function, and its derivative in the y direction on the boundary $(y = 0)$ of $\mathbb{R}^2_+$ is given by

$$K(x, 0; \xi, \eta) := \frac{\eta}{\pi[(x - \xi)^2 + \eta^2]} \qquad (x, 0) \in \partial\mathbb{R}^2_+, \ (\xi, \eta) \in \mathbb{R}^2_+ \qquad (8.24)$$

(see Exercise 8.5).

8.3 Neumann's function in the plane

We move on to present an integral representation for solutions of the Neumann problem for the Poisson equation:

$$\begin{aligned} \Delta u &= f && D, \\ \partial_n u &= g && \partial D, \end{aligned} \qquad (8.25)$$

where D is a smooth bounded domain. The first difficulty that arises is the nonuniqueness of the problem, which implies that it is impossible to find a unique integral formula. Furthermore, we should recall the solvability condition (7.9) for the Neumann problem. Nevertheless, the derivation of the integral representation for the Neumann problem is basically similar to the procedure for the Dirichlet problem.

Recall that the Green representation formula (8.5) enables us to reproduce the value of an arbitrary smooth function u at any point (ξ, η) in D provided that Δu is given in D, and u and $\partial_n u$ are given on ∂D. For the Neumann problem, u is not known on ∂D. We proceed now with almost the same idea that was used for the Dirichlet problem. Let $h(x, y; \xi, \eta)$ be a solution (depending on the parameter (ξ, η)) of the following Neumann problem:

$$\begin{aligned} \Delta h(x, y; \xi, \eta) &= 0 && (x, y) \in D, \\ \partial_n h(x, y; \xi, \eta) &= \partial_n \Gamma(x, y; \xi, \eta) + 1/L && (x, y) \in \partial D, \end{aligned} \qquad (8.26)$$

where L is the length of ∂D. Substituting $u = 1$ into the Green representation formula (8.5) implies that

$$\int_{\partial D} \partial_n \Gamma(x, y; \xi, \eta) ds = -1. \qquad (8.27)$$

Therefore, the boundary condition in (8.26) satisfies the solvability condition (7.9). It is known that (7.9) is not only a necessary condition but also a sufficient condition for the solvability of the problem.

Definition 8.16 A *Neumann function* for a domain D and the Laplace operator is the function

$$N(x, y; \xi, \eta) := \Gamma(x, y; \xi, \eta) - h(x, y; \xi, \eta) \quad (x, y), (\xi, \eta) \in D, (x, y) \neq (\xi, \eta), \qquad (8.28)$$

where $h(x, y; \xi, \eta)$ is a solution of (8.26).

In other words, a Neumann function satisfies

$$\Delta N(x, y; \xi, \eta) = -\delta(x - \xi, y - \eta) \quad (x, y) \in D,$$
$$\partial_n N(x, y; \xi, \eta) = -\tfrac{1}{L} \qquad\qquad (x, y) \in \partial D. \tag{8.29}$$

Therefore,

$$u(\xi, \eta) = \int_{\partial D} N(x, y; \xi, \eta) \partial_n u(x, y)\, ds$$
$$- \int_D N(x, y; \xi, \eta) \Delta u(x, y)\, dx dy + \frac{1}{L} \int_{\partial D} u\, ds. \tag{8.30}$$

Substituting the given data into (8.30), we obtain the following representation formula for solutions of the Neumann problem.

Theorem 8.17 *Suppose that $u \in C^2(\bar{D})$ is a solution of the Neumann problem*

$$\Delta u = f \quad D,$$
$$\partial_n u = g \quad \partial D. \tag{8.31}$$

Then

$$u(\xi, \eta) = \int_{\partial D} N(x, y; \xi, \eta) g(x, y)\, ds - \int_D N(x, y; \xi, \eta) f(x, y)\, dx dy + \frac{1}{L} \int_{\partial D} u\, ds. \tag{8.32}$$

Remark 8.18 (a) The kernel N is not called the Green function of the problem, since N does not satisfy the corresponding homogeneous boundary condition. There is no kernel function that satisfies

$$\Delta G(x, y; \xi, \eta) = -\delta(x - \xi, y - \eta) \quad (x, y) \in D,$$
$$\partial_n N(x, y; \xi, \eta) = 0 \qquad\qquad (x, y) \in \partial D. \tag{8.33}$$

(b) The Neumann function is determined up to an additive constant. In order to uniquely define N it is convenient to use the normalization

$$\int_{\partial D} N(x, y; \xi, \eta)\, ds = 0. \tag{8.34}$$

(c) The third term in the representation formula (8.32) is $(1/L) \int_{\partial D} u\, ds$, the average of u on the boundary, which is not given. But since the solution is determined up to an additive constant, it is convenient to add the condition

$$\int_{\partial D} u(x, y)\, ds = 0, \tag{8.35}$$

and then the problem is uniquely solved, and the corresponding integral representation uniquely determines the solution.

8.4 The heat kernel

Consider (again) the homogeneous heat problem with the Dirichlet condition

$$
\begin{aligned}
u_t - k u_{xx} &= 0 & & 0 < x < L,\ t > 0, \\
u(0, t) = u(L, t) &= 0 & & t \geq 0, \\
u(x, 0) &= f(x) & & 0 \leq x \leq L,
\end{aligned}
\tag{8.36}
$$

that was solved in Section 5.2. Using the separation of variables method, we found that the solution of the problem is of the form

$$
u(x, t) = \sum_{n=1}^{\infty} B_n \sin \frac{n \pi x}{L} e^{-k(\frac{n\pi}{L})^2 t},
\tag{8.37}
$$

where B_n are the Fourier coefficients

$$
B_n = \frac{2}{L} \int_0^L \sin \frac{n \pi y}{L} f(y)\, dy \qquad n = 1, 2, \ldots.
\tag{8.38}
$$

For fixed $t > \varepsilon > 0$ and $0 < x < L$, the series

$$
\frac{2}{L} \sum_{n=1}^{\infty} \left(e^{-k(\frac{n\pi}{L})^2 t} \sin \frac{n \pi x}{L} \right) \sin \frac{n \pi y}{L} f(y)
$$

converges uniformly (as a function of y). Therefore, we may integrate term by term and hence

$$
u(x, t) = \sum_{n=1}^{\infty} e^{-k(\frac{n\pi}{L})^2 t} \sin \frac{n \pi x}{L} \left(\frac{2}{L} \int_0^L \sin \frac{n \pi y}{L} f(y)\, dy \right)
$$

$$
= \int_0^L \left(\frac{2}{L} \sum_{n=1}^{\infty} e^{-k(\frac{n\pi}{L})^2 t} \sin \frac{n \pi x}{L} \sin \frac{n \pi y}{L} \right) f(y)\, dy.
$$

The function

$$
K(x, y, t) := \frac{2}{L} \sum_{n=1}^{\infty} e^{-k(\frac{n\pi}{L})^2 t} \sin \frac{n \pi x}{L} \sin \frac{n \pi y}{L}
\tag{8.39}
$$

is called the *heat kernel* of the initial boundary condition (8.36). The reader can verify that for every fixed y the kernel K is a solution of the heat equation and that it satisfies the Dirichlet conditions for $t > 0$. In addition, K is symmetric, i.e. $K(x, y, t) = K(y, x, t)$.

To summarize, we have obtained the following simple integral representation:

$$u(x, t) = \int_0^L K(x, y, t) f(y) \, dy, \tag{8.40}$$

for the solution of the initial boundary value problem (8.36).

Consider now the nonhomogeneous problem

$$\begin{aligned} u_t - k u_{xx} &= F(x, t) & 0 < x < L, \ t > 0, \\ u(0, t) = u(L, t) &= 0 & t \geq 0, \\ u(x, 0) &= f(x) & 0 \leq x \leq L. \end{aligned} \tag{8.41}$$

We apply the Duhamel principle (see Exercise 5.14). Let $v(x, t, s)$ be the solution of the following initial homogeneous problem (which depends on the parameter s)

$$\begin{aligned} v_t - k v_{xx} &= 0 & 0 < x < L, \ t > s, \\ v(0, t, s) = v(L, t, s) &= 0 & t \geq s, \\ v(x, s, s) &= F(x, s) & 0 \leq x \leq 1. \end{aligned}$$

Using the integral representation (8.40), we can express $v(x, t, s)$ in the form

$$v(x, t, s) = \int_0^L K(x, y, t - s) F(y, s) \, dy.$$

Therefore, by the Duhamel principle and the superposition principle, the solution of problem (8.41) is given by the integral representation

$$u(x, t) = \int_0^L K(x, y, t) f(y) \, dy + \int_0^t \int_0^L K(x, y, t - s) F(y, s) \, dy \, ds. \tag{8.42}$$

The significance of (8.39) and (8.42) is that they are valid in a much broader context (see also Section 9.12).

Remark 8.19 From the exponential decay in (8.39), it follows that the heat kernel is a smooth function for $0 < x < L$, $t > 0$. On the other hand, the heat kernel is singular at $t = 0$, for $x = y$. As for the Green function, the precise character of this singularity is explained rigorously by the theory of distributions. It turns out that for a fixed $0 < y < L$, the heat kernel $K(x, y, t)$ is a distribution with a support at $\Omega = [0, L] \times \mathbb{R}$ that solves the problem

$$\begin{aligned} K_t - k K_{xx} &= \delta(x - y) \delta(t) & 0 < x < L, \ -\infty < t < \infty, \\ K(x, y, t) &= 0 & t < 0, \\ K(0, y, t) = K(L, y, t) &= 0 & t > 0. \end{aligned} \tag{8.43}$$

In other words, for any smooth function φ with a compact support in Ω, we have

$$\varphi(y, 0) = \int_\Omega K(x, y, t)[-\partial_t \varphi(x, t) - \partial_{xx} \varphi(x, t)] \, dx \, dt,$$

and for any smooth function ψ with a compact support in $\{(x, t) \mid 0 \le x \le L, \ t < 0\}$, we have

$$\int_\Omega K(x, y, t) \psi(x, t) \, dx \, dt = 0.$$

The following is an alternative but equivalent characterization of the heat kernel. For any fixed $0 < y < L$ and $t > 0$, the heat kernel $K(x, y, t)$ is a distribution with a compact support in $[0, L]$ that satisfies

$$K_t - k\Delta K = 0 \quad 0 < x < L, \ 0 < t,$$
$$K(x, y, 0) = \delta(x - y),$$
$$K(0, y, t) = K(L, y, t) = 0 \quad t > 0.$$

In the latter formulation, t is considered as a parameter, and the precise meaning is that for any smooth function $\phi(x)$ with a compact support in $[0, L]$ we have

$$\begin{cases} \dfrac{\partial}{\partial t} \displaystyle\int_0^L K(x, y, t)\phi(x) \, dx - \int_0^L K(x, y, t)\partial_{xx}\phi(x) \, dx = 0 \qquad \forall t > 0, \\[2ex] \displaystyle\lim_{t \to 0_+} \int_0^L K(x, y, t)\phi(x) \, dx = \phi(y). \end{cases}$$

8.5 Exercises

8.1 (a) Show that the function that is defined in (8.22) is indeed the Green function in B_R, and that its radial derivative on the circle is the Poisson kernel which was derived in Section 7.8.

(b) Evaluate $\lim_{R \to \infty} G_R(x, y; \xi, \eta)$.

8.2 Prove that the Neumann function for the Poisson equation is symmetric, i.e.

$$N(x, y; \xi, \eta) = N(\xi, \eta; x, y),$$

for all $(x, y), (\xi, \eta) \in D$ such that $(x, y) \ne (\xi, \eta)$.

Hint The proof is similar to the proof of Theorem 8.10.

8.3 (a) Derive an explicit formula for the Green function of a disk as an infinite series, using (7.76) which is a formula for the solution of the Dirichlet problem for the Poisson equation.

(b) Calculate the sum of the above series and obtain the explicit formula (8.22) for the Green function of the disk.

8.4 (a) Write the Green function of (8.22) in polar coordinates.

(b) Using a reflection principle and part (a) find the Green function of half of a disk.

8.5 (a) Show that the function which is defined in (8.23) is indeed the Green function in $\mathbb{R}_+^2$, and that its derivative in the y direction for $y = 0$ is the Poisson kernel which is given by (8.24).

(b) Using a reflection principle and part (a) find the Green function of the positive quarter plane $x > 0$, $y > 0$.

8.6 Let $\mathbb{R}_+^2$ be the upper half-plane. Find the Neumann function of $\mathbb{R}_+^2$.

8.7 (a) Prove (8.9).

(b) Find the constant c in (8.10), and verify directly that ρ_ε is an approximation of the delta function.

8.8 Let $k \neq 0$. Show that the function $G_k(x, \xi) = e^{-k|x-\xi|}/2k$ is a fundamental solution of the equation

$$-u'' + k^2 u = 0 \qquad -\infty < x < \infty.$$

Hint Use one of Green's identities.

8.9 Show that the *Gaussian kernel*

$$K(x, y, t) := \begin{cases} \dfrac{1}{(4\pi kt)^{1/2}} e^{-\frac{(x-y)^2}{4kt}} & t > 0, \\ 0 & t < 0. \end{cases} \qquad (8.44)$$

is the heat kernel for the Cauchy problem

$$u_t - k u_{xx} = 0 \qquad -\infty < x < \infty, \ t > 0,$$
$$u(x, 0) = f(x) \qquad -\infty < x < \infty,$$

where f is a bounded continuous function on $\mathbb{R}$.

8.10 Use a reflection principle and the (Gaussian) heat kernel (8.44) to obtain the heat kernel for the problem

$$u_t - k u_{xx} = 0 \qquad 0 < x < \infty, \ t > 0,$$
$$u(0, t) = 0 \qquad t \geq 0,$$
$$u(x, 0) = f(x) \qquad 0 \leq x \leq \infty.$$

8.11 Let $D_R := \mathbb{R}^2 \setminus B_R$ be the exterior of the disk with radius R centered at the origin. Find the (Dirichlet) Green function of D_R.

8.12 (a) Use a reflection principle and the (Gaussian) heat kernel (8.44) to obtain the following alternative representation of the heat kernel for the initial boundary value problem (8.36):

$$K(x, y, t) = \frac{1}{(4\pi kt)^{1/2}} \sum_{n=-\infty}^{n=\infty} \left[e^{-\frac{(x-y-2Ln)^2}{4kt}} - e^{-\frac{(x+y-2Ln)^2}{4kt}} \right] \qquad t > 0. \qquad (8.45)$$

(b) Use (8.45) to show that the exact short time behavior ($t \to 0_+, x \neq y$) of the heat kernel for the problem (8.36) is given by (8.44).

(c) Use (8.39) to show that the exact large time behavior ($t \to \infty$) of the heat kernel for the problem (8.36) is given by

$$K(x, y, t) \approx \frac{2}{L} e^{-k(\frac{\pi}{L})^2 t} \sin \frac{\pi x}{L} \sin \frac{\pi y}{L} .$$

8.13 Let B_R be the disk with radius R centered at the origin. Find the Neumann function of B_R.

9

Equations in high dimensions

9.1 Introduction

To simplify the presentation we have concentrated so far mainly on equations involving two independent variables. In this chapter we shall extend the discussion to equations in higher dimensions. A considerable part of the theoretical and practical aspects that we studied for equations in two variables can be extended at once to higher dimensions. Nevertheless, we shall see that there are sometimes significant differences between problems in different dimensions.

9.2 First-order equations

The general first-order quasilinear equation for a function u in n variables is

$$\sum_{i=1}^{n} a_i(x_1, x_2, \ldots, x_n, u)u_{x_i} = c(x_1, x_2, \ldots, x_n, u). \tag{9.1}$$

The method of characteristics that we developed in Chapter 2 is also valid for (9.1). The initial condition for (9.1) is an $(n-1)$-dimensional surface Γ in the Euclidean space $\mathbb{R}^{n+1}$. We write Γ parameterically:

$$x_{0,i} = x_{0,i}(s_1, s_2, \ldots, s_{n-1}) \quad i = 1, 2, \ldots, n, \tag{9.2}$$

$$u_0 = u_0(s_1, s_2, \ldots, s_{n-1}). \tag{9.3}$$

Similarly to the two-dimensional case we write the characteristic equations

$$\frac{\partial x_i}{\partial t} = a_i(x_1, x_2, \ldots, x_n, u) \quad i = 1, 2, \ldots, n, \tag{9.4}$$

$$\frac{\partial u}{\partial t} = c(x_1, x_2, \ldots, x_n, u). \tag{9.5}$$

Solving the system of ODEs (9.4)–(9.5) with the initial data (9.2)–(9.3) at $t = 0$, we generate the solution $u(x_1, x_2, \ldots, x_n)$ of (9.1) as a parametric n-dimensional hypersurface

$$x_i = x_i(t, s_1, s_2, \ldots, s_{n-1}) \quad i = 1, 2, \ldots, n,$$
$$u = u(t, s_1, s_2, \ldots, s_{n-1}).$$

The transversality condition that we introduced in Chapter 2 takes now the form

$$J\big|_\Gamma = \begin{pmatrix} \dfrac{\partial x_{0,1}}{\partial s_1} & \dfrac{\partial x_{0,2}}{\partial s_1} & \cdots & \dfrac{\partial x_{0,n}}{\partial s_1} \\[2mm] \dfrac{\partial x_{0,1}}{\partial s_2} & \dfrac{\partial x_{0,2}}{\partial s_2} & \cdots & \dfrac{\partial x_{0,n}}{\partial s_2} \\[2mm] \vdots & \vdots & & \vdots \\[2mm] \dfrac{\partial x_{0,1}}{\partial s_{n-1}} & \dfrac{\partial x_{0,2}}{\partial s_{n-1}} & \cdots & \dfrac{\partial x_{0,n}}{\partial s_{n-1}} \\[2mm] a_1 & a_2 & \cdots & a_n \end{pmatrix} \neq 0. \tag{9.6}$$

When this condition holds, the parametric representation we obtained indeed provides the (locally) unique solution to (9.1). Generalizing the existence and uniqueness statement of Theorem 2.10 and the discussion that follows it to the n-dimensional case is straightforward.

Example 9.1 Solve the linear equation

$$xu_x + yu_y + zu_z = 4u,$$

subject to the initial condition $u(x, y, 1) = xy$.

The characteristic equations are

$$x_t = x, \quad y_t = y, \quad z_t = z, \quad u_t = 4u,$$

and the initial conditions can be written parametrically as

$$x(0, s_1, s_2) = s_1, \quad y(0, s_1, s_2) = s_2, \quad z(0, s_1, s_2) = 1, \quad u(0, s_1, s_2) = s_1 s_2.$$

The transversality condition can easily be seen to hold. Solving the characteristic equations and substituting in the initial condition yields:

$$x = s_1 e^t, \quad y = s_2 e^t, \quad z = e^t, \quad u = s_1 s_2 e^{4t}.$$

Therefore, the solution is given by $u(x, y, z) = xyz^2$.

Consider now the general first-order equation in n independent variables:

$$F(x_1, x_2, \ldots, x_n, u, u_{x_1}, u_{x_2}, \ldots, u_{x_n}) = 0. \tag{9.7}$$

The associated Cauchy problem consists of (9.7) and the initial condition provided as an $(n - 1)$-dimensional surface Γ in the Euclidean space $\mathbb{R}^{n+1}$, as described above in (9.2)–(9.3). The method of characteristic strips that we developed in Chapter 2 is also valid in the higher-dimensional case. The strip equations are given by

$$\frac{\partial x_i}{\partial t} = \frac{\partial F}{\partial p_i} \qquad i = 1, 2, \dots, n,$$

$$\frac{\partial u}{\partial t} = \sum_{i=1}^{n} p_i \frac{\partial F}{\partial p_i}, \qquad\qquad (9.8)$$

$$\frac{\partial p_i}{\partial t} = -\frac{\partial F}{\partial x_i} - p_i \frac{\partial F}{\partial u} \qquad i = 1, 2, \dots, n,$$

where we used the notation $p_i = \partial u / \partial x_i$. To obtain a unique solution we must supply appropriate initial conditions. One such condition is given by the initial surface Γ. The additional conditions (for p_i) are determined in a similar way to (2.89)–(2.90). We therefore write the initial conditions for p_i as

$$p_i(0, s_1, \dots, s_{n-1}) = p_{0,i}(s_1, s_2, \dots, s_{n-1}).$$

The functions $p_{0,i}$ are determined from the equations

$$\frac{\partial u_0}{\partial s_i} = \sum_{i=1}^{n} p_{0,i} \frac{\partial x_{0,i}}{\partial s_i} \qquad i = 1, 2, \dots, n - 1, \qquad\qquad (9.9)$$

and

$$F(x_{0,1}, x_{0,2}, \dots, x_{0,n}, u_0, p_{0,1}, p_{0,2}, \dots, p_{0,n}) = 0, \qquad\qquad (9.10)$$

provided that an appropriate transversality condition holds true.

9.3 Classification of second-order equations

In Chapter 3 we classified second-order equations in two independent variables into three categories. A similar classification in higher dimensions is more intricate. Let $u(x_1, x_2, \dots, x_n)$ be a function satisfying a second-order equation whose principal part is of the form

$$L_0[u] = \sum_{i,j=1}^{n} a_{ij} \frac{\partial^2 u}{\partial x_i \partial x_j}, \qquad\qquad (9.11)$$

where $a_{ij} = a_{ij}(x_1, x_2, \dots, x_n)$. Since the mixed derivatives are invariant under a change in the order of differentiation, we can assume without loss of generality that the coefficient matrix $\mathbf{A} = (a_{ij})$ is symmetric. Thus the principal part can be considered as

$$L_0 = (\vec{\nabla}_x)^t \mathbf{A} \vec{\nabla}_x, \qquad\qquad (9.12)$$

where $\vec{\nabla}_x$ denotes the gradient operator with respect to the variables $(x_1, \ldots, x_n)$. In the special case considered earlier in Chapter 3 we had $\mathbf{A} = \begin{pmatrix} A & B \\ B & C \end{pmatrix}$. In the process of changing the notation (9.11) to the notation (9.12) we may have generated first-order derivatives of u, but such terms have no effect on the principal part.

In order to obtain a classification scheme for equations in arbitrarily many variables, it is beneficial to review some of the fundamental issues that we saw during our analysis of equations in two variables. We recall the definition of characteristic curves provided in Chapter 2. A fundamental property of these curves is that when the initial condition is provided on such a curve, the associated Cauchy problem does not have a unique solution. Actually a characteristic curve can be defined as a curve satisfying this property. Later, in Chapter 3, we saw another important property of these curves: they are exactly the curves along which singularities propagate. It turns out that these two properties of the characteristic curves are related to each other. We shall analyze this relation, and elaborate on its significance to the classification of equations in n dimensions.

We start by defining the Cauchy problem for an equation in n variables.

Definition 9.2 Cauchy problem Find a function $u(x_1, x_2, \ldots, x_n)$ in the space C^2 satisfying a given second-order PDE whose principal part is given by (9.11), such that u and all its first derivatives are provided on a hypersurface Γ that is given parametrically by $\phi(x_1, x_2, \ldots, x_n) = 0$.

A necessary condition for a solution for a Cauchy problem to exist is that the mixed derivatives be compatible (in the sense that the mixed derivative does not depend on the order of differentiation). We assume that this condition holds (otherwise the problem is not meaningful). Formally, to find the solution of a Cauchy problem in a neighborhood of the initial surface we have to compute the second-order derivatives of u from the PDE itself and from the initial data. Differentiating the equation will then enable us to find the third-order derivatives and so forth. This process fails if we cannot eliminate some second-order derivative from the condition of the Cauchy problem. We thus *define* characteristic surfaces as follows.

Definition 9.3 A surface Γ will be called a *characteristic surface* with respect to a second-order PDE if it is not possible to eliminate at least one second derivative of u from the conditions of the Cauchy problem.

Example 9.4 Consider the hyperbolic equation in two variables

$$u_{\eta_1 \eta_2} = 0. \tag{9.13}$$

We try to solve the Cauchy problem consisting of (9.13) and the initial data

$$u(\eta_1, 0) = f(\eta_1), \quad u_{\eta_2}(\eta_1, 0) = g(\eta_1). \tag{9.14}$$

Recalling that the general solution to (9.13) is of the form $u(\eta_1, \eta_2) = F(\eta_1) + G(\eta_2)$, we see at once that the problem cannot, in general, be solved. The reason is that the problem does not contain a term involving the second derivative with respect to η_2, and, therefore, it is not possible to "get off" the initial surface (the η_1 axis) in the normal direction.

Having used the formulation of the Cauchy problem to define characteristic surfaces, we shall show that these are the *only* surfaces along which the solution can be singular.

Lemma 9.5 *Let $u(x_1, x_2, \ldots, x_n)$ be a solution of a Cauchy problem. Assume that $u \in C^1$ in some domain Ω, and, furthermore, $u \in C^2$ in Ω except for a surface Γ. Then Γ is a characteristic surface.*

Proof Suppose by contradiction that Γ is not characteristic. Then knowing the values of u and its derivatives on Γ and using the PDE, we can eliminate the second derivatives of u on both sides of Γ. But then the continuity of u and its first derivatives imply that the second derivatives are also continuous, which contradicts the assumptions.   □

We shall use Definition 9.3 to derive an analytic criterion for the existence of characteristic surfaces, and even to compute such surfaces. In Example 9.4 we could have verified that the surface $\eta_2 = 0$ is characteristic since the equation had no second derivative with respect to η_2. In general, we have to find out whether there exist surfaces such that the equation effectively has no second derivatives in the direction normal to them. For this purpose, let Γ be a surface (our candidate for a characteristic surface) described parametrically as $\phi_1(x_1, x_2, \ldots, x_n) = 0$. Consider an invertible change of variables from $(x_1, x_2, \ldots, x_n)$ to $(\eta_1, \eta_2, \ldots, \eta_n)$ given by

$$\eta_i = \phi_i(x_1, x_2, \ldots, x_n) \quad i = 1, 2, \ldots, n.$$

To express the principal part (9.11) in terms of the new variables, let us write

$$u(x_1, x_2, \ldots, x_n) = w(\phi_1(x_1, x_2, \ldots, x_n), \ldots, \phi_n(x_1, x_2, \ldots, x_n)).$$

We thus obtain

$$L_0 = \sum_{i,j=1}^{n} \alpha_{ij} \frac{\partial^2 w}{\partial \eta_i \partial \eta_j}, \quad \alpha_{ij} = \sum_{k,l=1}^{n} a_{kl} \frac{\partial \phi_i}{\partial x_k} \frac{\partial \phi_j}{\partial x_l}. \tag{9.15}$$

The condition for Γ to be a characteristic surface is therefore equivalent to asking that the coefficient of $\partial^2 w / \partial \eta_1^2$ should vanish. In other words, the quadratic form defined by the matrix $\mathbf{A}$ should vanish for the vector $\vec{\nabla}_x \phi_1$:

$$\alpha_{11} = 0 \Rightarrow (\vec{\nabla}_x \phi_1)^t \mathbf{A} \vec{\nabla}_x \phi_1 = 0. \tag{9.16}$$

There are degenerate cases where one of the variables, say, x_1, does not appear at all in the principal part, namely $a_{1j} = 0$ for all j. Obviously, in such a case the surface $x_1 = c$, for some constant c, is a characteristic surface in the sense defined above. However this is an "uninteresting" case, since in this case we cannot provide the first derivative with respect to x_1 on Γ, and thus the Cauchy problem should be reformulated. We therefore define the classification of second-order equations in the following way:

Definition 9.6 A PDE is called *elliptic* if it has no characteristic surfaces; it is called *parabolic* if there exists a coordinate system, such that at least one of the independent variables does not appear at all in the principal part of the operator, and the principal part is elliptic relative to the variables that do appear in it; all other equations are called *hyperbolic*.

Let us reexamine the transformation (9.15) for the principal part. If the principal part has no mixed derivatives we shall say that it is a *canonical form*. Notice that this definition is somehow different from the one we introduced in Chapter 3, but, in fact, it is equivalent to it. We saw in Chapter 3 that in addition to the classification scheme, any equation can be transformed to an appropriate canonical form. In the elliptic case, for instance, we converted the principal part into the form $\partial^2/\partial x_1^2 + \partial^2/\partial x_2^2$, while in the hyperbolic case the principal part was converted into the form $\partial^2/\partial x_1^2 - \partial^2/\partial x_2^2$. It is remarkable that while the classification we just described is valid in any dimension, it is not always possible to convert a given equation into a canonical form. The reason is basically combinatoric. A transformation into a canonical form requires equating all the mixed derivatives to zero. However as the dimension grows linearly, the number of mixed derivatives grows quadratically. Thus, when the dimension is 3, we have three functions ϕ_i, $i = 1, 2, 3$ at our disposal to set three terms (the three mixed derivatives) to zero. In dimension 4, however, the mission is, in general, impossible, since we are to set to zero the six coefficients of the mixed derivatives using only four degrees of freedom (ϕ_i, $i = 1, 2, 3, 4$). The surplus of equations over unknowns becomes even worse with increasing dimension.

Fortunately, in the special but frequent case of equations with constant coefficients we can transform the equation into a canonical form regardless of the dimension. To consider this case in some detail, we assume that $\mathbf{A}$ is a constant matrix. Notice that the principal part is, in fact, expressed as a quadratic form relative to $\mathbf{A}$. To study the quadratic form observe that since $\mathbf{A}$ is symmetric it is diagonalizable.

We thus write

$$\mathbf{Q}^t \mathbf{A} \mathbf{Q} = \mathbf{D} = \begin{pmatrix} \lambda_1 & 0 & \cdot & \cdot & \cdot & 0 \\ 0 & \lambda_2 & 0 & \cdot & \cdot & 0 \\ \cdot & \cdot & \cdot & \cdot & \cdot & \cdot \\ 0 & \cdot & \cdot & \cdot & 0 & \lambda_n \end{pmatrix}, \tag{9.17}$$

where $\mathbf{Q}$ is the diagonalizing matrix of $\mathbf{A}$, and $\{\lambda_i\}$ are the real eigenvalues of $\mathbf{A}$. The classification scheme we introduced earlier can now be readily implemented with respect to the quadratic form (9.17). For example, an equation is elliptic if the quadratic form is strictly positive or strictly negative. More generally, we write the full classification scheme in terms of the spectrum of $\mathbf{A}$.

Definition 9.7 Let $\mathbf{A}$ be the (constant) matrix forming the principal part of a second-order PDE. The equation is called *hyperbolic* if at least one of the eigenvalues is positive and one is negative; it is called *elliptic* if all the eigenvalues are of the same sign; it is called *parabolic* if at least one eigenvalue vanishes, and all the eigenvalues that do not vanish are of the same sign.

The spectral decomposition induced by $\mathbf{Q}$ provides us with a natural tool for transforming the principal part to a canonical form. Denote the ith column of $\mathbf{Q}$ by $\vec{q}_i$, and define the canonical variables

$$\xi_i = \vec{q}_i{}^t \cdot \vec{x}. \tag{9.18}$$

The new variables satisfy $\vec{\nabla}_\xi = \mathbf{Q}^t \vec{\nabla}_x$; thus it follows from (9.15) that the principal part relative to the variables ξ takes the form

$$L_0[u] = \sum_{i=1}^{n} \lambda_i \frac{\partial^2 v}{\partial \xi_i^2},$$

where we used the notation

$$u(x_1, x_2, \ldots, x_n) = v(\xi_1(x_1, x_2, \ldots, x_n), \ldots, \xi_n(x_1, x_2, \ldots, x_n)).$$

Example 9.8 Consider the Poisson equation in $\mathbb{R}^3$:

$$\Delta u = u_{x_1 x_1} + u_{x_2 x_2} + u_{x_3 x_3} = F(x_1, x_2, x_3).$$

The matrix $\mathbf{A}$ corresponding to the principal part is the identity matrix in $\mathbb{R}^3$. Therefore the equation is elliptic. Alternatively, using (9.16) the equation for the characteristic surface is

$$\phi_{x_1}^2 + \phi_{x_2}^2 + \phi_{x_3}^2 = 0.$$

Clearly, this equation has no nontrivial solution.

Example 9.9 The heat equation in a three-dimensional spatial domain is given by

$$u_t = k\Delta u.$$

The variable t does not show up at all in the principal part, while the reduction of the principal part to the other variables (x_1, x_2, x_3) is elliptic according to the previous example. Thus the equation is parabolic.

Example 9.10 The *Klein–Gordon equation* for a function $u(x_1, x_2, x_3, t)$ in four-dimensional space-time has the form

$$u_{tt} - c^2(u_{x_1x_1} + u_{x_2x_2} + u_{x_3x_3}) = V(x_1, x_2, x_3, u). \tag{9.19}$$

This is one of the fundamental equations of mathematical physics. Although the equation is nonlinear, we shall classify it according to the criteria we developed above, since the principal part *is* linear, and it is this part that determines the nature of the equation. The matrix associated with the principal part is

$$A = \begin{pmatrix} -c^2 & 0 & 0 & 0 \\ 0 & -c^2 & 0 & 0 \\ 0 & 0 & -c^2 & 0 \\ 0 & 0 & 0 & 1 \end{pmatrix}. \tag{9.20}$$

Therefore the equation is hyperbolic. The equation for the characteristic surfaces is

$$\phi_t^2 = c^2(\phi_{x_1}^2 + \phi_{x_2}^2 + \phi_{x_3}^2). \tag{9.21}$$

This is a generalization of the eikonal equation that we discussed in Chapters 1 and 2. As a matter of fact, we are interested in the level sets $\phi = $ constant. If we write the level sets as

$$\omega t = kS(x_1, x_2, x_3),$$

we find that S satisfies the same eikonal equation derived in Chapter 1. We point out, though, that there is a fundamental difference between the derivation of the eikonal equation in Chapter 1 and the one given in this chapter. In Chapter 1 we derived the eikonal equation as an asymptotic limit for large wave numbers; here, on the other hand, we obtained it as the *exact* equation for the characteristic surfaces of the wave operator!

Example 9.11 In dimension 4 or more there exist equations of types that we have not (fortunately...) encountered yet. For example, consider the equation

$$u_{x_1x_1} + u_{x_2x_2} - u_{x_3x_3} - u_{x_4x_4} = 0.$$

Heuristically speaking, this is a wave equation where the dimension of "time" is 2!

9.4 The wave equation in $\mathbb{R}^2$ and $\mathbb{R}^3$

We developed in Chapter 4 the d'Alembert formula for the solution $u(x, t)$ of the wave equation in dimensions $1 + 1$ (i.e. one space dimension and one time dimension). We also studied the way in which waves propagate according to this formula. In particular we observed two basic phenomena:

(1) Suppose that the initial data $u(x, 0)$ have a compact support. Then the support propagates with the speed of the wave while preserving its initial shape. This seems to contradict our daily experience that indicates that waves decay as they propagate.
(2) When the initial velocity $u_t(x, 0)$ is different from zero we observed an even more bizarre effect. Suppose that $u_t(x, 0)$ is compactly supported, and assume for simplicity that $u(x, 0) = 0$. Let x_0 be an arbitrary point along the x axis. Denote by l the distance between x_0 and the farthest point in the support of $u_t(x, 0)$. D'Alembert's formula implies that $u(x_0, t) = (1/2c) \int_{-\infty}^{\infty} u_t(x, 0) dx$ for all $t > l/c$, where c is the speed of the wave. Had we been living in a world in which sound waves behaved in this manner, we would be subjected to an unbearable noise!

Our experience shows, however, that there are here and there calm places and quiet moments in our turbulent world. Therefore the waves described by d'Alembert's formula do not provide a realistic description of actual waves. It turns out that the source of the difficulty is in the reduction to one space dimension. We shall demonstrate in this section that the wave equation in *three* space dimensions does not suffer from any of the difficulties we just pointed out. It is remarkable that three is a magical number in this respect. It is the only (!) dimension in which waves propagate while maintaining their original shape on the one hand, but decay in amplitude and do not leave a trace behind them on the other hand. In other words, it is the only dimension in which it is possible to use waves to transmit meaningful information. Is it a coincidence that we happen to live in such a world?

9.4.1 Radially symmetric solutions

The case of radially symmetric problems in dimension $3 + 1$ turns out to be particularly simple. We seek solutions $u(x_1, x_2, x_3, t)$ to the wave equation

$$u_{tt} - c^2 \Delta u = 0 \qquad (x_1, x_2, x_3) \in \mathbb{R}^3, \quad -\infty < t < \infty, \qquad (9.22)$$

that are of the form $u = u(r, t)$, where $r = \sqrt{x_1^2 + x_2^2 + x_3^2}$. In Exercise 9.4 the reader will show that the radial part of the Laplace operator in three dimensions is

$$\frac{\partial^2}{\partial r^2} + \frac{2}{r} \frac{\partial}{\partial r}$$

(see also Subsection A.5). Thus $u(r, t)$ satisfies the equation

$$u_{tt} - c^2 \left(\frac{\partial^2 u}{\partial r^2} + \frac{2}{r} \frac{\partial u}{\partial r} \right) = 0. \tag{9.23}$$

Defining $v(r, t) = ru(r, t)$, we observe that v satisfies $v_{tt} - c^2 v_{rr} = 0$. This is exactly the one-dimensional wave equation! Therefore the general radial solution for (9.23) can be written as

$$u(r, t) = \frac{1}{r}[F(r + ct) + G(r - ct)]. \tag{9.24}$$

Moreover, we can use the same strategy to solve the Cauchy problem that consists of (9.23) for $t > 0$ and the initial conditions

$$u(r, 0) = f(r), \quad u_t(r, 0) = g(r) \quad 0 \le r \le \infty. \tag{9.25}$$

In light of the equation for the auxiliary function v that we defined above, we can use d'Alembert's formula to write down an explicit solution for u. There is one obstacle, though; the initial conditions are only given along the ray $r \ge 0$, and not for all values of r. To resolve this difficulty we observe that if a radial function $h(r)$ is of the class C^1, then it must satisfy $h'(0) = 0$. In order for u to be a classical solution of the problem we shall assume that indeed f and g are continuously differentiable. Thus $f'(0) = g'(0) = 0$. We can therefore apply the method we introduced in Chapter 4 (see Exercise 4.4) to solve the one-dimensional wave equation over the ray $r > 0$. For this purpose we extend f and g to the whole line $-\infty < r < \infty$ by defining them to be the even extensions of the given f and g. Hence, the initial conditions for v are odd functions, and therefore the solution $v(r, t)$ is odd, which implies that u is an even function. We thus obtain the following radially symmetric solution for the three-dimensional (radial) wave equation:

$$u(r, t) = \frac{1}{2r} \left[(r + ct)\tilde{f}(r + ct) + (r - ct)\tilde{f}(r - ct) \right] + \frac{1}{2cr} \int_{r-ct}^{r+ct} s\tilde{g}(s)ds, \tag{9.26}$$

where $\tilde{f}$ and $\tilde{g}$ are the even extensions of f and g, respectively.

In spite of the similarity between (9.26) and the one-dimensional d'Alembert formula that was introduced in Chapter 4, they are, in fact, quite different from each other. Let us consider a few examples to demonstrate these differences.

Example 9.12 Let $u(r, t)$ be the radial solution to the Cauchy problem (9.22) for $c = 1$ and the initial conditions

$$u(r, 0) = 0, \quad u_t(r, 0) = \begin{cases} 1 & r \le 1, \\ 0 & r > 1. \end{cases}$$

Compute $u(2, \frac{1}{2})$, $u(2, \frac{3}{2})$, and $u(2, 4)$.

Substituting the problem's data into (9.26) and using the even extension principle of the initial data, we obtain:

$$u(2, \tfrac{1}{2}) = \tfrac{1}{4} \int_{3/2}^{5/2} s\tilde{g}(s)ds = 0,$$

$$u(2, \tfrac{3}{2}) = \tfrac{1}{4} \int_{1/2}^{7/2} s\tilde{g}(s)ds = \tfrac{1}{4} \int_{1/2}^{1} s\,ds = \tfrac{3}{32},$$

$$u(2, 4) = \tfrac{1}{4} \int_{-2}^{6} s\tilde{g}(s)ds = \tfrac{1}{4} \int_{-1}^{1} s\,ds = 0.$$

More generally, one can verify that for short time intervals the perturbation originating in the domain $r \leq 1$ does not influence the sphere $r = 2$ at all. After one unit of time the perturbation does reach that sphere, and after two units of time it reaches its maximum there. After this time the wave on the sphere $r = 2$ decays, and it vanishes completely after some finite time. This picture should be contrasted with the one-dimensional case, where we saw that the influence of initial data consisting of a compactly supported wave's speed never disappears.

Example 9.13 Let $u(r, t)$ be the radial solution of the Cauchy problem (9.22) with $c = 1$ and the initial data

$$u(r, 0) = f(r) = \begin{cases} 1 & r \leq 1 \\ 0 & r > 1 \end{cases}, \qquad u_t(r, 0) = 0.$$

Let us compute $u(r, t)$ for a sphere of radius $r > 1$. We obtain

$$u(r, t) = \frac{1}{2r}(r - t)\tilde{f}(r - t).$$

Notice that the solution is zero outside the shell $t - 1 \leq r \leq t + 1$; moreover, $\max_{\{r>0\}} |u(r, t)|$ decays like $1/r$ (see Figure 9.1).

9.4.2 The Cauchy problem for the wave equations in three-dimensional space

Consider the general Cauchy problem in $3 + 1$ dimensions consisting of

$$u_{tt} - c^2 \Delta u = 0 \quad (x_1, x_2, x_3) \in \mathbb{R}^3, \; 0 < t < \infty, \tag{9.27}$$

together with the initial conditions

$$u(x_1, x_2, x_3, 0) = f(x_1, x_2, x_3), \, u_t(x_1, x_2, x_3, 0) = g(x_1, x_2, x_3) \quad (x_1, x_2, x_3) \in \mathbb{R}^3. \tag{9.28}$$

We shall first show that it is enough to solve a simpler problem in which $f(\vec{x}) \equiv 0$. This simplification is a consequence of the following claim and the superposition principle.

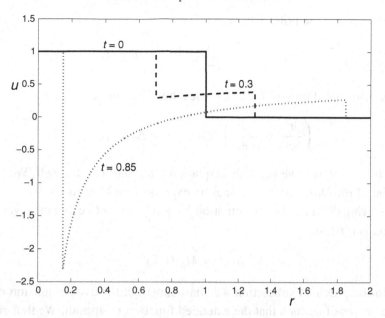

Figure 9.1 The solution of the problem in Example 9.13 for $t = 0$, 0.3, 0.85. We observe the propagation of the wave to the right (the domain $r > 1$), the reduction of the amplitude of the forward propagating wave, and the approach to the singularity at $r = 0$ that will occur at $t = 1$.

Proposition 9.14 *Let* $u(\vec{x}, t)$ *be the solution of the Cauchy problem (9.27)–(9.28) with the initial data*

$$u(x_1, x_2, x_3, 0) = 0, \quad u_t(x_1, x_2, x_3, 0) = g(x_1, x_2, x_3). \tag{9.29}$$

Then $v(\vec{x}, t) := u_t(\vec{x}, t)$ *solves the Cauchy problem (9.27) with the initial data*

$$v(x_1, x_2, x_3, 0) = g(x_1, x_2, x_3), \quad v_t(x_1, x_2, x_3, 0) = 0. \tag{9.30}$$

Proof Since (9.27) is an equation with constant coefficients, it is clear that if u is a solution, then so is $v = u_t$. Hence v solves the Cauchy problem (9.27) with the initial data

$$v(x_1, x_2, x_3, 0) = u_t(x_1, x_2, x_3, 0) = g(x_1, x_2, x_3),$$
$$v_t(x_1, x_2, x_3, 0) = u_{tt}(x_1, x_2, x_3, 0) = c^2 u_{xx}(x_1, x_2, x_3, 0) = 0.$$

□

We shall use an interesting observation due to the French mathematician Gaston Darboux (1842–1917) to solve the Cauchy problem (9.27) and (9.29). Let h be a differentiable function in $\mathbb{R}^3$. We define its *spherical mean* $M_h(a)$ over the sphere of radius a around the point $\vec{x}$ to be

$$M_h(a, \vec{x}) = \frac{1}{4\pi a^2} \int_{|\vec{\xi} - \vec{x}| = a} h(\vec{\xi}) ds_\xi. \tag{9.31}$$

Darboux discovered that $M_h(a, \vec{x})$ satisfies the differential equation

$$\left(\frac{\partial^2}{\partial a^2} + \frac{2}{a} \frac{\partial}{\partial a} \right) M_h(a, \vec{x}) = \Delta_x M_h(a, \vec{x}). \tag{9.32}$$

For an obvious reason we name this equation after Darboux himself. We leave the derivation of the *Darboux equation* as an exercise (see Exercise 9.7).

Considering (9.31) as a transformation $h \to M_h$, we notice that the inverse transformation is obvious:

$$h(\vec{x}) = M_h(0, \vec{x}). \tag{9.33}$$

Just as in the previous subsection we shall construct the even extension of M_h to negative values of a, such that the extended function is smooth. We thus require

$$\frac{\partial}{\partial a} M_h(0, \vec{x}) = 0. \tag{9.34}$$

This extension conforms with the definition of M_h: write M_h as

$$M_h(a, \vec{x}) = \frac{1}{4\pi} \int_{|\vec{\eta}| = 1} h(\vec{x} + a\vec{\eta}) ds_\eta, \tag{9.35}$$

where we have applied the change of variables $\vec{\xi} = \vec{x} + a\vec{\eta}$, and $\vec{\eta}$ varies over the unit sphere. The symmetry of the unit sphere now implies that M_h is an even function of a.

Equations (9.33) and (9.34) provide "initial" conditions for the Darboux equation. To connect the notion of spherical means and the wave equation, set M_u to be the spherical mean of $u(\vec{x}, t)$, where u is the solution of the Cauchy problem (9.27) and (9.29). We prove the following statement.

Proposition 9.15 $M_u(a, \vec{x}, t)$ *satisfies the radially symmetric wave equation (9.23).*

Proof The Darboux equation, the representation (9.31) and the wave equation imply

$$c^2 \left(\frac{\partial^2}{\partial a^2} + \frac{2}{a} \frac{\partial}{\partial a} \right) M_u(a, \vec{x}, t) = c^2 \Delta_x M_u(a, \vec{x}, t)$$

$$= \frac{1}{4\pi} \int_{|\vec{\eta}| = 1} c^2 \Delta_x u(\vec{x} + a\vec{\eta}) ds_\eta = \frac{\partial^2}{\partial t^2} M_u(a, \vec{x}, t).$$

Notice that the variables (x_1, x_2, x_3) are merely parameters in this equation. The initial conditions are

$$M_u(a, \vec{x}, 0) = M_f(a, \vec{x}) = 0, \qquad \frac{\partial}{\partial t} M_u(a, \vec{x}, 0) = M_g(a, \vec{x}).$$

□

Using the formula we derived in the previous subsection for the radial solution of the wave equation in three space dimensions we infer

$$M_u(a, \vec{x}, t) = \frac{1}{2ca} \int_{a-ct}^{a+ct} s M_g(s, \vec{x}) ds = \frac{1}{2ca} \int_{ct-a}^{ct+a} s M_g(s, \vec{x}) ds, \qquad (9.36)$$

where in the last equality we used the evenness of M. To eliminate $u(\vec{x}, t)$ we let a approach zero in (9.36). We obtain

$$u(\vec{x}, t) = t M_g(ct, \vec{x}). \qquad (9.37)$$

Thanks to formula (9.37) and to Proposition 9.14, we can now write a formula for the general solution of the Cauchy problem:

$$u(\vec{x}, t) = t M_g(ct, \vec{x}) + \frac{\partial}{\partial t}[t M_f(ct, \vec{x})], \qquad (9.38)$$

or, upon substituting the formula for the spherical means,

$$u(\vec{x}, t) = \frac{1}{4\pi c^2 t} \int_{|\vec{\xi} - \vec{x}| = ct} g(\vec{\xi}) ds_\xi + \frac{\partial}{\partial t}\left[\frac{1}{4\pi c^2 t} \int_{|\vec{\xi} - \vec{x}| = ct} f(\vec{\xi}) ds_\xi \right]. \qquad (9.39)$$

To understand the significance of the representation (9.39) we shall analyze separately the contributions of f and of g. Assume first that both f and g are compactly supported. The contribution to the first term in (9.39) is only from the *sphere* $|\vec{x} - \vec{\xi}| = ct$. Let $\vec{x}$ be outside the support of g. For sufficiently small times there is no contribution to the solution at $\vec{x}$, since the sphere is fully outside the support. There is a first time t_0 at which the sphere $|\vec{x} - \vec{\xi}| = ct$ intersects the support of g. Then we shall, in general, get a contribution to $u(\vec{x})$. On the other hand, when t is sufficiently large, the sphere $|\vec{x} - \vec{\xi}| = ct$ has expanded so much that it no longer intersects the support of g, and from that time on g will have no impact on the value of $u(\vec{x})$. This behavior is in marked contrast to the bizarre phenomenon we mentioned above in the one-dimensional case. The contribution of f to the solution at a point $\vec{x}$ outside the support of f is also felt only after an initial time period (the distance between $\vec{x}$ and the support of f divided by c), and, here, too, the perturbation proceeds without leaving a trace in $\vec{x}$. Hadamard called such a phenomenon *Huygens' principle in the narrow sense*. Huygens' principle is graphically depicted in Figure 9.2.

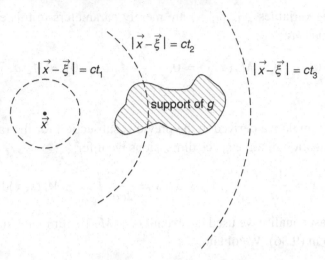

Figure 9.2 Wave propagation in $\mathbb{R}^3$.

Another feature of the wave equation that distinguishes the three-dimensional case from the one-dimensional case is the loss of regularity. We proved in Chapter 4 that if the initial data satisfy $f \in C^2$ and $g \in C^1$, then the solution is classical, namely, $u \in C^2$. Moreover, even when the initial data *do* have singular points, such as points of nondifferentiability, or even discontinuity, the solution is singular in exactly the same way, and the singularity propagates (while preserving its nature) along the characteristic curves. The situation is different in the three-dimensional case, as smooth initial data might develop singularities in finite time. To analyze this phenomenon, let us reexamine the radial case. We assume that the initial condition g vanishes identically and compute the solution at the origin. In considering the limit $r \to 0$ in (9.26) we recall that f is an even function, and, therefore, f' is odd. We obtain

$$u(0, t) = f(ct) + ctf'(ct), \tag{9.40}$$

Indeed the expression for $u(0, t)$ depends not only on f itself, but also on its derivative. Therefore even if $f \in C^2$, the solution may not be classical at the origin. Moreover, if f has discontinuities, the solution may be even unbounded. For example, let us look again at Example 9.13. Formula (9.40) implies that the solution blows up at $t = 1$, which is exactly the time it takes the singularity to travel from its original location $r = 1$ to the origin. The reason behind the spontaneous creation of singularities is geometric. The initial data in Example 9.13 are discontinuous on the unit sphere. As the wave propagates towards the origin (cf. Figure 9.1) it shrinks until it collapses at the origin to a point. In other words, the singularity that started its life as a two-dimensional object (a sphere) later turned into a zero-dimensional object (a point). This shrinking implies that the singularity concentrates (or focuses) and its nature worsens.

9.4.3 The Cauchy problem for the wave equation in two-dimensional space

Equipped with the solution of the wave equation in n spatial dimensions, we can solve the equation in a smaller number of dimensions by freezing one of the variables. We shall now demonstrate this method, called *Hadamard's method of descent*, to derive a formula for the solution of the Cauchy problem for the wave equation in $2 + 1$ dimensions:

$$v_{tt} - c^2(v_{x_1x_1} + v_{x_2x_2}) = 0 \qquad (x_1, x_2) \in \mathbb{R}^2, t > 0, \quad (9.41)$$
$$v(x_1, x_2, 0) = f(x_1, x_2), \quad v_t(x_1, x_2, 0) = g(x_1, x_2) \qquad (x_1, x_2) \in \mathbb{R}^2. \quad (9.42)$$

We substitute the initial conditions into (9.39). Since the problem does not depend on the variable x_3, we shall evaluate the solution at a point $(x_1, x_2, 0)$ in the (x_1, x_2) plane. To compute the surface integral we use the relation

$$\xi_3 = \sqrt{(ct)^2 - (\xi_1 - x_1)^2 - (\xi_2 - x_2)^2}$$

to express the integral in terms of (ξ_1, ξ_2), where these two variables vary over the disk $(\xi_1 - x)^2 + (\xi_2 - x)^2 \le (ct)^2$, i.e. the projection of the sphere over the plane. Note that the point $(\xi_1, \xi_2, -\xi_3)$ contributes to the integral the same as (ξ_1, ξ_2, ξ_3). Using the formula $ds_\xi = |ct/\xi_3|d\xi_1 d\xi_2$ for the surface element of the sphere, expressed in Cartesian coordinates, we obtain

$$v(x_1, x_2, 0, t) = \frac{1}{2\pi c} \int_{r \le ct} \frac{g(\xi_1, \xi_2)}{\sqrt{(ct)^2 - r^2}} d\xi_1 d\xi_2$$
$$+ \frac{\partial}{\partial t} \left[\frac{1}{2\pi c} \int_{r \le ct} \frac{f(\xi_1, \xi_2)}{\sqrt{(ct)^2 - r^2}} d\xi_1 d\xi_2 \right], \qquad (9.43)$$

where we have written $r = \sqrt{(\xi_1 - x_1)^2 + (\xi_2 - x_2)^2}$. By construction, $v(x_1, x_2, 0, t)$ is a solution the Cauchy problem in the plane. Thus we can omit x_3 from the list of variables of v.

There is a fundamental difference between the solution in two dimensions (9.43) and that in three dimensions (9.39). In the former case the integration is over a planar domain, while in the latter case the integration is over a boundary of a three-dimensional domain. Therefore, in the two-dimensional case even if the initial data have a compact support, once the initial perturbation has reached a planar point $\vec{x}$ outside the support, it will leave some trace there for all later times, since if $t_2 > t_1$, then the domain of integration at time t_2 includes the domain of integration at time t_1.

9.5 The eigenvalue problem for the Laplace equation

We have applied in Chapters 5–7 the method of separation of variables to solve a variety of canonical problems. The basic tool employed in this method was the solution of an appropriate eigenvalue problem. For example, when we dealt with equations with constant coefficients, we typically solved Sturm–Liouville problems like (5.9)–(5.10). The main difference between the method of separation of variables for equations in two variables and equations in more than two variables is that in the latter case the eigenvalue problem itself might be a PDE. We point out that the method is not circular. The PDE one needs to solve as part of the eigenvalue problem is of a lower dimension (it involves a smaller number of variables), and thus is simpler, than the underlying PDE we are solving. Since we can solve PDEs explicitly only for a small number of simple canonical domains, and since these domains occur in a large variety of canonical problems, we shall limit the discussion to rectangles, prisms, disks, balls and cylinders. Nevertheless, we start with a general discussion on the eigenvalues of the Laplace operator that applies to any smooth domain.

Let Ω be a bounded domain in $\mathbb{R}^2$ or in $\mathbb{R}^3$. We define the following inner product in the space of continuous functions in $\bar{\Omega}$:

$$\langle u, v \rangle = \int_\Omega u(\vec{x}) v(\vec{x}) \, d\vec{x}. \tag{9.44}$$

The following problem generalizes the Sturm–Liouville problem (5.9)–(5.10):

$$-\Delta u = \lambda u \quad \vec{x} \in \Omega, \tag{9.45}$$

$$u = 0 \quad \vec{x} \in \partial\Omega. \tag{9.46}$$

We call this problem a Dirichlet eigenvalue problem. The set of eigenvalues λ is called the *spectrum* of the Dirichlet problem. It can be shown that under certain smoothness assumptions on the domain Ω, there exists a discrete infinite sequence of eigenvalues $\{\lambda_n\}$ and eigenfunctions $\{u_n(\vec{x})\}$ solving (9.45)–(9.46). We show in the next subsection that many of the properties we presented in Chapter 6 for the eigenvalues and eigenfunctions of the Sturm–Liouville problem are also valid for the problem (9.45)–(9.46). We then proceed to compute the spectrum of the Laplacian in several canonical domains. One can similarly formulate the eigenvalue problem for the Laplace operator under the Neumann boundary condition

$$\partial_n u = 0 \quad \vec{x} \in \partial\Omega, \tag{9.47}$$

or even for the problem of the third kind.

9.5.1 Properties of the eigenfunctions and eigenvalues
of the Dirichlet problem

We review the ten properties that were presented in Chapter 6 for the Sturm–Liouville problem, and examine the analogous properties in the case of (9.45)–(9.46). We assume throughout that Ω is a sufficiently smooth (bounded) domain such that the eigenfunctions belong the class $C^2(\bar{\Omega})$ there. We also refer to the scalar product defined in (9.44).

1 Symmetry Using an integration by parts (Green's formula) we see that for any two functions satisfying the Dirichlet boundary conditions

$$\int_\Omega v \Delta u \, d\vec{x} = -\int_\Omega \vec{\nabla} v \cdot \vec{\nabla} u \, d\vec{x} = \int_\Omega u \Delta v \, d\vec{x}.$$

This verifies the symmetry of the Laplace operator.

2 Orthogonality

Proposition 9.16 *Eigenfunctions associated with different eigenvalues are orthogonal to each other.*

Proof Let v_n, v_m be two eigenfunctions associated with the eigenvalues $\lambda_n \neq \lambda_m$, respectively; namely,

$$-\Delta v_n = \lambda_n v_n, \tag{9.48}$$
$$-\Delta v_m = \lambda_m v_m. \tag{9.49}$$

The symmetry property implies

$$(\lambda_n - \lambda_m) \int_\Omega v_n v_m \, d\vec{x} = 0,$$

hence the orthogonality. □

3 The eigenvalues are real The proof is the same as the proof of Proposition 6.21.

4 The eigenfunctions are real Here the claim is identical to Proposition 6.22 and the related discussion in Chapter 6.

5 Multiplicity of the eigenvalues One of the main differences between the one-dimensional Sturm–Liouville problem and the multi-dimensional case we consider here involves multiplicity. In the multi-dimensional case (9.45)–(9.46) the multiplicity might be larger than 1 (but it is always finite!). This fact is of great physical significance. We shall demonstrate this property in the sequel through specific examples.

6 There exists a sequence of eigenvalues converging to ∞ We formulate the following proposition.

Proposition 9.17 *(a) The set of eigenvalues for the problem (9.45)–(9.46) consists of a monotone nondecreasing sequence converging to* ∞.

(b) The eigenvalues are all positive and have finite multiplicity.

Proof We only prove the statement that all the eigenvalues are positive. In the process of doing so, we shall discover an important formula for the characterization of the eigenvalues. Multiply (9.45) by u and integrate by parts over Ω. We obtain

$$\lambda = \frac{\int_\Omega |\nabla u|^2 \, d\vec{x}}{\int_\Omega u^2 \, d\vec{x}}. \tag{9.50}$$

Since the function $u = $ constant is not an eigenfunction, it follows that $\lambda > 0$. $\quad \square$

7 Generalized Fourier series Let $\{\lambda_n\}$ be the eigenvalue sequence for the Dirichlet problem, written in a nondecreasing order. Denote by V_n the subspace spanned by the eigenfunctions associated with the eigenvalue λ_n. We have shown that eigenfunctions belonging to different subspaces V_n are orthogonal to each other. We now select for each eigenspace V_n an orthonormal basis. We have thus constructed an orthonormal set of eigenfunctions $\{v_n(\vec{x})\}$. It is known that the sequence is *complete* with respect to the norm induced by the inner product (9.44). Thus we can formally expand smooth functions defined in Ω into a generalized Fourier series

$$f(\vec{x}) = \sum_{m=0}^{\infty} \alpha_m v_m(\vec{x}). \tag{9.51}$$

Due to the completeness of the orthonormal system $\{v_m\}$, the series is converging on average, and the generalized Fourier coefficients are given by

$$\alpha_m = \langle f(\vec{x}), v_m(\vec{x}) \rangle. \tag{9.52}$$

We shall demonstrate several such Fourier expansions in the next few subsections, although we shall not analyze their convergence in detail.

8 An optimization problem for the first eigenfunction We developed in (9.50) an integral formula for the eigenvalues. Denote the smallest eigenvalue (called the principal eigenvalue) by λ_0. Using a proof that is similar to that for Proposition 6.37, the following proposition can be shown.

Proposition 9.18 The Rayleigh–Ritz formula

$$\lambda_0 = \inf_{v \in V} \frac{\int_\Omega |\nabla v|^2 dx}{\int_\Omega v^2 dx}, \tag{9.53}$$

where

$$V = \{v \in C^2(\Omega) \cap C(\bar{\Omega}) \mid v \neq 0, \ v \mid_{\partial\Omega} = 0\}.$$

Moreover, λ_0 is a simple eigenvalue, and the infimum is only achieved for the associated eigenfunction.

9 Zeros of the eigenfunctions The zero set of a scalar function is generically a codimension one manifold (lines in the plane; surfaces in space). These sets are called *nodal surfaces*. The nodal surfaces can take quite intricate shapes. An interesting application of the shape of the nodal surfaces of the eigenfunctions for the Laplace operator is in the theory of Turing instability. This theory, proposed by the British mathematician Alan Mathison Turing (1912–1954) explains the spontaneous creation of patterns in chemical and biological systems. It is argued, for example, that the specific patterns of the zebra's stripes or the giraffe's spots can be explained with the aid of the nodal surfaces of certain eigenfunctions of the Laplacian [12].

10 Asymptotic behavior of the eigenvalues λ_n when $n \to \infty$ It can be shown in analogy to formula (6.76) that for $\Omega \subseteq \mathbb{R}^j$ the nth eigenvalue associated with (9.45)–(9.46) has the following asymptotic behavior in the limit $n \to \infty$:

$$\lambda_n \sim 4\pi^2 \left(\frac{n}{\omega_j |\Omega|}\right)^{\frac{2}{j}} \qquad j = 1, 2, 3 \ldots . \tag{9.54}$$

This formula is called *Weyl's asymptotic formula*. We have used here the notation ω_j to denote the volume of the unit ball in $\mathbb{R}^j$. For example, $\omega_1 = 2$, $\omega_2 = \pi$, $w_3 = 4\pi/3$.

9.5.2 The eigenvalue problem in a rectangle

Let Ω be the rectangle $\{0 < x < a, \ 0 < y < b\}$. We want to compute the eigenvalues of the Laplace operator in Ω:

$$\begin{aligned} u_{xx} + u_{yy} &= -\lambda u & 0 < x < a, \ 0 < y < b, \\ u(0, y) = 0, \ u(a, y) &= 0 & 0 < y < b, \\ u(x, 0) = 0, \ u(x, b) &= 0 & 0 < x < a. \end{aligned} \tag{9.55}$$

We use the symmetry of the rectangle to construct separable solutions of the form $u(x, y) = X(x)Y(y)$. We obtain two Sturm–Liouville problems

$$Y''(y) + \mu Y(y) = 0, \tag{9.56}$$
$$Y(0) = Y(b) = 0, \tag{9.57}$$

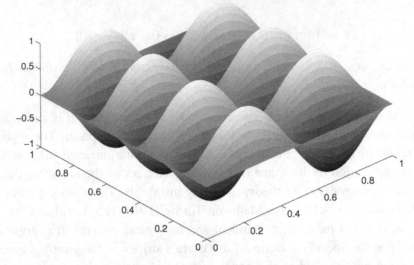

Figure 9.3 The (7,2) mode $u_{7,2}(x, y) = \sin(7\pi x)\sin(2\pi y)$.

and

$$X''(x) + (\lambda - \mu)X(x) = 0, \tag{9.58}$$
$$X(0) = X(a) = 0. \tag{9.59}$$

We have already solved such systems in Chapter 6. The solutions are

$$\lambda_{n,m} = \pi^2\left(\frac{n^2}{a^2} + \frac{m^2}{b^2}\right) \qquad m, n = 1, 2, \ldots, \tag{9.60}$$

$$u_{n,m}(x, y) = X_n(x)Y_m(y) = \sin\frac{n\pi x}{a}\sin\frac{m\pi y}{b} \qquad m, n = 1, 2, \ldots. \tag{9.61}$$

The graph of $u_{7,2}$ is depicted in Figure 9.3. Notice that the eigenvalue μ that appears in (9.56)–(9.59) is merely a tool in the computation, and it does not show up in the final answer.

The generalized Fourier expansion of a function in two variables $f(x, y)$ in the rectangle Ω by the system $\{u_{n,m}\}$ can be written as

$$f(x, y) = \sum_{n,m=1}^{\infty} A_{n,m}\sin\frac{n\pi x}{a}\sin\frac{m\pi y}{b}, \tag{9.62}$$

where the generalized Fourier coefficients are given by

$$A_{n,m} = \frac{4}{ab}\int_\Omega f(x, y)\sin\frac{n\pi x}{a}\sin\frac{m\pi y}{b}\,\mathrm{d}x\mathrm{d}y. \tag{9.63}$$

It is straightforward to find the corresponding eigenvalues and eigenfunctions for the Neumann problem in a rectangle. This is left to the reader as an exercise.

One of the important issues in the analysis of eigenvalues is their multiplicity. We saw in Chapter 6 that all the eigenvalues in a regular Sturm–Liouville problem are simple. In higher dimensions, though, some eigenvalues might have a multiplicity larger than 1. When this happens, we say that the problem has *degenerate states*. We prove now that the eigenvalue problem for Laplace equation in the unit square is degenerate.

Proposition 9.19 *There are infinitely many eigenvalues for the Dirichlet problem in the unit square that are not simple.*

Proof An eigenvalue λ is degenerate if there are two different pairs of positive integers (m, n) and (p, q) such that

$$p^2 + q^2 = m^2 + n^2.$$

Equations of this type appear frequently in number theory, where they are called *Diophantine equations*. To prove that this Diophantine equation has infinitely many two pairs of solutions we choose $p = m + 1$. The equation takes the form

$$2m + 1 = n^2 - q^2,$$

namely, there exists a solution for each choice of n and q, provided they have a different parity. There also exist 'trivial' solutions such as $(m, n) = (q, p)$; furthermore, if a pair of solutions is multiplied by an integer, one obtains a new solution.
□

9.5.3 The eigenvalue problem in a disk

Let Ω be the disk $\{0 \leq r < a, \ 0 \leq \theta \leq 2\pi\}$. We want to compute the eigenvalues and eigenfunctions of the Laplace equation there. Using a polar coordinate system the problem is written as:

$$u_{rr} + \frac{1}{r}u_r + \frac{1}{r^2}u_{\theta\theta} = -\lambda u \qquad 0 < r < a, \ 0 \leq \theta \leq 2\pi, \qquad (9.64)$$

$$u(a, \theta) = 0 \qquad 0 \leq \theta \leq 2\pi. \qquad (9.65)$$

Just like in Chapter 7 we construct separable solutions of the form $u(r, \theta) = R(r)\Phi(\theta)$. We use the standard arguments to obtain two systems of Sturm–Liouville problems:

$$\Phi''(\theta) + \mu\Phi(\theta) = 0 \qquad 0 \leq \theta \leq 2\pi, \qquad (9.66)$$

$$\Phi(0) = \Phi(2\pi), \qquad \Phi'(0) = \Phi'(2\pi), \qquad (9.67)$$

and

$$R''(r) + \frac{1}{r}R'(r) + \left(\lambda - \frac{\mu}{r^2}\right)R(r) = 0 \quad 0 < r < a, \tag{9.68}$$

$$|\lim_{r \to 0} R(r)| < \infty, \quad R(a) = 0. \tag{9.69}$$

The solution to (9.66)–(9.67) is (see Chapter 7)

$$\Phi_n(\theta) = A_n \cos n\theta + B_n \sin n\theta, \quad \mu_n = n^2 \quad n = 0, 1, 2, \ldots . \tag{9.70}$$

Therefore, the radial problem (9.68)–(9.69) becomes

$$R''(r) + \frac{1}{r}R'(r) + \left(\lambda - \frac{n^2}{r^2}\right)R(r) = 0 \quad 0 < r < a, \quad |\lim_{r \to 0} R(r)| < \infty, \quad R(a) = 0.$$
$$\tag{9.71}$$

Applying the change of variables $s = \sqrt{\lambda}r$, (9.71) is transformed into the canonical form

$$\psi''(s) + \frac{1}{s}\psi'(s) + \left(1 - \frac{n^2}{s^2}\right)\psi(s) = 0 \quad 0 < s < \sqrt{\lambda}a, \tag{9.72}$$

together with the boundary conditions

$$|\lim_{s \to 0} \psi(s)| < \infty, \quad \psi(\sqrt{\lambda}a) = 0, \tag{9.73}$$

where we write $R(r) = \psi(\sqrt{\lambda}r)$. The system (9.72)–(9.73) forms a singular Sturm–Liouville problem, Indeed, we can also write (9.72) in the form (see Chapter 6, and in particular (6.24) there):

$$(s\psi')' + \left(s - \frac{n^2}{s}\right)\psi = 0 \quad 0 < s < \sqrt{\lambda}a.$$

We call (9.72) a *Bessel equation of order n*. Equations of this type can be solved by the Frobenius–Fuchs method (expansion into a power series). It is easy to verify that the point $s = 0$ is a regular singular point for all Bessel equations. Moreover, one of the independent solutions is singular at $s = 0$, while the other one is regular there. Since we are looking for regular solutions to (9.64)–(9.65), we shall ignore the singular solution. The regular solution for the Bessel equation is called the *Bessel function of order n of the first kind*. It is denoted by J_n in honor of the German mathematician and astronomer Friedrich Wilhelm Bessel (1784–1846) who was among the first to study these functions. There is also a singular solution Y_n for the Bessel equation that is called the Bessel function of order n of the second kind. There exist several voluminous books such as [21] summarizing the rich knowledge accumulated over the years on the many fascinating properties of Bessel functions.

We list here some of these properties that are of particular relevance to our study of the eigenvalues in a disk.

(1) For every nonnegative integer n the zeros of the Bessel function J_n form a sequence of real positive numbers $\alpha_{n,m}$ that diverge to ∞ as $m \to \infty$.

(2) The difference between two consecutive zeros converges to π in the limit $m \to \infty$. A full proof of this interesting property is difficult; instead we present the following heuristic argument. For large n the eigenvalues are determined by the form of the solution for large values of s (since $\sqrt{\lambda}a \gg 1$). To estimate the behavior of the solution ψ of (9.72) at large s, it is useful write $\psi = s^{-1/2}\chi$. A little algebra shows that χ satisfies the equation

$$\chi'' + \chi + s^{-2}\left(\frac{1}{4} - n^2\right)\chi = 0.$$

Therefore we expect that for large argument the Bessel function will be approximately proportional to $s^{-1/2}\cos(s + \gamma)$, where γ is an appropriate constant. It can be shown that this indeed is the asymptotic behavior of the Bessel functions, and that $\gamma = -\frac{1}{2}n\pi - \frac{1}{4}\pi$, where n is the order of the function. This justifies our claim about the difference between consecutive zeros.

(3) We pointed out that (9.72) possesses only one solution that is not singular at the origin. We shall select a certain normalization for that solution. In the case $n = 0$ it is convenient to select the normalization $J_0(0) = 1$. When $n > 0$, however, it follows from the series expansion of the solution to (9.72) that $J_n(0) = 0$. We thus search for another normalization. An elegant way to select a normalization is to construct an integral representation for J_n. For this purpose consider the differential equation

$$\Delta v + v = (\Delta + 1)v = 0. \tag{9.74}$$

Clearly the function $v(y) = e^{iy} = e^{ir \sin \theta}$ satisfies this equation. Let us expand this function into a classical Fourier series in the variable θ:

$$e^{ir \sin \theta} = \sum_{n=-\infty}^{\infty} \Psi_n(r)e^{in\theta}. \tag{9.75}$$

Operating over (9.75) with $\Delta + 1$ we find

$$0 = \sum_{n=-\infty}^{\infty}\left[\Psi_n'' + \frac{1}{r}\Psi_n' + \left(1 - \frac{n^2}{r^2}\right)\Psi\right]e^{in\theta}.$$

Therefore we can identify the coefficients Ψ_n in (9.75) with the Bessel functions J_n. The Fourier formulas now provide the important integral representation for Bessel

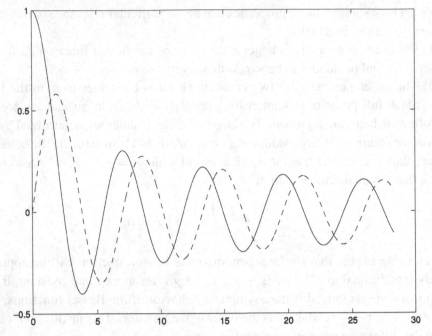

Figure 9.4 The Bessel functions J_0 (solid line) and J_1 (dashed line).

functions:

$$J_n(x) = \frac{1}{2\pi} \int_0^{2\pi} e^{ix \sin \theta} e^{-in\theta} \, d\theta. \tag{9.76}$$

Indeed the normalization we selected satisfies $J_0(0) = 1$. One of the applications of the integral representation (9.76) is the recursive formula

$$s J_{n+1}(s) = n J_n(s) - s J_n'(s). \tag{9.77}$$

We leave the proof of the recursive formula to Exercise 9.11. Notice that according to this formula it is enough to compute J_0, and then use this function to evaluate J_n for $n > 0$. The Bessel functions J_0 and J_1 are depicted in Figure 9.4.

(4) The following proposition is particularly useful for the expansion of functions defined over the disk in terms of Bessel functions.

Proposition 9.20 *Let n be a nonnegative integer. Then for all $m = 1, 2, \ldots$ we have*

$$\int_0^a r J_n^2 \left(\frac{\alpha_{n,m}}{a} r \right) \, dr = \frac{a^2}{2} J_{n+1}^2(\alpha_{n,m}), \tag{9.78}$$

where $\{\alpha_{n,m}\}$ are the zeros of J_n.

Proof Consider (9.72) for some eigenvalue $\lambda_{n,m}$. Multiplying the equation by $s^2 J_n'$ and integrating from 0 to $\sqrt{\lambda_{n,m}}a$, one obtains

$$\int_0^{\sqrt{\lambda_{n,m}}a} \left[s J_n'(s J_n')' + \left(1 - \frac{n^2}{s^2}\right) s^2 J_n J_n' \right] ds = 0.$$

Some of the terms in the integrand are complete derivatives. Performing the integrations we find

$$2 \int_0^{\sqrt{\lambda_{n,m}}a} s J_n^2 \, ds = \lambda_{n,m} a^2 [J_n'(\sqrt{\lambda_{n,m}}a)]^2.$$

Returning to the variable r we end up with

$$\int_0^s r J_n^2(\sqrt{\lambda_{n,m}}r) \, dr = \frac{a^2}{2} [J_n'(\sqrt{\lambda_{n,m}}a)]^2.$$

Observe that if the argument s in the recurrence formula (9.77) is a zero of J_n, then the formula reduces to $J_n'(s) = J_{n+1}(s)$. Since by assumption $\lambda_{n,m}$ is an eigenvalue, then $\sqrt{\lambda_{n,m}}a$ is indeed a zero of J_n, and the claim follows.   □

The eigenvalues of the Dirichlet problem in a disk are therefore given by the double index sequence

$$\lambda_{n,m} = \left(\frac{\alpha_{n,m}}{a}\right)^2 \qquad n = 0, 1, 2, \ldots, \qquad m = 1, 2, \ldots, \qquad (9.79)$$

while the eigenfunctions are

$$u_{n,m} = J_n\left(\frac{\alpha_{n,m}}{a}r\right)(A_{n,m} \cos n\theta + B_{n,m} \sin n\theta) \quad n = 0, 1, 2, \ldots, \quad m = 1, 2, \ldots.$$
$$(9.80)$$

This sequence forms a complete orthogonal system for the space of continuous functions in the disk of radius a with respect to the inner product

$$\langle f, g \rangle = \int_0^{2\pi} \int_0^a f(r, \theta) g(r, \theta) r \, dr \, d\theta. \qquad (9.81)$$

The Fourier–Bessel expansion for a function $h(r, \theta)$ over that disk is given by

$$h(r, \theta) = \sum_{n=0}^{\infty} \sum_{m=1}^{\infty} J_n\left(\frac{\alpha_{n,m}}{a}r\right)(A_{n,m} \cos n\theta + B_{n,m} \sin n\theta), \qquad (9.82)$$

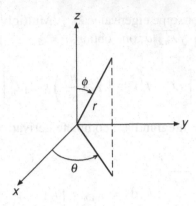

Figure 9.5 Our notation for the spherical coordinate system.

where according to Proposition 9.20 the Fourier–Bessel coefficients are

$$A_{n,m} = \frac{2}{\pi a^2 J_{n+1}(\alpha_{n,m})} \int_0^{2\pi} \int_0^a h(r,\theta) J_n\left(\frac{\alpha_{n,m}}{a} r\right) \cos n\theta \, r \, dr \, d\theta, \quad (9.83)$$

$$B_{n,m} = \frac{2}{\pi a^2 J_{n+1}(\alpha_{n,m})} \int_0^{2\pi} \int_0^a h(r,\theta) J_n\left(\frac{\alpha_{n,m}}{a} r\right) \sin n\theta \, r \, dr \, d\theta. \quad (9.84)$$

We end this subsection by pointing out that each eigenvalue (except for the case $n = 0$) is of multiplicity 2.

9.5.4 The eigenvalue problem in a ball

We solved the eigenvalue problem in a rectangle by writing the rectangle as a product of two intervals. Similarly we computed the eigenvalues in a disk using the observation that in polar coordinates the disk too can be written as a product of two intervals. We thus separated the eigenvalue problem in the disk into one eigenvalue problem on the unit circle (9.66)–(9.67), and another problem in the radial direction. Proceeding similarly in the case of the ball we define a spherical coordinate system $\{(r,\phi,\theta) | \, r > 0, \, 0 \le \phi \le \pi, \, 0 \le \theta \le 2\pi\}$ (see Figure 9.5), given by

$$\begin{aligned} x &= r \sin\phi \cos\theta, \\ y &= r \sin\phi \sin\theta, \\ z &= r \cos\phi. \end{aligned} \quad (9.85)$$

The reader will compute the Laplace operator in spherical coordinates in Exercise 9.4. Write

$$B_a := \{0 < r < a, 0 < \phi < \pi, 0 \le \theta \le 2\pi\}, \quad S^2 := \{0 \le \phi \le \pi, 0 \le \theta \le 2\pi\}.$$

The eigenvalue problem in a ball of radius a is given by

$$\frac{1}{r^2}\frac{\partial}{\partial r}\left(r^2\frac{\partial u}{\partial r}\right)+\frac{1}{r^2}\left[\frac{1}{\sin\phi}\frac{\partial}{\partial\phi}\left(\sin\phi\frac{\partial u}{\partial\phi}\right)+\frac{1}{\sin^2\phi}\frac{\partial^2 u}{\partial\theta^2}\right]$$

$$= -\lambda u \qquad (r,\phi,\theta)\in B_a, \tag{9.86}$$

$$u(a,\theta,\phi)=0 \qquad (\phi,\theta)\in S^2, \tag{9.87}$$

plus certain compatibility conditions to be presented later.

Writing u in the separable form $u(r,\theta,\phi)=R(r)Y(\phi,\theta)$, we obtain a system of two eigenvalue problems. One of them, defined over the unit sphere S^2, takes the form

$$\frac{1}{\sin\phi}\frac{\partial}{\partial\phi}\left(\sin\phi\frac{\partial Y}{\partial\phi}\right)+\frac{1}{\sin^2\phi}\frac{\partial^2 Y}{\partial\theta^2}=-\mu Y \qquad (\phi,\theta)\in S^2. \tag{9.88}$$

Equation (9.88) is subject to two conditions. The first condition is that the solution is periodic with respect to the variable θ:

$$Y(\phi,0)=Y(\phi,2\pi),\quad Y_\theta(\phi,0)=Y_\theta(\phi,2\pi). \tag{9.89}$$

The other condition is that Y is bounded everywhere on the unit sphere, and, in particular in the two poles $\phi=0$ and $\phi=\pi$, where the coefficients of (9.88) are not bounded.

The second problem for the radial function $R(r)$ consists of the equation

$$\frac{1}{r^2}\frac{\partial}{\partial r}\left(r^2\frac{\partial R}{\partial r}\right)=\left(\frac{\mu}{r^2}-\lambda\right)R \qquad 0<r<a, \tag{9.90}$$

the boundary condition

$$R(a)=0, \tag{9.91}$$

and the requirement that R is bounded at the origin (where (9.90) is singular).

We shall perform an extensive analysis of the eigenvalue problem (9.88)–(9.89). We seek eigenfunctions Y in a separable form $Y(\phi,\theta)=\Phi(\phi)\Theta(\theta)$. Substituting this form of Y into (9.88) gives rise to two equations:

$$\Theta''(\theta)+\nu\Theta(\theta)=0 \quad 0<\theta<2\pi, \tag{9.92}$$

$$\sin\phi\frac{\partial}{\partial\phi}\left[\sin\phi\frac{\partial\Phi(\phi)}{\partial\phi}\right]+(\mu\sin^2\phi-\nu)\Phi(\phi)=0 \quad 0<\phi<\pi. \tag{9.93}$$

The periodicity condition (9.89) implies the following eigenfunctions and eigenvalues for (9.92):

$$\nu_m=m^2,\quad \Theta_m(\theta)=A_m\cos m\theta+B_m\sin m\theta \qquad m=0,1,2,\dots, \tag{9.94}$$

Substituting the eigenvalues ν_m into (9.93), performing the change of variables $t = \cos\phi$, using $\sin\phi \, d/d\phi = -\sin^2\phi \, d/dt$, and setting $P(t) = \Phi(\phi(t))$, we obtain for $P(t)$ a sequence of eigenvalue problems

$$(1 - t^2)\frac{d}{dt}\left[(1 - t^2)\frac{dP}{dt}\right] + \left[(1 - t^2)\mu - m^2\right]P = 0 \quad -1 < t < 1,$$

$$m = 0, 1, 2, \ldots . \tag{9.95}$$

Equation (9.95) is a linear second-order ODE. It is a regular equation except at the end points $t = \pm 1$ which are regular singular points. We recall that we are looking for solutions that are bounded everywhere, *including* the singular points (the poles of the original unit sphere). It it convenient to consider first the case $m = 0$, and to proceed later to the cases where $m > 0$.

When $m = 0$ we obtain

$$\frac{d}{dt}\left[(1 - t^2)\frac{dP}{dt}\right] + \mu P = 0 \quad -1 < t < 1. \tag{9.96}$$

This equation is called the *Legendre equation* after the French mathematician Adrien-Marie Legendre (1752–1833). The problem of finding bounded solutions to this equation is called the Legendre eigenvalue problem. It is a singular Sturm–Liouville problem. Since the eigenvalue problem for the Laplace equation in a ball is important in many applications such as electromagnetism, quantum mechanics, gravitation, hydrodynamics, etc., the Legendre equation has been studied extensively. The following property of it is very useful for our purposes.

Proposition 9.21 *A solution of the Legendre equation is bounded at the end points $t = \pm 1$ if and only if the eigenvalues are $\mu_n = n(n + 1)$ for $n = 0, 1, \ldots$. Moreover, in this case the solution for μ_n is a polynomial of degree n that is called the Legendre polynomial. We denote this polynomial by $P_n(t)$.*

Proof We outline the main steps in the proof. The Legendre equation is solved by the series expansion (Frobenius–Fuchs) method. For example, we expand around the regular singular point $t = 1$. The solution takes the form

$$P = (t - 1)^\gamma \sum_{k=0}^{\infty} a_k(t - 1)^k.$$

Substituting the series into the equation, we find that the indicial equation for γ is $\gamma^2 = 0$. Thus $\gamma = 0$ is a double root. Hence there exists one solution that is regular at $t = 1$, while the other solution has a logarithmic singularity there. We observe that if $P(t)$ is a solution, then $P(-t)$ is a solution too. This implies that also at $t = -1$ there is one regular solution and one singular solution. Therefore we need to check whether the regular solution at $t = 1$ connects to the regular solution at $t = -1$.

For this purpose we compute the recursive formula for the coefficients a_k:

$$\frac{a_{k+1}}{a_k} = \frac{\mu - k(k+1)}{2(k+1)^2}.$$

Therefore, if μ is not of the form $k(k+1)$, the ratio between two consecutive terms in the series at $t = -1$ satisfies $[k(k+1) - \mu]/(k+1)^2 = O(1 - 1/k)$, and thus the series diverges there, i.e. the regular solution at $t = 1$ is in fact singular at $t = -1$.

It follows that the only way to obtain a solution that is regular at the two end points is to impose that the series is not infinite but rather terminates at some point and the solution is then a polynomial. This requires $\mu = k(k+1)$ for some positive integer k (an alternative proof of this result will be outlined in Exercise 9.13). Furthermore, the recurrence formula we wrote can be integrated to provide an explicit formula for the Legendre polynomial (the regular solution is normalized by $P_n(1) = 1$):

$$P_n(t) = \sum_{k=0}^{n} \frac{(n+k)!}{(n-k)!(k!)^2 2^k}(t-1)^k. \tag{9.97}$$

For example, the first few polynomials are

$$P_0(t) = 1, \tag{9.98}$$
$$P_1(t) = t, \tag{9.99}$$
$$P_2(t) = \frac{3}{2}t^2 - \frac{1}{2}. \tag{9.100}$$

□

Let us return now to the general case in which $m > 0$. Equation (9.95) is called the *associated Legendre equation* of order m. The structure of the eigenvalues and eigenfunctions of the associated Legendre equation is provided by the following proposition.

Proposition 9.22 *Fix* $m \in \mathbb{N}$. *The associated Legendre equation (9.95) has solutions that are bounded everywhere if and only if the eigenvalues* μ *are of the form* $\mu_n = n(n+1)$ *for* $n = 0, 1, \dots$. *Moreover, the eigenfunction* $P_n^m(t)$ *associated with such an eigenvalue* μ_n *can be expressed as*

$$P_n^m(t) = (1 - t^2)^{m/2}\frac{d^m P_n}{dt^m}. \tag{9.101}$$

Proof We first verify that indeed (9.101) satisfies (9.95). For this purpose we let P be some solution of the Legendre equation (9.96). Differentiating the equation *m* times we obtain

$$(1 - t^2)\frac{d^{m+2}P}{dt^{m+2}} - 2(m+1)t\frac{d^{m+1}P}{dt^{m+1}} + [\mu - m(m+1)]\frac{d^m P}{dt^m} = 0.$$

Substituting

$$L(t) = (1 - t^2)^{m/2}\frac{d^m P(t)}{dt^m},\tag{9.102}$$

we observe that L satisfies

$$(1 - t^2)\frac{d}{dt}\left[(1 - t^2)\frac{dL}{dt}\right] + [(1 - t^2)\mu - m^2]L = 0.\tag{9.103}$$

It follows that each solution of the associated Legendre equation is of the form (9.102). Clearly, if we now select $\mu = n(n+1)$ for a positive integer n, we shall obtain a solution to (9.95) that is bounded in both end points (since in this case P is a polynomial). We have thus shown that each function of the form (9.101) is indeed a valid solution of our problem. It remains to show that there are no further solutions.

Since each solution of the associated Legendre equation is of the form (9.102), we have to show that if $\mu \neq n(n+1)$, then L is singular at least at one end point. This can be proved by the same method as in the proof of Proposition 9.21; namely, if $\mu \neq n(n+1)$, then the solution $L(t)$ that is regular at $t = 1$ is singular at $t = -1$, and the solution that is regular at $t = -1$ is singular at $t = 1$. □

Since P_n is a polynomial of degree n, $P_n^m \equiv 0$ for $m > n$. We have thus established that the eigenvalues and eigenfunctions for the problem (9.88)–(9.89) are given by

$$\mu_n = n(n+1) \qquad\qquad n = 0, 1, \dots, \tag{9.104}$$

$$Y_{n,m}(\phi, \theta) = \{\cos m\theta\, P_n^m(\cos\phi), \sin m\theta\, P_n^m(\cos\phi)\} \quad n = 0, 1, \dots,$$

$$m = 0, 1, \dots, n. \tag{9.105}$$

In particular, μ_n is an eigenvalue with a multiplicity $n + 1$. The functions $Y_{n,m}$ are called *spherical harmonics* of order n. They can also be written in a complex form:

$$Y_{n,m}(\phi, \theta) = e^{im\theta}\, P_n^m(\cos\phi) \quad n = 0, 1, \dots, \quad m = -n, -n+1, \dots, n-1, n.$$

We turn our attention to the radial problem (9.90)–(9.91). Fix a nonnegative integer n. Let us substitute $\mu_n = n(n+1)$, and $R(r) = \rho(r)/\sqrt{r}$. Equation (9.90) now becomes a Bessel equation of order $n + \frac{1}{2}$:

$$\rho_n''(r) + \frac{1}{r}\rho_n'(r) + \left[\lambda - \frac{(n+\frac{1}{2})^2}{r^2}\right]\rho_n(r) = 0.\tag{9.106}$$

Under the change of variables $s = \sqrt{\lambda}r$ (see the previous section) we obtain from the boundary condition (9.91) (similarly to (9.79) and (9.80)) that the radial solution is of the form

$$R_{n,l}(r) = \frac{J_{n+\frac{1}{2}}(\sqrt{\lambda_{n,l}}r)}{\sqrt{r}}, \qquad \lambda_{n,l} = \left(\frac{\alpha_{n,l}}{a}\right)^2 \qquad l = 1, 2, \ldots ,$$

where $\alpha_{n,l}$ denote the zeros of the Bessel function $J_{n+\frac{1}{2}}$. We have thus shown that the eigenfunctions of the Dirichlet problem in the ball are

$$U_{n,m,l}(r, \phi, \theta) = \frac{1}{\sqrt{r}} J_{n+\frac{1}{2}}\left(\frac{\alpha_{n,l}r}{a}\right) Y_{n,m}(\phi, \theta) \quad n = 0, 1, \ldots ,$$

$$m = 0, 1, \ldots , n, \quad l = 1, 2, \ldots , \tag{9.107}$$

while the eigenvalues are

$$\lambda_{n,l} = \left(\frac{\alpha_{n,l}}{a}\right)^2 \quad n = 0, 1, \ldots , \quad l = 1, 2, \ldots . \tag{9.108}$$

Two important conclusions stemming from the calculations performed in this section are worthwhile mentioning.

Corollary 9.23 *The eigenvalue problem in a rectangle may or may not be degenerate. The eigenvalue problem in a disk is always degenerate, and the multiplicity is exactly 2. The degeneracy of the eigenvalue problem in the ball is even greater. For $n \geq 0$ and $l \geq 1$, the eigenvalue $\lambda_{n,l}$ has a multiplicity of $2n + 1$, since each such eigenvalue is associated with $2n + 1$ spherical harmonics.*

Corollary 9.24 *Let $Q_n(x, y, z)$ be a homogeneous harmonic polynomial of degree n in $\mathbb{R}^3$, i.e.*

$$Q_n(x, y, z) = \sum_{p+q+s=n} \alpha_{p,q,s} x^p y^q z^s.$$

Expressing Q_n in the spherical coordinate system we obtain $Q_n = r^n F(\phi, \theta)$. If we substitute Q_n into (9.86), we find that F is a spherical harmonic of degree n. Conversely, every function of the form $r^n Y_{n,m}(\phi, \theta)$ is a homogeneous harmonic polynomial (the proof is given as an exercise; see Exercise 9.15). It follows that the dimension of the space of all homogeneous harmonic polynomials of degree n in $\mathbb{R}^3$ is $2n + 1$.

One of the important applications of eigenfunctions is as a means for expanding functions into generalized Fourier series. For instance, the classical Fourier series can be derived as an expansion in terms of the eigenfunctions of the Laplacian on the unit circle. Similarly we use spherical harmonics, i.e. the eigenfunctions of the Laplacian on the unit sphere S^2 ((9.88) and the conditions that followed it), to expand functions f depending on the spherical variables ϕ, θ. We thus consider the space $C(S^2)$ of continuous functions over the unit sphere. For each pair of functions f and g in this space we define an inner product:

$$\langle f, g \rangle = \int_0^{2\pi} \int_0^{\pi} f(\phi, \theta) g(\phi, \theta) \sin \phi \, d\phi d\theta.$$

We write the following expansion for any function $f \in C(S^2)$

$$f(\phi, \theta) = \sum_{n=0}^{\infty} \left\{ \frac{1}{2} A_{n,0} P_n(\cos \phi) + \sum_{m=1}^{n} \left[A_{n,m} \cos m\theta P_n^m(\cos \phi) \right. \right.$$

$$\left. \left. + B_{n,m} \sin m\theta P_n^m(\cos \phi) \right] \right\}. \tag{9.109}$$

To find the coefficients $A_{n,m}$ and $B_{n,m}$ we need to compute the inner product between each pair of spherical harmonics. From the construction of the spherical harmonics, and from the general properties of the eigenfunctions of the Laplacian it follows that different spherical harmonics are orthogonal to each other. It remains to find the norms of the spherical harmonics.

Proposition 9.25 *The associated Legendre functions satisfy the identity*

$$\int_0^{\pi} \left(P_n^m(\cos \phi) \right)^2 \sin \phi \, d\phi = \frac{2}{2n+1} \frac{(n+m)!}{(n-m)!}. \tag{9.110}$$

The proof is relegated to Exercise 9.17.

We thus obtain the following formulas for the coefficients of the expansion (9.109):

$$A_{n,m} = \frac{(2n+1)(n-m)!}{2\pi(n+m)!} \int_0^{2\pi} \int_0^{\pi} f(\phi, \theta) \cos m\theta P_n^m(\cos \phi) \sin \phi \, d\phi d\theta, \tag{9.111}$$

$$B_{n,m} = \frac{(2n+1)(n-m)!}{2\pi(n+m)!} \int_0^{2\pi} \int_0^{\pi} f(\phi, \theta) \sin m\theta P_n^m(\cos \phi) \sin \phi \, d\phi d\theta. \tag{9.112}$$

9.6 Separation of variables for the heat equation

Let Ω be a bounded domain in $\mathbb{R}^n$, and let $u(\vec{x}, t)$ be the solution to the heat problem

$$u_t - \Delta u = F(\vec{x}, t) \qquad \vec{x} \in \Omega, \quad t > 0, \tag{9.113}$$

$$u(\vec{x}, t) = 0 \qquad \vec{x} \in \partial\Omega, \tag{9.114}$$

$$u(\vec{x}, 0) = f(\vec{x}) \qquad \vec{x} \in \Omega. \tag{9.115}$$

Denote by $\{\lambda_m, v_m(\vec{x})\}_{m=1}^{\infty}$ the spectrum of the Laplace equation (Dirichlet problem) in Ω. We solve the problem (9.113)–(9.115) by expanding u, F, and f into a formal series of eigenfunctions (see (9.51)–(9.52)):

$$u = \sum_{m=1}^{\infty} T_m(t) v_m(\vec{x}), \quad F = \sum_{m=1}^{\infty} F_m(t) v_m(\vec{x}), \quad f = \sum_{m=1}^{\infty} f_m v_m(\vec{x}). \quad (9.116)$$

Substituting the expansion (9.116) into (9.113)–(9.115) we obtain a system of ODEs for $\{T_n(t)\}$:

$$T'_m(t) + \lambda_m T_m(t) = F_m(t), \quad T(0) = f_m \quad m = 1, 2, \ldots. \quad (9.117)$$

Example 9.26 Solve the following heat problem for $u(x, y, t)$:

$$u_t = \Delta u \quad 0 < x, y < \pi, \ t > 0, \quad (9.118)$$

$$u(0, y, t) = u(\pi, y, t) = u(x, 0, t) = u(x, \pi, t) = 0 \quad 0 \le x, y \le \pi, t \ge 0, \quad (9.119)$$

$$u(x, y, 0) = 1 \quad 0 \le x, y \le \pi. \quad (9.120)$$

The eigenvalues and eigenfunctions of the Laplacian in this rectangle are given by $\{m^2 + n^2, \sin mx \sin ny\}$. Therefore, the solution of (9.118), subject to the boundary condition (9.119), is of the form:

$$u(x, y, t) = \sum_{n,m=1}^{\infty} A_{n,m} \sin mx \ \sin ny \ e^{-(m^2+n^2)t}.$$

Substituting the initial conditions and computing the generalized Fourier coefficients $A_{n,m}$, we obtain

$$A_{n,m} = \frac{8}{\pi^2} \begin{cases} \dfrac{1}{nm} & m = 2k + 1, \ n = 2l + 1, \\ 0 & \text{otherwise.} \end{cases}$$

Therefore the solution can be written as

$$u(x, y, t) = \frac{8}{\pi^2} \sum_{k,l=0}^{\infty} \frac{1}{(2k+1)(2l+1)} \sin[(2k+1)x] \sin[(2l+1)y] e^{-[(2k+1)^2+(2l+1)^2]t}.$$

9.7 Separation of variables for the wave equation

The basic structure of the solution of the wave equation in a bounded domain Ω in $\mathbb{R}^n$ is similar to the corresponding solution of the heat equation that we presented in the preceding section. Let $u(\vec{x}, t)$ be the solution of the wave problem

$$u_{tt} - c^2 \Delta u = F(\vec{x}, t) \qquad \vec{x} \in \Omega, \ t > 0, \qquad (9.121)$$

$$u(\vec{x}, t) = 0 \qquad \vec{x} \in \partial\Omega, \ t > 0, \qquad (9.122)$$

$$u(\vec{x}, 0) = f(\vec{x}), \quad u_t(\vec{x}, 0) = g(\vec{x}) \qquad \vec{x} \in \Omega. \qquad (9.123)$$

Denote again the spectrum of the Dirichlet problem for the Laplace equation in Ω by $\{\lambda_m, v_m(\vec{x})\}_{m=1}^{\infty}$. We expand the solution, the initial condition, and the forcing term F into a generalized Fourier series in $\{v_m\}$, just like in (9.116). Similarly to (9.117) we obtain

$$T_m''(t) + c^2 \lambda_m T_m(t) = F_m(t), \quad T(0) = f_m, \ T'(0) = g_m \quad m = 1, 2, \ldots.$$
$$(9.124)$$

Example 9.27 Vibration of a circular membrane Denote by $u(r, \theta, t)$ the amplitude of a membrane with a circular cross section. Then u satisfies the following problem:

$$\frac{\partial^2 u}{\partial t^2} - c^2 \left(\frac{\partial^2 u}{\partial r^2} + \frac{1}{r} \frac{\partial u}{\partial r} + \frac{1}{r^2} \frac{\partial^2 u}{\partial \theta^2} \right) = F(r, \theta, t) \qquad 0 < r < a, \ 0 \le \theta \le 2\pi, \ t > 0,$$
$$(9.125)$$

$$u(r, \theta, 0) = f(r, \theta), \quad \frac{\partial u}{\partial t}(r, \theta, 0) = g(r, \theta) \qquad 0 < r < a, \ 0 \le \theta \le 2\pi, \quad (9.126)$$

$$u(a, \theta, t) = 0 \qquad 0 \le \theta \le 2\pi, \ t \ge 0. \qquad (9.127)$$

The system (9.125)–(9.127) models, for example, the vibrations of a drum or of a trampoline, where the forcing term F is determined by the beating of the drummer or by the forces exerted by the people jumping on the trampoline.

To solve (9.125)–(9.127) we expand u, F, f, and g into the eigenfunctions of the Laplace equation in a disk. To fix ideas, we shall consider a specific physical problem in which the trampoline starts from a horizontal rest position (i.e. $f = g = 0$), and the people jumping on it do so rhythmically with a constant frequency, namely, $F(r, \theta, t) = F_0(r, \theta) \sin \omega t$. We use (9.79) and (9.80) to expand

$$u(r, \theta, t) = \sum_{n=0}^{\infty} \sum_{m=1}^{\infty} J_n\left(\frac{\alpha_{n,m}}{a} r\right) [A_{n,m}(t) \cos n\theta + B_{n,m}(t) \sin n\theta], \quad (9.128)$$

$$F(r, \theta, t) = \sin \omega t \sum_{n=0}^{\infty} \sum_{m=1}^{\infty} J_n\left(\frac{\alpha_{n,m}}{a} r\right) (C_{n,m} \cos n\theta + D_{n,m} \sin n\theta). \quad (9.129)$$

Substituting the expansion (9.129) into (9.125) provides a set of ODEs for the coefficients $\{A_{n,m}, B_{n,m}\}$:

$$A''_{n,m}(t) + c^2\lambda_{n,m}A_{n,m}(t) = C_{n,m}\sin\omega t, \quad A_{n,m}(0) = A'_{n,m}(0) = 0, \quad (9.130)$$
$$B''_{n,m}(t) + c^2\lambda_{n,m}B_{n,m}(t) = D_{n,m}\sin\omega t, \quad B_{n,m}(0) = B'_{n,m}(0) = 0. \quad (9.131)$$

We obtain

$$u = \sum_{n=0,m=1}^{\infty} (c^2\lambda_{n,m} - \omega^2)^{-1} J_n\left(\frac{\alpha_{n,m}}{a}r\right)(C_{n,m}\cos n\theta + D_{n,m}\sin n\theta)$$
$$\times \left(\sin\omega t - \frac{\omega\sin c\sqrt{\lambda_{n,m}}\,t}{c\sqrt{\lambda_{n,m}}}\right). \quad (9.132)$$

Clearly the solution (9.132) is valid only if we are careful to stay away from the resonance condition

$$\omega^2 - c^2\lambda_{n,m} = 0. \quad (9.133)$$

Notice that, in general, obtaining an equality between two real numbers is unlikely. However even if condition (9.133) holds only approximately for some values of n, m, then one of the terms in the series (9.132) would have a very small denominator. When the system is in a state of resonance, or near a resonance, the trampoline's amplitude becomes very large, and it might collapse. We comment, though, that when the amplitude is large the linear model (9.125)–(9.127) is no larger valid.

9.8 Separation of variables for the Laplace equation

We present in this section two additional examples for solving the Laplace equation in multi-dimensional domains.

Example 9.28 Laplace equation in a cylinder Let u be a harmonic function in a cylinder with radius a and height h that is a solution of the Dirichlet problem (we employ a cylindrical coordinate system (r, θ, z)):

$$\frac{\partial^2 u}{\partial r^2} + \frac{1}{r}\frac{\partial u}{\partial r} + \frac{1}{r^2}\frac{\partial^2 u}{\partial\theta^2} + \frac{\partial^2 u}{\partial z^2} = 0 \quad 0 < r < a, \; 0 < \theta < 2\pi, \; 0 < z < h,$$
$$(9.134)$$
$$u(r, \theta, 0) = u(r, \theta, h) = 0 \quad 0 \le \theta \le 2\pi, \; 0 < r < a, \quad (9.135)$$
$$u(a, \theta, z) = f(\theta, z) \quad 0 \le \theta \le 2\pi, \; 0 < z < h. \quad (9.136)$$

The cylinder Ω is the product of a disk and an interval:

$$\Omega = \{0 < r < a, \; 0 \le \theta \le 2\pi\} \times \{0 < z < h\}.$$

We construct accordingly separated solutions of the form $u = R(r)\Psi(\theta)Z(z)$. We obtain two eigenvalue problems for Z and for Ψ:

$$Z''(z) + \mu Z(z) = 0, \qquad Z(0) = Z(h) = 0, \tag{9.137}$$

$$\Psi''(\theta) + \nu\Psi = 0, \qquad \Psi(0) = \Psi(2\pi), \ \Psi'(0) = \Psi'(2\pi). \tag{9.138}$$

These problems are by now well known to us. The solutions are given by the sequences

$$Z_n(z) = \sin\frac{n\pi z}{h}, \quad \mu_n = \left(\frac{n\pi}{h}\right)^2 \quad n = 1, 2, \dots, \tag{9.139}$$

$$\Psi_m(\theta) = a_m \cos m\theta + b_m \sin m\theta, \quad \nu_m = m^2 \quad m = 0, 1, 2, \dots. \tag{9.140}$$

The equation for the doubly indexed radial component $R_{n,m}(r)$ is

$$R_{n,m}'' + \frac{1}{r}R_{n,m}' - \left(\frac{m^2}{r^2} + n^2\right) R_{n,m} = 0. \tag{9.141}$$

Under the transformation $r \to inr$ (9.141) becomes a Bessel equation. The Bessel function J_m with a complex argument is often denoted by I_m, i.e. $R_{n,m} = A_{n,m}I_m(nr)$. In order to satisfy the Dirichlet condition on the cylinder's envelope we write as usual a formal eigenfunction expansion. The structure of the series will be studied in Exercise 9.18.

Example 9.29 Laplace equation in a ball The Dirichlet problem in a ball $B_a = \{(r, \phi, \theta) | 0 < r < a, \ 0 \le \phi \le \pi, \ 0 \le \theta \le 2\pi\}$ is written as

$$\frac{1}{r^2}\frac{\partial}{\partial r}\left(r^2\frac{\partial u}{\partial r}\right) + \frac{1}{r^2}\left[\frac{1}{\sin\phi}\frac{\partial}{\partial\phi}\left(\sin\phi\frac{\partial u}{\partial\phi}\right) + \frac{1}{\sin^2\phi}\frac{\partial^2 u}{\partial\theta^2}\right] = 0 \quad (r, \phi, \theta) \in B_a,$$
$$\tag{9.142}$$

$$u(a, \phi, \theta) = f(\phi, \theta) \quad (\phi, \theta) \in \partial B_a. \tag{9.143}$$

It can easily be checked that the separated solutions for the problem are of the form

$$u(r, \phi, \theta) = r^n Y_{n,m}(\phi, \theta) \qquad n = 0, 1, \dots, \quad m = 0, 1, \dots,$$

where $Y_{n,m}$ are spherical harmonics. Therefore, Corollary 9.24 implies that the solution of (9.142)–(9.143) is a (maybe infinite) linear combination of homogeneous harmonic polynomials. To facilitate the process of eliminating the coefficients of this linear combination, we write it in the form

$$u = \sum_{n=0}^{\infty} \left(\frac{r}{a}\right)^n \left[\frac{A_{n,0}}{2}P_n(\cos\phi) + \sum_{m=1}^{n} P_n^m(\cos\phi)(A_{n,m}\cos m\theta + B_{n,m}\sin m\theta)\right].$$
$$\tag{9.144}$$

The coefficients $\{A_{n,m}, B_{n,m}\}$ are given by the Fourier formula (9.111)–(9.112).

We have pointed out on several occasions the analogy between solving the Dirichlet problems in the ball and in the disk. We saw in Chapter 7 that the generalized Fourier series in a disk can be summed up in the Poisson integral representation. A similar representation can also be obtained in the case of the ball. Instead of deriving it as the infinite sum of harmonic polynomials, we shall construct it later in this chapter using the theory of Green's functions.

9.9 Schrödinger equation for the hydrogen atom

In the second half of the nineteenth century it was already realized by scientists that the spectrum of the hydrogen atom (and other elements) consists of many discrete lines (values). The Swiss scientist Johann Jacob Balmer (1825–1898) discovered in 1885 that many of the spectral lines in the visible part of the spectrum obey the formula

$$\nu_k = R(2^{-2} - k^{-2}), \tag{9.145}$$

where ν_k is the frequency, $k = 3, 4, \ldots$, and R is a constant called the Rydberg constant. Balmer then made the bold guess that (9.145) is, in fact, a special case of a more general rule of the form

$$\nu_{k_1,k_2} = R(k_1^{-2} - k_2^{-2}), \tag{9.146}$$

where k_1 and k_2 are certain integers. Indeed such spectral lines were discovered in the invisible part of the spectrum! One of the earliest confirmations of the young quantum mechanics developed by the German physicists Werner Heisenberg (1901–1976) and Max Born (1882–1970) and by Schrödinger was the derivation of the Balmer's spectral formula from theoretical principles (we point out, though, that the formula was also derived earlier by Bohr using his 'old' formulation of quantum mechanics).

The hydrogen atom consists of a nucleus and an electron. Therefore, we can split its motion into a part related to the motion of the center of mass, and a part related to the relative motion between the electron and the nucleus. We shall ignore the motion of the center of mass of the atom, and concentrate on the energy levels resulting from the electric attraction between the nucleus and the electron. The Schrödinger equation for the motion of an electron in an electric field generated by a nucleus with exactly one proton takes the form [17]

$$-\left(\frac{\hbar}{2m}\Delta + \frac{e^2}{r}\right)u = Eu. \tag{9.147}$$

Here $\hbar$ is the Planck constant divided by 2π, m is the reduced mass of the atom (i.e. $m = m_1 m_2/(m_1 + m_2)$, where m_1 is the electron's mass, and m_2 is the nucleus's

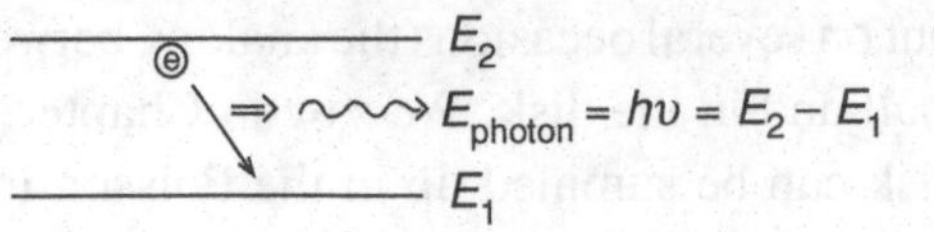

Figure 9.6 A sketch for the emission spectral lines of an atom. As an electron moves from the energy level E_2 to the energy level E_1, it emits a photon. The photon's energy is determined by the difference $E_2 - E_1$. Since the energy levels are discrete (as will be explained in the text), only discrete values are observed for the photon's energy. These photon's energy values determine the observed emission spectral lines.

mass; since $m_2 \gg m_1$, in practice $m \sim m_1$), r is the distance between the electron and the center of the nucleus, e is the electron's charge, E is the energy of the system, and $u(x, y, z)$ is the wave function. The standard physical interpretation of the wave function is that if we normalize u such that the integral of its square over the entire space is 1, then the square of u at any given point is the probability density for the electron to be at that point.

Equation (9.147) has the general form $L[u] = Eu$, i.e. it is an eigenvalue problem. Unlike the eigenvalue problems we have encountered so far, it is not formulated in a bounded domain with boundary conditions. Rather it is written for the entire space $\mathbb{R}^3$ under the condition that $|u|^2$ is integrable. It turns out that there is a discrete infinite set of negative eigenvalues. These are the bound states of the hydrogen. We remark that there is also a continuous part to the spectrum (see Chapter 6), but we shall not treat it here. The discrete eigenvalues are all negative, and the continuous part has no negative component. Note also that we keep the equation in a dimensional form to obtain a comparison between the theoretical results and the experimental observations mentioned above.

We exploit the radial symmetry of the Schrödinger equation (9.147) to seek solutions in the form $u(r, \theta, \phi) = R(r)Y_{l,m}(\theta, \phi)$, where $Y_{l,m}$ are the spherical harmonics. Substituting this form into (9.147), and using our results on the Laplacian's spectrum on the unit sphere, we obtain for R:

$$(r^2 R')' + \frac{2mr^2}{\hbar^2} \left(E + \frac{e^2}{r} \right) R - n(n+1)R = 0, \quad n = 0, 1, \ldots . \qquad (9.148)$$

Just as in our discussion of the Laplacian's spectrum in a disk, it is convenient to scale the eigenvalue E out of the problem. We thus introduce the new variable $\rho = \alpha r$, where $\alpha = 8m|E|/\hbar^2$. Equation (9.148) is now written as

$$\frac{1}{\rho^2}(\rho^2 R')' + \left[\frac{\lambda}{\rho} - \frac{1}{4} - \frac{n(n+1)}{\rho^2} \right] R = 0, \qquad (9.149)$$

where

$$\lambda = \frac{e^2 \sqrt{m}}{\hbar \sqrt{2|E|}}, \qquad (9.150)$$

and we use $R_n = R_n(\rho) = R(\rho)$, in spite of the mild abuse of notation.

To be able to normalize u, we look only for solutions that decay to zero as $\rho \to \infty$. Since the leading order terms (for large ρ) in (9.149) are $R'' - \frac{1}{4}R = 0$, we write R as $R = F_n(\rho)e^{-\frac{1}{2}\rho}$. We obtain that F_n satisfies

$$F_n'' + \left(\frac{2}{\rho} - 1\right) F_n' + \left[\frac{\lambda - 1}{\rho} - \frac{n(n+1)}{\rho^2}\right] F_n = 0. \qquad (9.151)$$

Applying the Frobenius–Fuchs theory, the solutions of (9.151) are of the form $F_n(\rho) = \rho^{\mu_n} \sum_{k=0}^{\infty} a_k^n \rho^k$. The indicial equation for μ_n is

$$\mu_n(\mu_n + 1) - n(n+1) = 0.$$

The only solution leading to a wave function that is bounded in the origin is $\mu_n = n$. It is convenient to write at this point $F_n(\rho) = \rho^n L_n(\rho)$. Under this transformation, the function L_n satisfies the *Laguerre equation*

$$\rho L_n'' + [2(n+1) - \rho] L_n' + (\lambda - n - 1)L_n = 0. \qquad (9.152)$$

From the analysis above, we know that (9.152) has an analytic solution of the form $L_n(\rho) = \sum_{k=0}^{\infty} a_k^n \rho^k$. Indeed we obtain by substitution a recursive equation of the form

$$a_{k+1} = \frac{k + n + 1 - \lambda}{(k+1)(k+2n+2)} a_k. \qquad (9.153)$$

A delicate analysis of the ODE (9.152) shows that the solutions for which the power series does not terminate after finitely many terms grow at ∞ at the rate of e^ρ. It follows that a necessary condition for the normalization of the wave function u is that the power series will be, in fact, a polynomial.

A necessary and sufficient condition for the power series to terminate after finitely many terms is $\lambda = k + n + 1$. We can thus derive from (9.150) the discrete energy levels:

$$E_j = -\frac{me^4}{2\hbar^2 j^2} \qquad j = 1, 2, \dots . \qquad (9.154)$$

To convert the formula for the energy levels into the observed spectral lines, we recall Planck's quantization rule stating that a wave with frequency ν carries an energy quanta of $h\nu$. Moreover, a hydrogen atom emits radiation when an electron 'jumps' from one bound state to another. In the transition process the electron emits radiation (photons). The frequency of the radiated energy is the difference between the energies in the states before and after the transition (see Figure 9.6). Therefore one expects to observe spectral lines with frequencies of the form (9.146) with

$$R = \frac{2\pi^2 m e^4}{h^3}. \tag{9.155}$$

Remark 9.30 (1) The $2n+1$ degeneracy of the spherical harmonics implies that each energy state of the hydrogen atom is associated with actual $2n+1$ distinct eigenfunctions (or eigenstates). The different $2n+1$ states can be distinguished from each other by subjecting the atom to an external magnetic field. This is precisely the *Zeeman effect*.

(2) To appreciate the quantities involved in the calculation we carried out, we recall that $\hbar = 1.054 \times 10^{-27}$ erg, $e = 4.8 \times 10^{-10}$ esu, $m = 0.911 \times 10^{-27}$ g, and $R \approx 3.3 \times 10^{15}$ s^{-1}.

(3) It is interesting to note that Balmer was not a 'professional' physicist. He was a mathematics teacher in a girls' school in Basel. His interest was in finding simple geometrical principles in the sciences and humanities. For example, he wrote an article on architectural interpretation of Ezekiel's prophecy. His belief in nature's harmony led him to argue that the spectral lines of the elements satisfy beautiful arithmetic relations. It seems at first sight that the spectral lines of the hydrogen atom prove his belief that the world is based on harmonic arithmetic rules. It turns out, however, that (9.146) is only approximately correct. The Schrödinger equation we wrote neglects relativistic effects. When one considers the relativistic Schrödinger equation (or the Dirac equation), it is discovered that each energy level we found has an inner structure (called the fine structure). The actual formula for the spectral lines is then found to be more involved and not as elegant and integral as Balmer believed.

9.10 Musical instruments

Musical instruments differ in shape, size, and mode of operation. Yet, they all share the same basic principle – they emit sound waves. We shall analyze a number of instruments in light of the eigenvalue theory we have developed. In particular, we consider instruments based on strings, air compression, and membranes. Some less common instruments, such as the gong, are based on plates (see the next chapter) and will not be studied here. We emphasize that our models are crude, and they aim only to capture basic mechanisms. Much more extensive analysis can be found in [5] and numerous websites.

String instruments Let us recall our solution (5.44) for the wave equation in a finite interval with homogeneous boundary conditions. To understand the physical interpretation of this solution, we notice that the vibration of the string is a superposition of fundamental vibrations that are called *normal modes* (or standing waves, or just harmonics) with natural frequencies $\omega_n = c\pi n/L$. The name harmonics (and hence also the notions of harmonic functions and harmonic analysis) was coined by the ancient Greeks, who discovered that the basic frequencies of a string are all integral multiples of a single fundamental frequency $\omega_1 = c\pi/L$. They noted that this phenomenon is pleasing to the ear, or, rather, to the brain.

Thus, the harmonics produced by a string instrument, such as a violin or a guitar, are determined by two parameters: c and L. The speed of sound c depends, as was explained in Chapter 1, upon the material composing the string; it is a fixed property of the instrument. The string's length L is adjusted by the musician, and provides a dynamical control of the basic tones.

The explanation above for the performance of a string instrument is oversimplified. Let us look, for example, at the guitar. In reality the motion of the guitar's string is too weak actually to move a large enough mass of air to create a sound wave that could be detected by our ears (to demonstrate this fact, one can try to play an electric guitar without an amplifier). What the string's oscillations really do is activate a complex vibrating system that consists also of the top plate of the guitar's sound box and of the air inside the sound box. We therefore have a coupled system of vibrating bodies: the string (modeled by the one-dimensional wave equation of Chapter 5), the top plate (whose wave equation is based on the energy associated with plates; see Chapter 10), and a mass of air (whose wave equation was derived in Chapter 1). The final outcome of this complex system is an air pressure wave of considerable amplitude that is radiated from the sound box through its sound hole. Another interesting point is that a guitar string actually generates a combination of several basic modes. The combination depends not only on the string's length and material, but also on the playing technique. When the guitarist strikes the string, we should consider the wave equation with zero initial amplitude and some nonzero initial velocity; on the other hand, when the guitarist plucks the string, the initial velocity is zero, but the initial amplitude is different from zero. Different initial conditions give rise, of course, to different linear combinations of the basic modes.

Wind instruments In a wind instrument the sound wave is generated directly. To create a musical sound, the player blows air into the instrument. Therefore we use the acoustics equations (1.22). We consider first long and narrow instruments with a uniform cross section, such as the flute or the clarinet. Because of their geometry, we can assume for simplicity that the pressure depends only on the longitudinal direction that we denote by z. We thus write the wave equation for the pressure:

$$p_{tt} - c^2 p_{zz} = 0 \quad 0 < z < L, \quad t > 0, \tag{9.156}$$

where L is the length of the instrument. The pressure p in (9.156) is actually the deviation of the pressure in the instrument from the bulk pressure in air. Therefore, in the flute, which is open to the outside air at both ends, we use the Dirichlet boundary condition $p = 0$ at $z = 0, L$. In the clarinet, for example, one end $z = 0$ is open, and we also use the homogeneous Dirichlet condition there, but the other end $z = L$ is closed. Hence the air velocity is constant (zero) there. Therefore the second equation of the pair (1.22) implies that we should use there a homogeneous Neumann condition, i.e. $p_z(L) = 0$. From our discussion of the wave equation in Chapter 5 we conclude (at least in our crude model) that the basic frequencies of the flute are determined by the eigenvalues of the Sturm–Liouville problem (5.9)–(5.10), while the basic frequencies of the clarinet are determined by the eigenvalues of the problem (6.44)–(6.45). It follows that the fundamental frequencies of the flute include all multiples of $\omega = c\pi/L$, while in the clarinet we expect to obtain only the odd multiples of ω. The flute and the clarinet have basically a pipe geometry. The organ consists of many pipes. Typically some of them will be open at both ends, and some only open at one end. Finally, the basic harmonics generated by the flute depend on the location of the open holes. Basically, the holes shorten the length of the flute, thus increasing the frequency.

We proceed to analyze elongated wind instruments with a varying cross section such as the horn or the oboe. We still use a one-dimensional wave equation like (9.156), but we need to take into account also the variations in the cross section $S(z)$. It can be shown that the appropriate model for waves in a narrow elongated structure with a varying cross section is given by

$$S(z)p_{tt} - c^2 (S(z)p_z)_z = 0 \qquad 0 < z < L, \ t > 0. \tag{9.157}$$

Equation (9.157) is known in the scientific music literature as *Webster's horn equation*. Separating variables, as in Chapter 5, we obtain that the basic frequencies are $c\lambda_n$, where λ_n are the eigenvalues of the Sturm–Liouville problem

$$\left(S(z)v'(z)\right)' + \lambda S(z)v(z) = 0 \qquad 0 < z < L. \tag{9.158}$$

We have not specified the boundary conditions, since they depend on the specific instrument.

So far we have considered an arbitrary cross section $S(z)$. In general there is no closed form solution to problem (9.158). Fortunately, the profile of many instruments can be well approximated by the formula

$$S(z) = b(z - z_0)^{-\gamma}, \tag{9.159}$$

where γ is a parameter, called *flare* in musical jargon. For example, the horn roughly corresponds to $\gamma = 2$, while the oboe corresponds to $\gamma = 7$. Obviously, the flute and the clarinet correspond to $\gamma = 0$. For this special family of cross sections, (9.158) becomes

$$z^2 v'' - \gamma z v' + \lambda z^2 v = 0 \quad L_1 < z < L_2, \tag{9.160}$$

where we have shifted the origin to simplify the equation and to fit the actual shape of the instrument. Applying the transformation $v(z) = z^{(1+\gamma)/2} u(z)$, one can check that u satisfies the Bessel equation

$$u''(z) + \frac{1}{z} u'(z) + \left\{ \lambda - \frac{[(\gamma+1)/2]^2}{z^2} \right\} u(z) = 0 \quad L_1 < z < L_2. \tag{9.161}$$

The fundamental system of solutions of (9.161) is the pair

$$\left(J_{(\gamma+1)/2}(\sqrt{\lambda}z), \, Y_{(\gamma+1)/2}(\sqrt{\lambda}z) \right),$$

where $J_{(\gamma+1)/2}$ is the Bessel function of the first kind of order $(\gamma + 1)/2$, and $Y_{(\gamma+1)/2}$ is the Bessel function of the second kind (which is the singular solution to the Bessel equation of order $(\gamma + 1)/2$). The eigenvalues λ_n can now be computed by the algebraic method of Proposition 6.18, once the boundary conditions at the end points have been specified. The freedom introduced by incorporating a profile function $S(z)$ can be used for design purposes to generate better acoustical instruments and devices.

Drums The drum is modeled as a circular membrane attached at its boundary to a fixed frame. We have already solved the equation for the vibrations of such membranes, and found that the normal modes are $J_n(\alpha_{n,m} r / a) \cos n\theta$ and $J_n(\alpha_{n,m} r / a) \sin n\theta$, where $\alpha_{n,m}$ is the mth root of the Bessel function J_n, and a is the drum's radius. The basic frequencies of the membrane are $\omega_{n,m} = c\alpha_{n,m}/a$. In a typical drum, though, the membrane vibrates over a bounded domain in space. Therefore the assumption of free vibrations is not accurate. This assumption is particularly problematic for the first mode $\omega_{0,0}$, since this mode (and only this mode!) corresponds to an eigenfunction with a constant sign. Thus, the integral under the membrane surface is not zero, implying that a considerable compression or expansion of the air inside the drum is required. Therefore this mode is often not heard.

9.11 Green's functions in higher dimensions

In this section we study (Dirichlet) Green's function for the Laplace equation

$$\begin{aligned} \Delta u &= f & D, \\ u &= g & \partial D, \end{aligned} \tag{9.162}$$

where D is a smooth bounded domain in $\mathbb{R}^N$, and $N \geq 3$. We show that most of the properties of Green's function for a planar domain are valid also for higher dimensions. As in the two-dimensional case, the fundamental solution of the Laplace equation plays an important role in the study of Green's function. The essential difference, which has many implications, is that in the two-dimensional case the fundamental solution changes sign and does not decay at infinity, while for $N \geq 3$ the fundamental solution is positive and decays to zero at infinity.

Just like in the two-dimensional case, the fundamental solution in higher dimensions is a radial symmetric harmonic function that is singular at the origin. Set

$$\vec{x} = (x_1, \ldots, x_N), \qquad r = |\vec{x}| = \left(\sum_{n=1}^{N} x_i^2 \right)^{1/2}.$$

The fundamental solution $\Gamma(\vec{x}; 0) = \Gamma(|\vec{x}|)$ with a pole at the origin satisfies the following Euler (equidimensional) equation:

$$u_{rr} + \frac{N-1}{r} u_r = 0 \qquad r > 0. \tag{9.163}$$

Therefore, for $N \geq 3$

$$\Gamma(\vec{x}, 0) = \frac{1}{N(N-2)\omega_N} |x|^{2-N}, \tag{9.164}$$

where ω_N is the volume of a unit ball N in $\mathbb{R}^N$. Hence, the fundamental solution with a pole at $\vec{y}$ is given by

$$\Gamma(\vec{x}; \vec{y}) = \Gamma(|\vec{x} - \vec{y}|).$$

The fact that Γ is indeed a fundamental solution of the Laplace equation will be proved later (see Corollary 9.31). One can verify that

$$\left| \frac{\partial \Gamma}{\partial x_i}(\vec{x}; \vec{y}) \right| \leq C_N |\vec{x} - \vec{y}|^{1-N} \qquad 1 \leq i \leq N, \tag{9.165}$$

$$\left| \frac{\partial^2 \Gamma}{\partial x_i \partial x_j}(\vec{x}; \vec{y}) \right| \leq C_N |\vec{x} - \vec{y}|^{-N} \qquad 1 \leq i, j \leq N. \tag{9.166}$$

Fixing $\vec{y}$, the function $\Gamma(\vec{x}; \vec{y})$ is harmonic as a function of $\vec{x}$ for all $\vec{x} \neq \vec{y}$. For $\varepsilon > 0$, we use the notation

$$B_\varepsilon := \{\vec{x} \mid |\vec{x} - \vec{y}| < \varepsilon\}, \qquad D_\varepsilon := D \setminus B_\varepsilon.$$

Let $u \in C^2(\bar{D})$. Use the second Green identity (7.19) on the domain D_ε where $v(\vec{x}) = \Gamma(\vec{x}; \vec{y})$ is harmonic to derive

$$\int_{D_\varepsilon} (\Gamma \Delta u - u \Delta \Gamma) d\vec{x} = \int_{\partial D_\varepsilon} (\Gamma \partial_n u - u \partial_n \Gamma) d\sigma.$$

Therefore,

$$\int_{D_\varepsilon} \Gamma \Delta u \, d\vec{x} = \int_{\partial D} (\Gamma \partial_n u - u \partial_n \Gamma) d\sigma + \int_{\partial B_\varepsilon} (\Gamma \partial_n u - u \partial_n \Gamma) d\sigma.$$

Letting $\varepsilon \to 0$, and recalling that the outward normal derivative on ∂B_ε is the (inward) radial derivative (see Figure 8.1), we find

$$\left| \int_{\partial B_\varepsilon} \Gamma \partial_n u \, d\sigma \right| \le C\varepsilon \sup |\nabla u| \to 0 \qquad \text{where } \varepsilon \to 0,$$

$$\int_{\partial B_\varepsilon} u \partial_n \Gamma \, ds = \frac{1}{N \omega_N \, \varepsilon^{N-1}} \int_{\partial B_\varepsilon} u \, d\sigma \to u(\vec{y}) \qquad \text{where } \varepsilon \to 0.$$

Thus,

$$u(\vec{y}) = \int_{\partial D} \left[\Gamma(\vec{x} - \vec{y}) \partial_n u - u \partial_n \Gamma(\vec{x} - \vec{y}) \right] d\sigma - \int_D \Gamma(\vec{x} - \vec{y}) \Delta u \, d\vec{x}. \tag{9.167}$$

Equation (9.167) is called *Green's representation formula*, and the function

$$\Gamma[f](\vec{y}) := - \int_D \Gamma(|\vec{x} - \vec{y}|) f(\vec{x}) \, d\vec{x}$$

is called the *Newtonian potential* of f.

As in the planar case, it follows that harmonic functions are smooth. Moreover, we have the following corollary.

Corollary 9.31 *If $u \in C_0^2(\mathbb{R}^N)$ (i.e. u has a compact support), then*

$$u(\vec{y}) = - \int_{\mathbb{R}^N} \Gamma(\vec{x}; \vec{y}) \Delta u(\vec{x}) \, d\vec{x}. \tag{9.168}$$

In other words, $-\Delta \Gamma(\vec{x}; \vec{y}) = \delta(\vec{x} - \vec{y})$, so Γ is a fundamental solution of the Laplace equation on $\mathbb{R}^N$.

Consider again the Dirichlet problem (9.162). Let $h(\vec{x}; \vec{y})$ be a solution (which depends on the parameter $\vec{y}$) of the following Dirichlet problem:

$$\Delta h(\vec{x}; \vec{y}) = 0 \quad \vec{x} \in D, \qquad h(\vec{x}; \vec{y}) = \Gamma(\vec{x} - \vec{y}) \quad \vec{x} \in \partial D. \tag{9.169}$$

We suppose that there exists a solution for (9.169), and use the following definition.

Definition 9.32 *(Dirichlet) Green's function for the Laplace equation on D is given* by

$$G(\vec{x}; \vec{y}) := \Gamma(\vec{x}; \vec{y}) - h(\vec{x}; \vec{y}) \qquad \vec{x}, \vec{y} \in D, \vec{x} \neq \vec{y}. \tag{9.170}$$

Hence, the Green function satisfies

$$\begin{aligned} \Delta G(\vec{x}; \vec{y}) &= -\delta(\vec{x} - \vec{y}) & \vec{x} \in D, \\ G(\vec{x}; \vec{y}) &= 0 & \vec{x} \in \partial D. \end{aligned} \tag{9.171}$$

As in the two-dimensional case, we can write

$$u(\vec{y}) = -\int_{\partial D} \partial_n G(\vec{x}; \vec{y}) u(\vec{x}) \, d\sigma - \int_D G(\vec{x}; \vec{y}) \Delta u(\vec{x}) \, d\vec{x}. \tag{9.172}$$

Substituting the given data into (9.172), we finally arrive at following integral representation formula for solutions of the Dirichlet problem for the Poisson equation.

Theorem 9.33 *Let $u \in C^2(\bar{D})$ be a solution of the Dirichlet problem*

$$\begin{aligned} \Delta u &= f & D, \\ u &= g & \partial D. \end{aligned} \tag{9.173}$$

Then

$$u(\vec{y}) = -\int_{\partial D} \partial_n G(\vec{x}; \vec{y}) g(\vec{x}) \, d\sigma - \int_D G(\vec{x}; \vec{y}) f(\vec{x}) \, d\vec{x}. \tag{9.174}$$

Theorem 9.33 enables us to solve the Dirichlet problem in a domain D provided that Green's function is known, and that it is a priori known that the solution is in $C^2(D) \cap C^1(\bar{D})$. This additional regularity is indeed ensured if f, g, and ∂D are sufficiently smooth.

There are two integral kernels in the representation formula (9.174):

(1) Green's function $G(\vec{x}; \vec{y})$, which is defined for all distinct points $\vec{x}, \vec{y} \in D$.
(2) $K(\vec{x}; \vec{y}) := -\partial_n G(\vec{x}; \vec{y})$, which is the outward normal derivative of Green's function on the boundary of the domain D. This kernel, which is naturally called the *Poisson kernel* on D, is defined for $\vec{x} \in \partial D$ and $\vec{y} \in D$.

It turns out that Green's function can be represented as an infinite series using the eigenfunction expansion. More precisely, let D be a smooth bounded domain in $\mathbb{R}^N$. Appealing to the results of Section 9.5 we recall that the Laplace operator with the Dirichlet boundary condition admits a complete orthonormal system (with respect to the inner product $\langle u, v \rangle = \int_D u(\vec{x}) v(\vec{x}) \, d\vec{x}$) of eigenfunctions $\{\phi_n(\vec{x})\}_{n=0}^\infty$. The corresponding eigenvalues $\{\lambda_n\}_{n=0}^\infty$ are listed in a nondecreasing order and are repeated according to their multiplicity.

Using the eigenfunction expansion, it (formally) follows that the solution of the Dirichlet problem

$$\Delta u = f(\vec{x}) \quad \vec{x} \in D,$$
$$u(\vec{x}) = 0 \quad \vec{x} \in \partial D$$

(9.175)

is given by

$$u(\vec{x}) = \sum_{n=0}^{\infty} B_n \phi_n(\vec{x}),$$

(9.176)

where B_n are the (generalized) Fourier coefficients

$$B_n = \frac{-1}{\lambda_n} \int_D \phi_n(\vec{y}) f(\vec{y}) \, d\vec{y} \quad n = 0, 1, \dots .$$

(9.177)

Interchanging (formally) the order of summation and integration, we obtain

$$u(\vec{x}) = -\sum_{n=0}^{\infty} \left[\frac{1}{\lambda_n} \int_D \phi_n(\vec{y}) f(\vec{y}) \, d\vec{y} \right] \phi_n(\vec{x})$$

$$= -\int_D \left[\sum_{n=0}^{\infty} \frac{\phi_n(\vec{x})\phi_n(\vec{y})}{\lambda_n} \right] f(\vec{y}) \, d\vec{y}.$$

Therefore,

$$G(\vec{x}, \vec{y}) := \sum_{n=0}^{\infty} \frac{\phi_n(\vec{x})\phi_n(\vec{y})}{\lambda_n}.$$

(9.178)

The following theorem summarizes some essential properties of (Dirichlet) Green's function.

Theorem 9.34 *Let D be a smooth bounded domain in $\mathbb{R}^N$, $N \geq 3$. Then*

(a) *The Laplace operator has a unique (Dirichlet) Green function on D.*
(b) *Green's function is symmetric, i.e.*

$$G(\vec{x}; \vec{y}) = G(\vec{y}; \vec{x}),$$

for all $\vec{x}, \vec{y} \in D$ such that $\vec{x} \neq \vec{y}$.
(c) *For a fixed $\vec{x} \in D$, the function $G(\vec{x}; \vec{y})$ is a positive harmonic function in $D \setminus \{\vec{x}\}$ and vanishes on ∂D.*
(d) *Fix $\vec{x} \in \partial D$. Then as a function of $\vec{y}$, the Poisson kernel $K(\vec{x}; \vec{y})$ is a positive harmonic function in D that vanishes on $\partial D \setminus \{\vec{x}\}$.*
(e) *Let D_1, D_2 be smooth bounded domains in $\mathbb{R}^N$, $N \geq 3$ such that $D_1 \subset D_2$. Then*

$$0 \leq G_1(\vec{x}; \vec{y}) \leq G_2(\vec{x}; \vec{y}) \leq \Gamma(|\vec{x} - \vec{y}|) \quad \vec{x}, \vec{y} \in D_1,$$

where G_i is the Green function in D_i for $i = 1, 2$.

Proof The proof of the theorem is similar to the proof for the two-dimensional case and therefore is left for the reader (see Exercise 9.20). Note that by part (e) Green's function is pointwise bounded from above by the fundamental solution Γ. In particular at each point there is an upper bound for the Green function that is independent of D. This property is not valid on the plane, where the sequence $\{G_n\}$ of Green's functions for B_n, the disks of radius n and a center at the origin, tends to infinity as $n \to \infty$. $\square$

Example 9.35 Let B_R be an open ball with radius R and center at the origin. We want to calculate Green's function and the Poisson kernel for B_R. For $\vec{x} \in B_R$, the point $\tilde{x} := (R^2/|\vec{x}|^2)\vec{x}$ is said to be the *inverse point* of $\vec{x}$ with respect to the sphere ∂B_R. It is convenient to define the ideal point ∞ as the inverse of the origin. Set

$$G(\vec{x}; \vec{y}) := \begin{cases} \Gamma(|\vec{x} - \vec{y}|) - \Gamma\left(\dfrac{|\vec{y}|}{R}|\tilde{x} - \vec{y}|\right) & \vec{y} \neq 0, \\ \Gamma(|\vec{x}|) - \Gamma(R) & \vec{y} = 0. \end{cases}$$

$$= \Gamma(\sqrt{|\vec{x}|^2 + |\vec{y}|^2 - 2\vec{x} \cdot \vec{y}}) - \Gamma(\sqrt{(|\vec{x}||\vec{y}|/R)^2 + R^2 - 2\vec{x} \cdot \vec{y}}).$$

$$(9.179)$$

An elementary calculation implies that G is indeed Green's function on B_R, and that the Poisson kernel is given by

$$K(\vec{x}; \vec{y}) = \frac{R^2 - |\vec{y}|^2}{N\omega_N R}|\vec{x} - \vec{y}|^{-N}.$$

$$(9.180)$$

As an immediate consequence of the explicit expression (9.180) for the Poisson kernel in a ball, we can derive the mean value principle (see Exercise 9.21).

Theorem 9.36 The mean value principle *Let D be a domain in $\mathbb{R}^N$, $N \geq 3$, and let u be a harmonic function in D. Let $\vec{x}_0$ be a point in D and let B_R be a ball of radius R around $\vec{x}_0$ that is fully included in D. Then the value of u at $\vec{x}_0$ is the average of the values of u on the sphere ∂B_R.*

Just like in the two-dimensional case, the mean value principle implies the strong and weak maximum principles.

Theorem 9.37 *(a) Let u be a harmonic function in a domain D (here we also allow for unbounded $D \subset \mathbb{R}^N$). If u attains it maximum (minimum) at an interior point of D, then u is constant.*

(b) Suppose that D is a bounded domain, and let $u \in C^2(D) \cap C(\bar{D})$ be a harmonic function in D. Then the maximum (minimum) of u in $\bar{D}$ is achieved on the boundary ∂D.

Example 9.38 Let $\mathbb{R}^N_+$ be the upper half-space. As in Example 8.15, one can use the *reflection principle* to obtain the corresponding Green's function and Poisson's kernel (see Exercise 9.23).

9.12 Heat kernel in higher dimensions

In this section we give a short survey of the generalization to higher dimensions of the results of Section 8.4 concerning the heat kernel and integral representations for the heat equation. For simplicity, we assume in the following discussion that the heat conduction coefficient k equals 1.

Let $D \subset \mathbb{R}^N$ be a smooth bounded domain in $\mathbb{R}^N$, $N \geq 2$. Recall from Section 9.5 that the Laplace operator with the Dirichlet boundary condition admits a complete orthonormal system (with respect to the inner product $\langle u, v \rangle = \int_D u(\vec{x})v(\vec{x})\,d\vec{x}$) of eigenfunctions $\{\phi_n(\vec{x})\}_{n=0}^\infty$, and eigenvalues $\{\lambda_n\}_{n=0}^\infty$. The eigenvalues are listed in a nondecreasing order and are repeated according to their multiplicity.

Using the eigenfunction expansion, it (formally) follows that the solution of the initial boundary value problem

$$
\begin{aligned}
u_t - \Delta u &= 0 & \vec{x} &\in D,\ t > 0, \\
u(\vec{x}, t) &= 0 & \vec{x} &\in \partial D,\ t \geq 0, \\
u(\vec{x}, 0) &= f(\vec{x}) & \vec{x} &\in D
\end{aligned}
\tag{9.181}
$$

is given by

$$
u(\vec{x}, t) = \sum_{n=0}^\infty B_n \phi_n(\vec{x}) e^{-\lambda_n t},
\tag{9.182}
$$

where B_n are the (generalized) Fourier coefficients

$$
B_n = \int_D \phi_n(\vec{y}) f(\vec{y})\,d\vec{y} \qquad n = 0, 1, \ldots .
\tag{9.183}
$$

It turns out that the above series converges uniformly for $t > \varepsilon > 0$; therefore we may interchange the order of summation and integration, and hence

$$
\begin{aligned}
u(\vec{x}, t) &= \sum_{n=0}^\infty \left[\int_D \phi_n(\vec{y}) f(\vec{y})\,d\vec{y} \right] \phi_n(\vec{x}) e^{-\lambda_n t} \\
&= \int_D \left[\sum_{n=0}^\infty e^{-\lambda_n t} \phi_n(\vec{x}) \phi_n(\vec{y}) \right] f(\vec{y})\,d\vec{y}.
\end{aligned}
$$

We have derived the following integral representation:

$$u(\vec{x}, t) = \int_D K(\vec{x}, \vec{y}, t) f(\vec{y}) \, d\vec{y}, \tag{9.184}$$

where K is the *heat kernel*:

$$K(\vec{x}, \vec{y}, t) := \sum_{n=0}^{\infty} e^{-\lambda_n t} \phi_n(\vec{x}) \phi_n(\vec{y}). \tag{9.185}$$

By the Duhamel principle, it follows that the solution of the initial boundary value problem

$$\begin{cases} u_t - \Delta u = F(\vec{x}, t) & \vec{x} \in D, \ t > 0, \\ u(\vec{x}, t) = 0 & \vec{x} \in \partial D, \ t \geq 0, \\ u(\vec{x}, 0) = f(\vec{x}) & \vec{x} \in D \end{cases} \tag{9.186}$$

is given by the following representation formula:

$$u(\vec{x}, t) = \int_D K(\vec{x}, \vec{y}, t) f(\vec{y}) \, d\vec{y} + \int_0^t \int_D K(\vec{x}, \vec{y}, t - s) F(\vec{y}, s) \, d\vec{y} ds. \tag{9.187}$$

The main properties of the heat kernel for the one-dimensional case are also valid in higher dimensions. The following theorem summarizes these properties and some other properties that were not stated for the one-dimensional case.

Theorem 9.39 *Let $K(\vec{x}, \vec{y}, t)$ be the heat kernel of problem (9.186). Then*

(a) *The heat kernel is symmetric, i.e. $K(\vec{x}, \vec{y}, t) = K(\vec{y}, \vec{x}, t)$.*
(b) *For a fixed $\vec{y}$ (or a fixed $\vec{x}$), the heat kernel K as a function of t and $\vec{x}$ (or $\vec{y}$) solves the heat equation for $t > 0$, and satisfies the Dirichlet boundary conditions.*
(c) *$K(\vec{x}, \vec{y}, t) \geq 0$.*
(d) *Suppose that $D_1 \subset D_2$, and let K_i be the heat kernel in $D_i, i = 1, 2$. Then $K_1(\vec{x}, \vec{y}, t) \leq K_2(\vec{x}, \vec{y}, t)$ for all $\vec{x}, \vec{y} \in D_1$ and $t > 0$.*
(e)

$$\int_D K(\vec{x}, \vec{y}, t) \, d\vec{y} \leq 1.$$

(f) *For all $t, s > 0$ the heat kernel satisfies the following semigroup property:*

$$K(\vec{x}, \vec{y}, t + s) = \int_D K(\vec{x}, \vec{z}, t) K(\vec{z}, \vec{y}, s) \, d\vec{z}.$$

(g) *The following* trace formula *holds:*

$$\int_D K(\vec{x}, \vec{x}, t) \, d\vec{x} = \sum_{n=0}^{\infty} e^{-\lambda_n t}.$$

(h) Let $G(\vec{x}, \vec{y})$ be (Dirichlet) Green's function of the Laplace equation on a smooth bounded domain D. Then

$$G(\vec{x}, \vec{y}) = \int_0^\infty K(\vec{x}, \vec{y}, t)\, dt.$$

Proof (a),(b) Formally these parts follow directly from (9.185), which defines the heat kernel. In order to justify the convergence and term-by-term differentiations one should use the exponential decay of the terms $e^{-\lambda_n t}$ and the bounds (which may depend on n) on the eigenfunctions and their derivatives.

(c) Suppose on the contrary that there exists $(\vec{x}_0, \vec{y}_0, t_0)$, $\vec{x}_0, \vec{y}_0 \in D$, $t_0 > 0$ such that $K(\vec{x}_0, \vec{y}_0, t_0) < 0$. Then $K(\vec{x}, \vec{y}, t)$ is negative in some neighborhood of $(\vec{x}_0, \vec{y}_0, t_0)$. Let u be a solution of problem (9.186) with $F = 0$ and with $f(\vec{y})$, which is a nonnegative function that is strictly positive only for $\vec{y}$ in a small neighborhood of $\vec{y}_0$. The representation formula (9.187) implies that u is negative at $(\vec{x}_0, t_0)$, but this contradicts the maximum principle for the heat equation.

(d) Let f be an arbitrary nonnegative smooth function with a compact support in D_1. For $i = 1, 2$ the functions

$$u_i(\vec{x}, t) = \int_{D_i} K_i(\vec{x}, \vec{y}, t) f(\vec{y})\, d\vec{y}$$

solve the problems

$$\begin{aligned}
(u_i)_t - \Delta u_i &= 0 & \vec{x} \in D_i,\ t > 0, \\
u_i(\vec{x}, t) &= 0 & \vec{x} \in \partial D_i,\ t \geq 0, \\
u_i(\vec{x}, 0) &= f(\vec{x}) & \vec{x} \in D_i.
\end{aligned}$$

On the other hand, by the maximum principle $0 \leq u_1(\vec{x}, t) \leq u_2(\vec{x}, t)$. Therefore,

$$\int_{D_1} [K_2(\vec{x}, \vec{y}, t) - K_1(\vec{x}, \vec{y}, t)] f(\vec{y})\, d\vec{y} \geq 0.$$

Now, since $f \geq 0$ is an arbitrary function, a similar argument to the one used in the proof of part (c) shows that $K_1(\vec{x}, \vec{y}, t) \leq K_2(\vec{x}, \vec{y}, t)$.

(e) Let $0 \leq f_n \leq 1$ be a sequence of compactly supported smooth functions that converge monotonically to the function 1. Then

$$u_n(\vec{x}, t) := \int_D K(\vec{x}, \vec{y}, t) f_n(\vec{y}) d\vec{y} \ \to\ u(\vec{x}, t) := \int_D K(\vec{x}, \vec{y}, t) d\vec{y}.$$

Now, u_n is a monotone sequence of solutions to the heat problem

$$\begin{aligned}
u_t - \Delta u &= 0 & \vec{x} \in D,\ t > 0, \\
u(\vec{x}, t) &= 0 & \vec{x} \in \partial D,\ t \geq 0, \\
u(\vec{x}, 0) &= f_n(\vec{x}) & \vec{x} \in D.
\end{aligned}$$

On the other hand, $v(\vec{x}, t) = 1$ solves the heat problem

$$\begin{aligned} v_t - \Delta v = 0 &\qquad \vec{x} \in D, \ t > 0, \\ v(\vec{x}, t) = 1 &\qquad \vec{x} \in \partial D, \ t \geq 0, \\ v(\vec{x}, 0) = 1 &\qquad \vec{x} \in D. \end{aligned}$$

The maximum principle implies that $u_n(\vec{x}, t) \leq v(\vec{x}, t)$, and therefore,

$$u(\vec{x}, t) = \int_D K(\vec{x}, \vec{y}, t) \, d\vec{y} \leq v(\vec{x}, t) = 1.$$

(f) Fix $s > 0$, and let f be a smooth function. Set

$$v(\vec{x}, t) := \int_D K(\vec{x}, \vec{y}, t + s) f(\vec{y}) \, d\vec{y}.$$

The function

$$\begin{aligned} u(\vec{x}, t) : &= \int_D \left[\int_D K(\vec{x}, \vec{z}, t) K(\vec{z}, \vec{y}, s) d\vec{z} \right] f(\vec{y}) d\vec{y} \\ &= \int_D K(\vec{x}, \vec{z}, t) \left[\int_D K(\vec{z}, \vec{y}, s) f(y) d\vec{y} \right] d\vec{z} \end{aligned}$$

is a solution of the problem

$$\begin{aligned} u_t - \Delta u = 0 &\qquad \vec{x} \in D, \ t > 0, \\ u(\vec{x}, t) = 0 &\qquad \vec{x} \in \partial D, \ t \geq 0, \\ u(\vec{x}, 0) = v(\vec{x}, 0) &\qquad \vec{x} \in D. \end{aligned}$$

On the other hand, $v(\vec{x}, t)$ is also a solution of the same problem. Thanks to the uniqueness theorem $u = v$, hence

$$\int_D \left[\int_D K(\vec{x}, \vec{z}, t) K(\vec{z}, \vec{y}, s) \, d\vec{z} - K(\vec{x}, \vec{y}, t + s) \right] f(\vec{y}) \, d\vec{y} = 0.$$

Since f is an arbitrary function, it follows that

$$K(\vec{x}, \vec{y}, t + s) = \int_D K(\vec{x}, \vec{z}, t) K(\vec{z}, \vec{y}, s) \, d\vec{z}.$$

(g) The trace formula follows directly from the orthonormality of the sequence $\{\phi_n(\vec{x})\}_{n=0}^\infty$ of all eigenfunctions (see Exercise 9.22). The proof of the Weyl asymptotic formula (9.54) relies on this trace formula.

(h) Follows from the expansion formulas (9.185) and (9.178), and integration with respect to t. $\qquad\qquad\square$

9.13 Exercises

9.1 (a) Generalize the characteristic method for the eikonal equation (see Chapter 2) to solve the eikonal equation in three space dimensions.

(b) Let $u(x, y, z)$ be a solution to the eikonal equation in $\mathbb{R}^3$ (homogeneous medium). Assume

$$u(0, 0, 0) = u_x(0, 0, 0) = u_y(0, 0, 0) = 0.$$

Show that $(\partial^n u/\partial z^n)|_{(0,0,0)} = 0$, for all $n \geq 2$.

9.2 Solve the equation $u_x^2 + u_y^2 + u_z^2 = 4$ subject to the initial condition $u(x, y, 1 - x - y) = 3$.

9.3 Prove formula (9.26).

9.4 Derive the formulation of the Laplace equation in a spherical coordinate system (r, θ, ϕ).

9.5 Find the radial solution to the Cauchy problem (9.22) under the initial conditions

$$u(r, 0) = 2, \quad u_t(r, 0) = 1 + r^2.$$

9.6 Find the radial solution to the Cauchy problem (9.22), with $c = 1$ subject to the initial conditions

$$u(r, 0) = ae^{-r^2}, \quad u_t(r, 0) = be^{-r^2}.$$

9.7 Derive the Darboux equation (9.32).

9.8 Find the eigenfunctions, eigenvalues, and the generalized Fourier formula for the Laplace operator in the rectangle $0 < x < a$, $0 < y < b$ subject to Neumann boundary conditions.

9.9 Find the eigenfunctions, eigenvalues, and the generalized Fourier formula for the Laplace operator in the three-dimensional box

$$0 < x < a, \ 0 < y < b, \ 0 < z < c$$

under the Dirichlet boundary conditions.

9.10 (a) Prove that the Dirichlet eigenvalue problem for the Laplace equation in the unit square has infinitely many eigenvalues with multiplicity three or more.

(b) Let Ω be a rectangle with sides a and b, such that the ratio a^2/b^2 is not a rational number. Show that the Dirichlet eigenvalue problem in Ω is not degenerate.

9.11 Use the representation (9.76) to derive the recurrence formula (9.77).

9.12 The Bessel functions share many common properties with the classical trigonometric functions $\sin nx$ and $\cos nx$. Some of these properties were discussed in Subsection 9.5.3. Let us consider two additional properties:

(a) Show that, like $\sin nx$ and $\sin(n + 1)x$, the Bessel functions $J_n(x)$ and $J_{n+1}(x)$ do not vanish at the same point.

(b) The formula relating $\sin(\alpha + \beta)$ with $\sin \alpha$, $\cos \alpha$, $\sin \beta$ and $\cos \beta$ is in general taught in high school trigonometry classes. Show that for Bessel functions there exist similar formulas, except that they now involve an infinite series:

$$J_n(\alpha + \beta) = \sum_{m=-\infty}^{\infty} J_m(\alpha)J_{n-m}(\beta).$$

9.13 (a) Let v_1 and v_2 be two smooth solutions to the Legendre equation (9.96) in $[-1,1]$ associated with different coefficients μ_1 and μ_2. Show that $\int_{-1}^{1} v_1(s)v_2(s)ds = 0$.
(b) Use the result in part (a), the Weierstrass approximation theorem and the Legendre polynomials constructed in this chapter to prove that if μ is not of the form $\mu = k(k+1)$ for some integer k, then the Legendre equation has no smooth solutions on $[-1, 1]$.

9.14 Find the general solution of the wave equation in a cube under the Dirichlet boundary value conditions.

9.15 Prove that every function of the form $Q(r, \phi, \theta) = r^n Y_{n,m}(\phi, \theta)$ is a homogeneous harmonic polynomial.

9.16 Find the general solution of the heat equation in a disk under the Neumann boundary conditions.

9.17 (a) Prove the *Rodriguez formula*:

$$P_n(t) = \frac{1}{2^n n!} \frac{d^n}{dt^n}(t^2 - 1)^n. \tag{9.188}$$

(b) Prove Proposition 9.25.

9.18 Write the solution of the Dirichlet problem for the Laplace equation on a cylinder as a generalized Fourier series, and find the corresponding formula for the generalized Fourier coefficients.

9.19 Prove formulas (9.179)–(9.180) for the Green function and Poisson kernel in the ball B_R.

9.20 Complete the proof of Theorem 9.34 (see the corresponding proof for the two-dimensional case).

9.21 (a) Use the explicit formula for the Poisson kernel in a ball (formula (9.180)) to prove the mean value principle (Theorem 9.36).
(b) Provide an alternative proof of the same theorem that relies on the proof method of Theorem 7.7.
(c) Prove the strong and weak maximum principles for harmonic functions on a domain D in $\mathbb{R}^N$.

9.22 Complete the proof of Theorem 9.39.

9.23 Using the reflection method, find explicit formulas for the Green function and Poisson kernel in the half space $\mathbb{R}_+^N$.
Hint See Example 8.15.

9.24 Let $D \subset \mathbb{R}^N$ be a smooth bounded domain. Let $\{\phi_n(\vec{x})\}_{n=0}^{\infty}$ and $\{\lambda_n\}_{n=0}^{\infty}$ be the orthonormal sequence of eigenfunctions and the corresponding eigenvalues for the Laplace operator with the Dirichlet boundary condition. Let $\lambda \neq \lambda_n$ be a real number.

(a) Find the eigenfunction expansion of the Green function $G_\lambda(\vec{x}; \vec{y})$ for the Dirichlet problem in D for the Helmholtz equation $\Delta u + \lambda u = 0$. So, G_λ satisfies

$$\Delta G_\lambda(\vec{x}; \vec{y}) + \lambda G_\lambda(\vec{x}; \vec{y}) = -\delta(\vec{x} - \vec{y}) \quad \vec{x} \in D,$$

$$G_\lambda(\vec{x}; \vec{y}) = 0 \qquad\qquad\qquad \vec{x} \in \partial D.$$

(b) Calculate $\lim_{\lambda \to \lambda_0}(\lambda_0 - \lambda)G_\lambda(\vec{x}, \vec{y})$.

(c) Find the large time asymptotic formula for the heat kernel.

Hint Calculate $\lim_{t \to \infty} e^{\lambda_0 t} K(\vec{x}, \vec{y}, t)$.

(d) Compare your answers to parts (b) and (c), and try to explain your result.

9.25 (a) Find the eigenfunction expansion of the (Dirichlet) Green function in the rectangle $\{(x, y) \mid 0 < x < a, \ 0 < y < b\}$.

(b) Find the eigenfunction expansion of the (Dirichlet) Green function in the disk $\{(x, y) \mid x^2 + y^2 < R^2\}$.

10

Variational methods

The PDEs we have considered so far were derived by modeling a variety of phenomena in physics, engineering, etc. In this chapter we shall derive PDEs from a new perspective. We shall show that many PDEs are related to optimization problems. The theory that associates optimization with PDEs is called the *calculus of variations* [20]. It is an extremely useful theory. On the one hand, we shall be able to solve many optimization problems by solving the corresponding PDEs. On the other hand, sometimes it is simpler to study (and solve) certain optimization problems than to study (and solve) the related PDE. In such cases, the calculus of variations is an indispensable theoretical and practical tool in the study of PDEs. The calculus of variations can be used for both static problems and dynamic problems. The dynamical aspects of this theory are based on the Hamilton principle that we shall derive below. In particular, we shall show how to apply this principle for wave propagation in strings, membranes, etc.

We shall see that the connection between optimization problems and the associated PDEs is based on the a priori assumption that the solution to the optimization problem is smooth enough for the PDE to make sense. Can we justify this assumption? In many cases we can. Moreover, even if the solution is not smooth, we would like to define an appropriate concept of weak solutions as we already did earlier in this book in different contexts. How should we define them? To answer these questions we need to introduce special inner product function spaces (see Chapter 6). After introducing these spaces, we shall be able to define a natural notion of weak solutions for a large variety of PDEs.

10.1 Calculus of variations

Let Γ be a simple closed curve in $\mathbb{R}^3$. A surface whose boundary is Γ is said to be *spanned* by Γ. To define the concept of *minimal surfaces* let us consider for the moment a surface $S = S(u)$, characterized by a graph of a function $u(x, y)$ defined over a region D in $\mathbb{R}^2$, such that the boundary ∂D is mapped by u to $\partial S = \Gamma$ (in particular $S(u)$ is spanned by Γ). The surface area A of S is given by

$$E(u) := A(S(u)) = \int_D \sqrt{1 + u_x^2 + u_y^2}\, dxdy. \tag{10.1}$$

A surface S is called a (local) *minimal* if its surface area is smaller than the surface area of all other surfaces spanned by Γ that are close to S in an appropriate sense. More precisely, a function v is called *admissible* if the surface $S(v)$ is spanned by Γ, v is continuously differentiable in D (to guarantee that the local surface element is defined), and v is continuous in the closure of D. A function u is a (local) *minimizer* for the surface area problem if u is an admissible function and

$$E(u) \le E(v), \tag{10.2}$$

for every admissible functions v (that are close to u).

The problem of characterizing and computing minimal surfaces has been considered by many mathematicians since the middle of the eighteenth century, with major contributions to the subject provided by Lagrange and Laplace. The richness of the problem was not realized, however, until the soap film experiments performed by the Belgian physicist Joseph Antoine Plateau (1801–1883) around 1870.[1]

The problem of minimizing E is analogous to the problem of minimizing a differentiable function $f : \mathbb{R} \to \mathbb{R}$. As we recall from calculus, a necessary condition for a point $x \in \mathbb{R}$ to be a local minimizer is that the derivative of f is zero at x. The function $E(u)$ defined above is a mapping that associates a real number with a *function u*. Such objects are called *functionals*. There is a trick that enables us to use the theory of optimizing real functions to optimize functionals. The idea is to consider a fixed function u which is our candidate for a minimizer. We then introduce a real parameter ε and represent any admissible function v as $v = u + \varepsilon\psi$. This construction implies that ψ must belong to the space of functions

$$\mathcal{A} = \{\psi \in C^1(D) \cap C(\bar{D}), \quad \psi(x, y) = 0 \text{ for } (x, y) \in \partial D\}. \tag{10.3}$$

We rewrite (10.2) in the form $E(u) \le E(u + \varepsilon\psi)$ for small $|\varepsilon|$ and for all $\psi \in \mathcal{A}$. Considering $E(u + \varepsilon\psi)$ as a real function of ε (with u and ψ fixed), we apply the standard argument from calculus to require the necessary condition

$$\frac{d}{d\varepsilon} E(u + \varepsilon\psi)\bigg|_{\varepsilon=0} = 0, \tag{10.4}$$

---

[1] Incidentally, while such experiments are now frequently performed by children in science museums around the world, Plateau himself did not see a single minimal surface! He was blinded early in his scientific career as a result of looking directly at the sun while performing optical experiments.

The expression on the left hand side of (10.4) is called *the first variation of E* at u. It is denoted by $\delta E(u)(\psi)$. The somewhat unusual notation indicates that the first variation depends on u, and it is a functional (in fact, a linear functional) of ψ. We shall demonstrate in the sequel explicit computations of first variations. To avoid too cumbersome notation, we shall often denote the first variation for short by $\delta E(u)$.

Before demonstrating the implications of (10.4) for our model problem of minimal surfaces, let us look at a simpler problem. If we assume that the minimal surface u has small derivatives, we can approximate the functional $E(u)$ by a simpler functional. Using the approximation $\sqrt{1+x} \sim 1 + \frac{1}{2}x + \cdots$, we expand $E(u) = |D| + \frac{1}{2}\int_D \left(u_x^2 + u_y^2\right) dxdy + \cdots$, where $|D|$ denotes the area of D. Neglecting high order terms, we replace the problem of minimizing $E(u)$ with the problem of minimizing the functional

$$G(u) = \frac{1}{2}\int_D \left(u_x^2 + u_y^2\right) dxdy = \frac{1}{2}\int_D |\nabla u|^2 \, dxdy. \tag{10.5}$$

The functional G is called the *Dirichlet functional* or *Dirichlet integral*. It plays a prominent role in many branches of science and engineering.

We are now ready to use (10.4) to derive an equation for the local minimizers of G. Let us compute in detail the differentiation in (10.4): it is easy to check that

$$G(u + \varepsilon\psi) = G(u) + \varepsilon\int \nabla u \cdot \nabla\psi \, dxdy + \varepsilon^2 G(\psi). \tag{10.6}$$

Thus,

$$\delta G(u) = \frac{d}{d\varepsilon}G(u + \varepsilon\psi)\bigg|_{\varepsilon=0} = \int_D \nabla u \cdot \nabla\psi \, dxdy. \tag{10.7}$$

Therefore, a necessary condition for u to be a local minimizer is that it satisfies

$$\int_D \nabla u \cdot \nabla\psi \, dxdy = 0 \qquad \forall\psi \in \mathcal{A}. \tag{10.8}$$

To derive an explicit equation for u we integrate the last integral by parts. Using Green's identity (7.20) and the condition on ψ at the boundary ∂D, we obtain

$$\int_D \Delta u\psi \, dxdy = 0 \qquad \forall\psi \in \mathcal{A}. \tag{10.9}$$

At this point we invoke Lemma 1.1. Thanks to this lemma we conclude that if Δu is continuous, then (10.9) implies

$$\Delta u = 0 \quad \text{in } D. \tag{10.10}$$

By construction, u must satisfy the boundary condition

$$u(x, y) = g(x, y) \qquad (x, y) \in \partial D, \tag{10.11}$$

where g is the (given) graph of u over ∂D. We have therefore proved that a necessary condition for a smooth function u to minimize the Dirichlet functional G is that u is a solution of the Dirichlet problem for the Laplace equation. The PDE that is obtained by equating the first variation of a functional to zero is called the *Euler–Lagrange equation*. We can therefore say that "Laplace $=$ Euler–Lagrange of Dirichlet"....

We now return to our original problem of minimal surfaces. It is convenient to derive the minimal surface equation as a special case of a more general equation that is valid for any functional that depends on a function and its derivatives. Consider for this purpose a function $F(x_1, x_2, \ldots, x_n, L_1(u), L_2(u), \ldots, L_m(u))$, where L_i is a linear operator (such as a differential operator or the identity operator). For instance, the integrand in the Dirichlet integral is

$$F = \left(\frac{\partial}{\partial x} u\right)^2 + \left(\frac{\partial}{\partial y} u\right)^2.$$

To compute the first variation of

$$K(u) = \int_D F \, dx_1 dx_2 \ldots dx_n, \tag{10.12}$$

we expand F into a Taylor series around a base function u. Using

$$F(u + \varepsilon\psi) = F(u) + \varepsilon \sum_{i=1}^{m} \frac{\partial F}{\partial L_i}(u)L_i(\psi) + O(\varepsilon^2), \tag{10.13}$$

and equating the first variation to zero, we obtain the Euler–Lagrange equation

$$\delta K(u) = \int_D \sum_{i=1}^{m} \frac{\partial F}{\partial L_i}(u)L_i(\psi) \, dx_1 \ldots dx_n = 0. \tag{10.14}$$

The reader is encouraged to use (10.14) for an alternative derivation of (10.7). We now use (10.14) to derive a number of further examples of Euler–Lagrange equations:

Example 10.1 The minimal surface equation In the minimal surface problem, $F(u) = \sqrt{1 + u_x^2 + u_y^2}$. Therefore,

$$\delta A(S(u)) = \int_D \frac{1}{\sqrt{1 + u_x^2 + u_y^2}} \nabla u \cdot \nabla \psi \, dxdy. \tag{10.15}$$

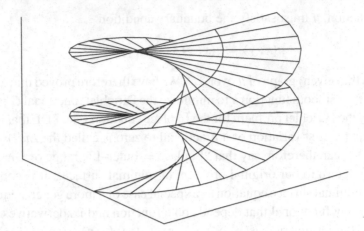

Figure 10.1 The helicoid.

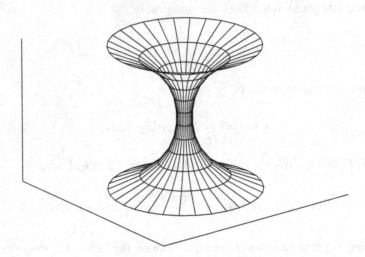

Figure 10.2 The catenoid.

Integrating by parts with the aid of the divergence theorem, we obtain the minimal surface equation:

$$\nabla \cdot \left(\frac{1}{\sqrt{1 + u_x^2 + u_y^2}} \nabla u \right) = \frac{\partial}{\partial x} \left(\frac{u_x}{\sqrt{1 + u_x^2 + u_y^2}} \right) + \frac{\partial}{\partial y} \left(\frac{u_y}{\sqrt{1 + u_x^2 + u_y^2}} \right) = 0.$$

(10.16)

Examples of minimal surfaces are depicted in Figure 10.1 (the helicoid) and in Figure 10.2 (the catenoid). Notice, though, that the surfaces in these examples (as well as the examples in Figure 10.3) cannot be represented as global graphs; rather they can be written explicitly in a parametric form. For example, the helicoid is expressed as

$$x = \rho \cos \theta, \quad y = \rho \sin \theta, \quad z = d\rho, \tag{10.17}$$

where $\theta \in [0, 2\pi]$, and $\rho \in (a, b)$ and a, b, d are some fixed parameters. Similarly, the catenoid is written as

$$x = d \cosh \frac{\rho}{d} \cos \theta, \quad y = d \cosh \frac{\rho}{d} \sin \theta, \quad z = \rho. \tag{10.18}$$

In the two examples we have analyzed so far, the values of the unknown function u were known at the boundary. We consider now the problem of minimizing functionals without constraints at the boundary. The calculation of the first variation is the same as in the case of problems *with* boundary constraints. The difference is in the last step where the first variation is used to obtain a PDE. We demonstrate the derivation of the Euler–Lagrange equation in such a case through a variant of the Dirichlet integral.

Example 10.2 Reconstruction of a function from its gradient Many applications in optics and other image analysis problems require a surface $u(x, y)$ to be computed from measurements of its gradient. This procedure is particularly useful in determining the phase of light waves or sound waves. If the measurement is exact, the solution is straightforward. Since, however, there is always an experimental error, the measurement can be considered at best as an approximation of the gradient. Denote the measured vector that approximates the gradient by $\vec{f}^t = (f_1, f_2)$. Typically, a given vector field is *not* a gradient of a scalar function u. To be a gradient $\vec{f}$ must satisfy the compatibility condition $\partial f_1/\partial y = \partial f_2/\partial x$. If this condition indeed holds, we can find u (locally) through a simple integration. Since we expect generically that measurement errors will corrupt the compatibility condition, we seek other means for estimating the phase u. One such estimate is provided by the *least squares approximation*:

$$\min \ K(u) := \int_D |\nabla u - \vec{f}|^2 \, dxdy, \tag{10.19}$$

where D is the domain where the gradient is measured. To apply (10.14), we write

$$F(u_x, u_y) = |\nabla u - \vec{f}|^2 = |\nabla u|^2 - 2\nabla u \cdot \vec{f} + |\vec{f}|^2.$$

The differentiation is simple to perform and we obtain

$$\delta K(u) = 2 \int_D \left(\nabla u - \vec{f} \right) \cdot \nabla \psi \, dxdy = 0. \tag{10.20}$$

Integrating (as usual) by parts, we get

$$\frac{1}{2}\delta K(u) = \int_D \left(-\Delta u + \vec{\nabla}\cdot\vec{f}\right)\psi \; dxdy + \int_{\partial D}\left(\partial_n u - \vec{f}\cdot\hat{n}\right)\psi \; ds = 0,$$
(10.21)

where $\hat{n}$ is the unit outer normal vector to ∂D. Since the first variation must vanish, in particular for functions ψ that are identically zero at ∂D, we must equate the first integral (10.21) to zero to obtain the Euler–Lagrange equation

$$\Delta u = \vec{\nabla}\cdot\vec{f} \quad (x, y) \in D.$$
(10.22)

Then (10.21) reduces to

$$\int_{\partial D}\left(\partial_n u - \vec{f}\cdot\hat{n}\right)\psi \; ds = 0.$$
(10.23)

Now, taking advantage of the fact that this relation holds for ψ that are nonzero on ∂D as well, we obtain

$$\partial_n u = \vec{f}\cdot\hat{n} \quad (x, y) \in \partial D.$$
(10.24)

We have demonstrated how to deduce appropriate boundary conditions from the optimization problem. Such boundary conditions, which are inherent to the variational problem (in contrast to being supplied from outside), are called *natural boundary conditions*.

Physical systems in equilibrium are often characterized by a function that is a local minimum of the potential energy of the system. This is one of the reasons for the great value of variational methods. In the next examples we consider two classical problems from the theory of elasticity.

Example 10.3 Equilibrium shape of a membrane under load Consider a thin membrane occupying a domain $D \subset \mathbb{R}^2$ when at a horizontal rest position, and denote its vertical displacement by $u(x, y)$. Assume that the membrane is subjected to a transverse force (called in elasticity *load*) $l(x, y)$ and constrained to satisfy $u(x, y) = g(x, y)$, for $(x, y) \in \partial D$. Since the membrane is assumed to be in equilibrium, its potential energy must be at a minimum. The potential energy consists of the energy stored in the stretching of the membrane and the work done by membrane against the load l. The local stretching of the membrane from its horizontal rest shape is given by $d(\sqrt{1 + u_x^2 + u_y^2} - 1)$, where d is the elasticity constant of the membrane. Assuming that the membrane's slopes are small, we approximate the local stretching by $\frac{1}{2}d(u_x^2 + u_y^2)$. The work against the load is $-lu$. Therefore, we have to minimize

$$Q(u) = \int_D \left(\frac{d}{2}(u_x^2 + u_y^2) - lu\right) dxdy.$$
(10.25)

The first variation is

$$\delta Q(u) = \int_D (d\nabla u \cdot \nabla \psi - l\psi)\, dxdy. \qquad (10.26)$$

Integrating the first term by parts, using the boundary condition $\psi = 0$ on ∂D, and equating the first variation to zero we obtain

$$\int_D (d\Delta u + l)\psi\, dxdy = 0.$$

Therefore, the Euler–Lagrange equation for the membrane is the Poisson equation:

$$d\Delta u = -l(x, y)\ (x, y) \in D, \qquad u(x, y) = g(x, y)\ (x, y) \in \partial D. \qquad (10.27)$$

Example 10.4 The plate equation Consider a thin plate under a load l whose amplitude with respect to a planar domain D is given by $u(x, y)$. Integration of the equations of elasticity leads to the following expression for the plate's energy:

$$P(u) = \int_D \left\{ \frac{d}{2}\left[(\Delta u)^2 + 2(1-\lambda)^{-1}\left(u_{xy}^2 - u_{xx}u_{yy}\right)\right] - lu \right\} dxdy, \qquad (10.28)$$

where λ is called the *Poisson ratio* and d is called the *flexural rigidity* of the plate. The Poisson ratio is a characteristic of the medium composing the plate. It measures the transversal compression of an element of the plate when it is stretched longitudinally. For example, $\lambda \approx 0.5$ for rubber, and $\lambda \approx 0.27$ for steel. The parameter d depends not only on the material constituting the plate, but also on its thickness.

To find the Euler–Lagrange equations for the plate we compute the first variation of P. To simplify the calculations we assume that the plate is clamped, i.e. both u and $\partial u/\partial n$ are given on ∂D. Computing the first variation of the first and third terms in (10.28) is straightforward:

$$\delta_1 = \int_D (d\Delta u\Delta\psi - l\psi)\, dxdy, \qquad (10.29)$$

where ψ is the variation, and the boundary conditions imply that ψ and $\partial\psi/\partial n = 0$ on ∂D. Integrating by parts twice the first integral in (10.29) and using the boundary conditions on ψ we obtain

$$\delta_1 = \int_D (d\Delta^2 u - l)\,\psi\, dxdy, \qquad (10.30)$$

where

$$\Delta^2 = \frac{\partial^4}{\partial^4 x} + 2\frac{\partial^4}{\partial^2 x\partial^2 y} + \frac{\partial^4}{\partial^4 y}$$

is called for obvious reasons the *biharmonic operator*.

We proceed to compute the first variation of the middle term in (10.28):

$$\delta_2 = \frac{d}{1-\lambda} \int_D (2u_{xy}\psi_{xy} - u_{xx}\psi_{yy} - u_{yy}\psi_{xx}) \, dxdy. \tag{10.31}$$

The computation is facilitated by the important observation that the integrand in (10.31) is the divergence of a certain vector:

$$2u_{xy}\psi_{xy} - u_{xx}\psi_{yy} - u_{yy}\psi_{xx} = \frac{\partial}{\partial x}(u_{xy}\psi_y - u_{yy}\psi_x) + \frac{\partial}{\partial y}(u_{xy}\psi_x - u_{xx}\psi_y). \tag{10.32}$$

Thanks to this identity and to the divergence theorem we can convert the variation δ_2 into a boundary integral. This integral involves the first derivatives of ψ. Since $\partial\psi/\partial n = \psi = 0$ at the boundary, both the normal *and* the tangential derivatives of ψ vanish there. Therefore, $\partial\psi/\partial x = \partial\psi/\partial y = 0$ on the boundary ∂D, and the boundary integral we derived for the variation δ_2 is identically zero. We finally obtain

$$\delta P(u) = \int_D \left(d\Delta^2 u - l\right) \psi \, dxdy = 0, \tag{10.33}$$

which implies that the Euler–Lagrange equation for thin plates is given by

$$d\Delta^2 u = l. \tag{10.34}$$

Another way to see that the middle term in the integral in (10.28) does not contribute to the Euler–Lagrange equation is to observe that the corresponding integrand is the Hessian $u_{xx}u_{yy} - u_{xy}^2$, which is a divergence of a vector field; i.e. it equals $\nabla \cdot (u_x u_{yy}, -u_x u_{xy})$.

Notice that the Poisson ratio does not play a role in the final equation! This does not mean, though, that clamped rubber plates and clamped steel plates bend in the same way under the same load: the coefficient d in (10.34) does depend on the material (in addition to its dependence on the plate's thickness). We *can* conclude the surprising fact that for any given steel plate there is a rubber plate that bends in exactly the same way. Equation (10.34) is the first fourth-order equation we have encountered so far in this book. As it turns out, fourth-order equations are rare in applications; among the exceptions are the plate equation we just derived, the equation for the vibrations of rods that we derive later in this chapter, and certain equations in lens design.

10.1.1 Second variation

It is well known that equating the first derivative of a real (scalar) function $f(x)$ to zero only provides a necessary condition for potential minimizers of f. To

determine whether a stationary point x_0 (where $f'(x_0) = 0$) is indeed a local minimizer, we have to examine higher derivatives of f. For example, if $f''(x_0) > 0$, we
can conclude that indeed x_0 is a local minimizer.

Similarly, to verify that a function u is a local minimum of some functional, we
must compute the *second variation* of the functional, and evaluate it at u. When
considering a general functional $Q(u)$, the first variation was defined as

$$\delta Q(u)(\psi) := \frac{d}{d\varepsilon} Q(u + \varepsilon\psi) \Big|_{\varepsilon=0}$$

for ψ in an appropriate function space. Similarly, if the first variation of Q at u is
zero, we define the second variation of Q there through

$$\delta^2 Q(u)(\psi) := \frac{d^2}{d\varepsilon^2} Q(u + \varepsilon\psi) \Big|_{\varepsilon=0}. \tag{10.35}$$

Just like the case of the first variation, the second variation is a functional of ψ that
depends on u.

For example, we consider the second variation of the Dirichlet functional G.
From (10.6) it follows at once that $\delta^2 G(v)(\psi) = G(\psi) > 0$ for *any functions* v and
ψ. Therefore, the harmonic function u that we identified above as a candidate for a
minimizer is indeed a local minimizer. In fact, it can be shown that u is the unique
minimizer of G. Notice, however, that the association between minimizers of G
and the harmonic function was contingent on the harmonic function being a smooth
function!

A functional Q such that $\delta^2 Q(v)(\psi) > 0$ for all appropriate v and ψ is called
(strictly) *convex*. Such functionals are particularly useful to identify since they have
a unique minimizer.

Is there always a unique minimum? This question has far reaching implications
in many branches of science and technology. In fact, it is also raised in unex-
pected disciplines such as philosophy and even theology. In contrast to the ethical
monotheism of the Prophets of Israel, the Hellenic monotheism was based on logi-
cal arguments, basically claiming that since God is the best, i.e. optimal, and since
the best must be unique, then there is only one god. This argument did not convince
the ancient Greeks (were they aware of the possibility of many local extrema?),
who stuck to their belief in a plurality of gods.

Indeed one of the intriguing questions raised by Plateau and many mathematicians after him was whether the minimal surface problem has a unique solution for
any given spanning curve Γ. The answer is no. In Figure 10.3 we depict an example
of a spanning curve Γ for which there exist more than one minimal surface.

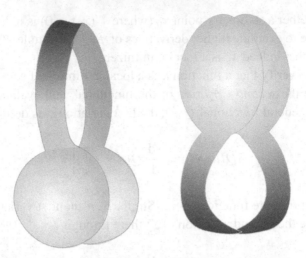

Figure 10.3 An example of two distinct minimal surfaces spanned by the same curve.

10.1.2 Hamiltonians and Lagrangians

Newton founded his theory of mechanics in the second part of the seventeenth century. The theory was based upon three laws postulated by him. The laws provided a set of tools for computing the motion of bodies, given their initial positions and initial velocities, by calculating the forces they exert on each other, and relating these forces to the acceleration of the bodies. Motivated by the introduction of steam machines towards the end of the eighteenth century and the beginning of the nineteenth century, scientists developed the theory of thermodynamics, and with it the important concept of energy. Then, in 1824 Hamilton started his systematic derivation of an axiomatic geometric theory of light. He realized that his theory is equivalent to a variational principle, called the *Fermat principle*, which states that light propagates so as to travel between two arbitrary points in minimal time. For example, the eikonal equation that we studied in Chapter 2 is related to the Euler–Lagrange equation associated with this principle. During his optics research, Hamilton observed that apparently different notions such as optical travel time and energy are in fact related by another physical object called *action*. Moreover, he showed that the entire theory of Newtonian mechanics can be formulated in terms of actions and energies, instead of in terms of forces and acceleration. Hamilton's new theory, now called *Hamilton's principle*, enabled the use of variational methods to study not just static equilibria, but also dynamical problems.

We first demonstrate Hamilton's principle by applying it to a standard one-dimensional problem in classical mechanics. Consider a discrete system of n interacting particles whose location at time t is given by $(x_1(t), x_2(t), \ldots, x_n(t))$. The kinetic energy is given by $E_k(x_1, \ldots, x_n) = \frac{1}{2} \sum_1^n m_i (dx_i/dt)^2$, while the potential energy is given by $E_p(x_1, x_2, \ldots, x_n)$. Since the force acting on particle i is $F_i = -\nabla_{x_i} E_p(x_1, \ldots, x_n)$, Newton's second law takes the form

$$\frac{\mathrm{d}}{\mathrm{d}t}\left(m_i \frac{\mathrm{d}x_i}{\mathrm{d}t}\right) = -\nabla_{x_i} E_p(x_1, \ldots, x_n) \quad i = 1, 2, \ldots, n. \tag{10.36}$$

To derive Hamilton's principle we define the total energy of the system (called the *Hamiltonian*) $E = E_k + E_p$. We also define the *Lagrangian* of the system $L = E_k - E_p$. The *action* in Hamilton's formalism is defined as

$$J = \int_{t_1}^{t_2} L \, \mathrm{d}t, \tag{10.37}$$

where t_1 and t_2 are two arbitrary points along the time axis. Hamilton postulated that a mechanical system evolves such that $\delta J = 0$, where the variation is taken with respect to all orbits $(y_1(t), \ldots, y_n(t))$ such that

$$y_i(t_1) = x_i(t_1), \quad y_i(t_2) = x_i(t_2), \quad i = 1, 2, \ldots, n. \tag{10.38}$$

Computing the first variation using (10.14), noting that the Lagrangian is a function of the form

$$L = L\left(x_1(t), \ldots, x_n(t), \frac{\mathrm{d}x_1}{\mathrm{d}t}, \frac{\mathrm{d}x_2}{\mathrm{d}t}, \ldots, \frac{\mathrm{d}x_m}{\mathrm{d}t}\right),$$

we write

$$\delta J = \int_{t_1}^{t_2}\left[\sum_1^n\left(m_i \frac{\mathrm{d}x_i}{\mathrm{d}t}\frac{\mathrm{d}\varphi_i}{\mathrm{d}t} - \frac{\partial E_p}{\partial x_i}\varphi_i\right)\right]\mathrm{d}t = \int_{t_1}^{t_2}\left[-\sum_1^n\left(m_i \frac{\mathrm{d}^2 x_i}{\mathrm{d}t^2} - \frac{\partial E_p}{\partial x_i}\right)\varphi_i\right]\mathrm{d}t = 0,$$

$$\tag{10.39}$$

where φ_i is the variation with respect to the particle x_i, and we have used the fact that the end point constraints (10.38) imply that $\varphi_i(t_1) = \varphi_i(t_2) = 0$. We have thus found that the Newton equations (10.36) are the Euler–Lagrange equations for the action functional.

The concept of the Lagrangian seems a bit odd at first sight. The sum of the kinetic and potential energies is the total energy, which is an intuitively natural physical object. But why should we consider their difference? To give an intuitive meaning to the difference $E_k - E_p$ it is useful to look a bit closer at the historical development of mechanics. Although Newton wrote clear laws for the dynamics of bodies, he and many other scientists looked for metaphysical principles behind them. As the mainstream philosophy of the eighteenth century was based on the idea of a single God, it was natural to assume that such a God would create a world that is 'perfect' in some sense. This prompted the French scientist Pierre de Maupertuis (1698–1759) to define the notion of action of a moving body. According to Maupertuis, the action of a body moving from a to b is $A = \int_a^b p \, dx$, where p is the particle's momentum. He then formulated his principle of least action, stating that the world is such that action is always minimized. Converting this definition of action to energy-related terms we write

$$A = \int_a^b p \, dx = \int_a^b m \frac{dx}{dt} \, dx = \int_{t_1}^{t_2} m \left(\frac{dx}{dt} \right)^2 dt = 2 \int_{t_1}^{t_2} E_k \, dt.$$

Here t_1 and t_2 are the initial and terminal times for the particle's path. The difficulty with this formula is that it only includes the kinetic energy, while the motion is determined by both the kinetic energy and the potential energy. Therefore, Lagrange used the identity $2E_k = E + L$ to write the action as

$$A = \int_{t_1}^{t_2} (E + L) \, dt.$$

Since the energy is a constant of the motion, extremizing $\int_{t_1}^{t_2} L \, dt$ is the same as extremizing $\int_{t_1}^{t_2} (E + L) \, dt$.

We proceed to demonstrate Hamilton's principle for a continuum. For this purpose we return to the problem of the elastic string. We consider a string clamped at the end points a and b, say $u(a, t) = u_a$, $u(b, t) = u_b$, where $u(x, t)$ is the string's deviation from the horizontal rest position. The kinetic energy of the string is given by $E_k = \frac{1}{2} \int_a^b \rho u_t^2 \, ds$, where $\rho(x, t)$ is the mass density, and $ds = \sqrt{1 + u_x^2} \, dx$ is a unit length element. The potential energy consists of the sum of the energy due to the stretching of the string, and the work done against a load $l(x, t)$:

$$E_p = \int_a^b \left(d \left(\sqrt{1 + u_x^2} - 1 \right) - l u \sqrt{1 + u_x^2} \right) dx,$$

where $d(x, t)$ is the string's elastic coefficient, and $l(x, t)$ is the load on the string. Notice that we allow the density, the elastic coefficient, and the load to depend on x and t. The action is thus given by

$$J = \int_{t_1}^{t_2} \int_a^b \left[\frac{1}{2} \sqrt{1 + u_x^2} \, \rho u_t^2 - d \left(\sqrt{1 + u_x^2} - 1 \right) + l u \sqrt{1 + u_x^2} \right] dx dt. \quad (10.40)$$

Consider variations $u + \varepsilon \psi$ such that ψ vanishes at the string's end points a and b, and also at the initial and terminal time points t_1 and t_2. Neglecting the term that is cubic in the derivatives u_x, u_t we get for the first variation:

$$\delta J = \int_{t_1}^{t_2} \int_a^b \left[\sqrt{1 + u_x^2} \, \rho u_t \psi_t - d(1 + u_x^2)^{-1/2} u_x \psi_x + l \sqrt{1 + u_x^2} \psi \right] dx dt.$$

$$(10.41)$$

We further integrate by parts the terms $u_t \psi_t$ (with respect to the t variable) and $u_x \psi_x$ (with respect to the x variable). The boundary conditions specified above for the variation ψ imply that all the boundary terms (both spatial and temporal) vanish. Therefore, equating the first variation to zero and integrating by parts we obtain

$$\delta J = \int_{t_1}^{t_2} \int_a^b \left\{ (-\sqrt{1 + u_x^2}\, \rho u_t)_t + [d(1 + u_x^2)^{-1/2} u_x]_x + l\sqrt{1 + u_x^2} \right\} \psi \, dx dt = 0.$$
(10.42)

The last equation implies the dynamical equation for the string's vibrations

$$(\rho u_t)_t - (1 + u_x^2)^{-1/2}[d(1 + u_x^2)^{-1/2} u_x]_x - l = 0. \tag{10.43}$$

If we assume that ρ and d are constants, and use the small slope approximation $|u_x| \ll 1$, we obtain again the one-dimensional wave equation.

Remark 10.5 The observant reader may have noticed that our nonlinear string model (10.43) is different from the string model (1.28) that we derived in Chapter 1. One difference is that in the current model we have allowed for variation of the mass density ρ and the elasticity coefficient d in space and time. A more subtle difference is in the form of the nonlinearity. The string model in Chapter 1 is based on the constitutive law (1.27). The model in this section is based on the action (10.40). It can be shown that this action is equivalent to a constitutive law of the form

$$\vec{T} = \hat{e}_\tau, \tag{10.44}$$

i.e. the tension is assumed to be uniform across the string. The model (1.27) can be called a "spring-like" string, while the model (10.44) can be called an "inextensible" spring.

Example 10.6 Vibrations of rods The potential energy of the string is stored in its stretching i.e. a string resists being stretched. We define a rod as an elastic body that also resists being bent. This means that we have to add to the elastic energy of the string a term that penalizes bending. The amount of bending of a curve $f(x)$ is measured by its *curvature*:

$$\kappa(x) = \frac{f_{xx}}{(1 + f_x^2)^{3/2}}.$$

Therefore, the Lagrangian for a rod under a load l can be written as

$$L = \int_a^b \left[\frac{1}{2}\sqrt{1 + u_x^2}\, \rho u_t^2 - \frac{d_1}{2} \frac{u_{xx}^2}{(1 + u_x^2)^2} - d_2 \left(\sqrt{1 + u_x^2} - 1 \right) + lu\sqrt{1 + u_x^2} \right] dx.$$
(10.45)

To simplify the computation in this example we introduce the small slopes ($|u_x| \ll$ 1) assumption at the outset. We thus approximate the action by

$$J = \int_{t_1}^{t_2} \int_a^b \left(\frac{1}{2} \rho u_t^2 - \frac{d_1}{2} u_{xx}^2 - \frac{d_2}{2} u_x^2 + lu \right) \, dx\, dt. \qquad (10.46)$$

Computing the first variation we find

$$\delta J = \int_{t_1}^{t_2} \int_a^b \left(\rho u_t \psi_t - d_1 u_{xx} \psi_{xx} - d_2 u_x \psi_x + l\psi \right) \, dx\, dt. \qquad (10.47)$$

In order to obtain the Euler–Lagrange equation we need to integrate the last integral by parts. Just as in the case of the plate, we assume that the rod is clamped, i.e. we specify u and u_x at the end points a and b. Therefore, the variation ψ vanishes at the spatial and temporal end points, and in addition, ψ_x vanishes at a and b. We thus obtain that the vibrations of rods are determined by the equation

$$(\rho u_t)_t - (d_2 u_x)_x + (d_1 u_{xx})_{xx} - l = 0. \qquad (10.48)$$

In Exercise 10.9 the reader will use the separation of variables method to solve the initial boundary value problem for equation (10.48).

10.2 Function spaces and weak formulation

In Chapter 6 we defined the notion of inner product spaces. We also introduced there concepts such as norms, complete orthonormal sets, and generalized Fourier expansions. In this section we take a few further steps in this direction. We shall introduce the concept of Hilbert spaces and show how to use it in the analysis of optimization problems and PDEs. We warn the reader that this is a small section dealing with an extensive subject. One of the main difficulties is that we are dealing with function spaces. Not only do these spaces turn out to be of infinite dimension, but the very meaning of "function" and "integral" must be very carefully examined. Therefore, our presentation will be minimal and confined to the basic facts that are essential in applications. We recommend [7] for more extensive exposition of the subject.

Let V be a (real) inner product space of functions defined over a domain $D \subset \mathbb{R}^n$. We have already seen in Chapter 9 that in such a space there is a well-defined (induced) norm, $\|f\| := \langle f, f \rangle^{1/2}$ for $f \in V$. We used the norm to define (Definition 6.9) convergence in the mean. We shall need later to define an alternative notion of convergence that is called *weak convergence*. To better distinguish between the different types of convergence, we shall replace 'convergence in the mean' by the title *strong convergence*. So, a sequence $\{v_n\}_{n=1}^\infty$ *converges strongly to* v, if $\lim_{n\to\infty} \|v_n - v\| = 0$.

A natural example for an inner product space is the space of all functions in a bounded domain D that are continuous in $\bar{D}$, equipped with the inner product

$$\langle f, g \rangle = \int_D f(\vec{x}) g(\vec{x}) \, d\vec{x}. \tag{10.49}$$

So far the definitions of an inner product space and the norms induced by the inner product are quite similar to the same notions in linear algebra. It is tempting, therefore, to proceed and borrow further ideas from linear algebra in developing the theory of function spaces. One of the most useful objects in linear algebra is a *basis* for a vector space. Indeed, intuitively, the generalized Fourier series we wrote in Chapter 6 looks like an expansion with respect to a basis that consists of the system of eigenfunctions of the given Sturm–Liouville problem. In order actually to define a basis in a function space we must overcome a serious obstacle, namely that the space must be 'complete' in an appropriate sense. To explain what we have in mind, recall again the example of the function space V consisting of the continuous functions over $\bar{D}$ under the inner product (10.49). Can we say that the eigenfunctions of the Laplace operator in D form a basis for this space? As a more concrete example, consider the function space V_E consisting of all continuous functions defined on $[0, \pi]$ that vanish at the end points. Can we say that the sequence $\{\sin nx\}$ is a basis for this space? To answer this question, recall from linear algebra that if B is a basis of an n-dimensional vector space V, then it spans V and, in particular, every linear combination of vectors in B is an element in V. Since we expect, from the examples above, that in the case of function spaces a basis will consist of infinitely many terms, we have to be slightly more careful and require that B be a linearly independent set that 'spans' V, and that each sequence in V whose terms are infinitesimally close (in norm) to each other, strongly converges to a vector in V. Note the latter condition is the completeness condition on the space V.

As it turns out, however, if we consider the space V_E above, and the candidate for a basis $B_E = \{\sin nx\}$, then there exist such linear combinations of functions in B_E that *do not* converge to a continuous function. In fact, this is not a surprise for us; we computed in Chapters 5 and 6 examples in which the Fourier series converged to discontinuous functions. This means that the function space V_E is not complete in some sense. Therefore, we now set about completing it.

For this purpose we recall from calculus the concept of a Cauchy sequence. A sequence of functions $\{f_n\}$ in an inner product space V is said to be a *Cauchy sequence* if for each $\varepsilon > 0$ there exists $N = N(\varepsilon)$ such that $\| f_n - f_m \| < \varepsilon$ whenever $n, m > N$. We note that every (strongly) converging sequence in V is a Cauchy sequence.

We proceed to construct out of an inner product space V a new space that consists of all the Cauchy sequences of vectors (functions) in V. Note that this space contains V since for any $f \in V$, the constant sequence $\{f\}$ is a Cauchy sequence. It turns out that the resulting space H, the *completion* of V, is an inner product space that has the property that every Cauchy sequence in H has a limit which is also an element of H. We can now introduce the following definition.

Definition 10.7 An inner product space in which every Cauchy sequence converges is said to be *complete*. Complete inner product spaces are called *Hilbert spaces* in honor of the German mathematician David Hilbert (1862–1943).[2]

In our example above, we constructed a Hilbert space out of the space V_E. The Hilbert space thus constructed is denoted $L_2[0, \pi]$ (or just L_2 if the domain under consideration is clear from the context). The construction implies in particular that every function in L_2 is either continuous or can be approximated (in the sense of strong convergence) to arbitrary accuracy by a continuous function.

Definition 10.8 A set W of functions in a Hilbert space H with the property that for every $f \in H$ and for every $\varepsilon > 0$ there exists a function $f_\varepsilon \in W$ such that $\| f - f_\varepsilon \| < \varepsilon$ is called a *dense set*.

Thus, the set of continuous functions on $[0, \pi]$ is dense in $L_2[0, \pi]$. Our discussion on the construction of Hilbert spaces has been heuristic. In particular it is not obvious that the space we completed out of V_E is still an inner product space. Nevertheless, it can be shown that essentially every inner product space V_I can be extended *uniquely* into a Hilbert space H_I such that V_I is dense in H_I, and such that the inner product $\langle f, g \rangle_{V_I}$ of V_I is extended into an inner product $\langle \phi, \psi \rangle_{H_I}$ for the elements of H_I, such that if $f, g \in V_I$, we have

$$\langle f, g \rangle_{V_I} = \langle f, g \rangle_{H_I}.$$

We are now ready to define a basis in a Hilbert space.

Definition 10.9 A set B of functions in a Hilbert space H is said to be a *basis* of H if its vectors are linearly independent and the set of finite linear combinations of functions from B is dense in H.

---

[2] If you find the concept of Hilbert space hard to grasp, then you are not alone. Hilbert was a professor at Göttingen University which was a center of mathematics research from the days of Gauss and Riemann up until the mid-1930s. The Hungarian–American mathematician John von Neumann (1903–1957), one of the founders of the modern theory of function spaces, visited Göttingen in the mid-1920s to give a lecture on his work. The legend is that shortly into the lecture Hilbert raised his hand and asked, "Herr von Neumann, could you explain to us *again* what a Hilbert space is?"

Example 10.10 In Chapter 6, we stated below Proposition 6.30 that for a given regular (or periodic) Sturm–Liouville problem on $[a, b]$, the eigenfunction expansion of a function u that satisfies $\int_a^b u^2(x)r(x)\,dx < \infty$ (r is the corresponding weight function) converges strongly to u. In other words, the orthonormal system of all eigenfunctions of a given regular (or periodic) Sturm–Liouville problem forms a basis in the Hilbert space of all functions such that the norm $\|u\|_r := (\int_a^b u^2(x)r(x)\,dx)^{1/2}$ is finite. In particular, the system $\{\sin nx\}_{n=1}^\infty$ (or $\{\cos nx\}_{n=0}^\infty$) is a basis of $L_2[0, \pi]$.

Remark 10.11 Since any orthonormal sequence is a linearly independent set, Proposition 6.13 implies that a complete orthonormal sequence in a Hilbert space is a basis.

As another example of a Hilbert space that is particularly useful in the theory of PDEs we consider the space $C^1(D)$ equipped with the inner product

$$\langle u, v \rangle = \int_D (uv + \nabla u \cdot \nabla v)\,d\vec{x}. \tag{10.50}$$

The special Hilbert space obtained from the completion process of the space above is called a *Sobolev space* after the Russian mathematician Sergei Sobolev (1908–1989). It is denoted by $H_1(D)$. Just like the case of the space L_2, the set of continuously differentiable functions in D is dense in $H_1(D)$. Other examples of Hilbert spaces are obtained for functions with special boundary behavior, for instance functions that vanish on the boundary of D.

What is the theory of Hilbert space we elaborated on good for? We shall now consider several applications of it.

10.2.1 Compactness

When we studied in calculus the problem of minimizing real valued functions, we had at our disposal a theorem that guaranteed that a continuous function in a closed bounded set K must achieve its maximum and minimum in K. Establishing a priori the existence of a minimizer for a functional is much harder. To understand the difficulty involved, let us recall from calculus that if A is a set of real numbers bounded from below, then it has a well-defined infimum. Moreover, there exists at least one sequence $a_n \subset A$ that converges to the infimum. Consider now, for example, the Dirichlet integral $G(u)$ defined over the functions in

$$\mathcal{B} = \{u \in C^1(D) \cap C(\bar{D}), \ u = g \ \ \vec{x} \in \partial D\}$$

for some domain D. Clearly G is bounded from below by 0. Therefore, there exists a sequence $\{u_k\}$ such that

$$\lim_{k\to\infty} G(u_k) = \inf_{u\in B} G(u).$$

Such a sequence $\{u_k\}$ is called a *minimizing sequence*.

The trouble is that a priori it is not clear that the infimum is achieved, and in fact, it is not even clear that the minimizing sequence u_k has convergent subsequences in B. Achieving the infimum is not always possible even for a sequence of numbers (for example if they are defined over an open interval), but we do like to retain some sort of convergence. In $\mathbb{R}^n$ we know that any bounded sequence has at least one convergent subsequence. This is the *compactness* property of bounded sets in $\mathbb{R}^n$.

Is it also true for the space B? The answer is no. We have seen examples in which a Fourier series converges strongly to a discontinuous function. This is a case in which a sequence of functions in B – the partial sums of the Fourier series – *does not* have any subsequence converging to a function in B. In Exercise 10.11 the reader will show that any orthonormal (infinite) sequence in a given infinite-dimensional Hilbert space H is bounded, but does not admit any subsequence converging strongly to a function in H.

It turns out that, if we consider infinite bounded sequences of functions in Hilbert spaces, we can still maintain to some extent the property of compactness. Unfortunately we have to weaken the meaning of convergence.

Definition 10.12 A sequence of functions $\{f_n\}$ in a Hilbert space H is said to *converge weakly* to a function f in H if

$$\lim_{n\to\infty} \langle f_n, g \rangle = \langle f, g \rangle \quad \forall g \in H. \tag{10.51}$$

Note that by the Riemann–Lebesgue lemma (see (6.38)), any (infinite) orthonormal sequence in a given infinite-dimensional inner product space converges weakly to 0.

The following theorem explains why we call the property (10.51) weak convergence, and also provides the fundamental compactness property of Hilbert spaces.

Theorem 10.13 *Let H be a Hilbert space. The following statements hold:*

(a) *Every strongly convergent sequence $\{u_n\}$ in H also converges weakly. The converse is not necessarily true.*

(b) *If $\{u_n\}$ converges weakly to u, then*

$$\|u\| \le \liminf_{n\to\infty} \|u_n\|. \tag{10.52}$$

(c) *Every sequence $\{u_n\}$ in H that is bounded (in the sense that $\|u_n\|_H \le C$) has at least one weakly convergent subsequence.*

(d) Every weakly convergence sequence in H is bounded.

Proof We only prove parts (a) and (b):

(a) We need to show that if $\|u_n - u\| \to 0$ in H, then $\{u_n\}$ converges weakly to u. For this purpose we write for an arbitrary function $f \in H$

$$|\langle u_n, f \rangle - \langle u, f \rangle| = |\langle u_n - u, f \rangle| \le \|u_n - u\|^{1/2}\|f\|^{1/2}, \tag{10.53}$$

where the last step follows from the Cauchy–Schwartz inequality (see (6.8)).

The second part of (a) follows from a counterexample. Consider the sequence $\{\sin nx\} \subset L_2([0, \pi])$. Then by the Riemann–Lebesgue lemma $\{\sin nx\}$ converges weakly to 0, while $\|\sin nx\| = \sqrt{\pi/2}$ and therefore, $\{\sin nx\}$ does not converge strongly to 0.

(b) If $\{u_n\}$ converges weakly to u, then, in particular, $\langle u_n, u \rangle \to \|u\|^2$. By the Cauchy–Schwartz inequality, it follows that

$$\|u\|^2 = \lim_{n \to \infty} |\langle u_n, u \rangle| \le \|u\| \liminf_{n \to \infty} \|u_n\|. \tag{10.54}$$

$\square$

We have therefore shown that it is useful to work in Hilbert spaces to guarantee compactness in some sense. It still remains to show in applications that a given minimizing sequence is indeed bounded and thus admits a weakly convergence subsequence, and that at least one of its limits indeed achieves the infimum for the underlying functional. We shall demonstrate all of this through an example below.

10.2.2 The Ritz method

Consider the problem of minimizing a functional $G(u)$, where u is taken from some Hilbert space H. The Ritz method is based on selecting a basis B (preferably orthonormal) for H, and expressing the unknown minimizer u in terms of the elements ϕ_n of B:

$$u = \sum_{n=1}^{\infty} \alpha_n \phi_n. \tag{10.55}$$

The functional minimization problem has been transformed to an algebraic (albeit infinite-dimensional) minimization problem in the unknown coefficients α_n. This process is similar to our discussion after the introduction of the Rayleigh quotient in Chapters 6 and 9.

Practically, we can use the fact that since the series expansion for u is convergent, we expect the coefficients to decay as $n \to \infty$. We can therefore truncate the expansion at some finite term N and write

$$u \approx \sum_{n=1}^{N} \alpha_n \phi_n. \tag{10.56}$$

This approximation leads to a finite-dimensional algebraic system that can be handled by a variety of numerical tools as discussed in Chapter 11.

Remark 10.14 A very interesting question is: what would be an optimal basis? It is clear that some bases are superior to others. For example, the series (10.55) might converge much faster in one basis than in another basis. In fact, the series might even be finite if we are fortunate (or clever). For instance, suppose that we happened to choose a basis that contains the minimizing function u itself. Then the series expansion would consist of just one term!

At the other extreme, we might face the problem of not having *any* obvious candidate for a basis. This would happen when we consider a Hilbert space of functions defined over a general domain that has no symmetries. We shall address this question in Chapter 11.

Example 10.15 To demonstrate the Ritz method we return to the problem of phase reconstruction (Example 10.2). In typical applications D is the unit disk. We shall seek the minimizer of $K(u)$ in the space $H_1(D)$. What would be a good basis for this space? The first candidate that comes to mind is the basis

$$\left\{ J_n\left(\frac{\alpha_{n,m}}{a}r\right) \cos n\theta \right\} \cup \left\{ J_n\left(\frac{\alpha_{n,m}}{a}r\right) \sin n\theta \right\}$$

that we constructed in (9.80). While this basis would certainly do the work, it turns out that in practice physicists use another basis. Phase reconstruction is an important step in a process called adaptive optics, in which astronomers correct images obtained by telescopes. These images are corrupted by atmospheric turbulence (this is similar to scintillation of stars when they are observed by a naked eye). Thus astronomers measure the phase and use these measurements to adjust flexible mirrors to correct the image. The Dutch physicist Frits Zernike (1888–1966) proposed in 1934 to expand the phase in a basis in which he replaced the Bessel functions above by radial functions that are polynomials in r.

The *Zernike basis* for the space L_2 over the unit disk consists of functions that have the same angular form as the Bessel basis above. The radial Bessel functions, though, are replaced by orthogonal polynomials. Using complex number notation, we write the Zernike functions as

$$Z_n^m(r, \theta) = R_n^m(r)\, e^{im\theta}, \tag{10.57}$$

where the polynomials R_n^m are orthogonal over the interval $(0, 1)$ with respect to the inner product $\langle f(r), g(r) \rangle := \int_0^1 f(r)g(r)r\, dr$. For some reason Zernike did not choose the polynomials to be orthonormal, but rather set $\langle R_n^m, R_{n'}^m \rangle = [1/2(n+1)]\delta_{n,n'}$. In fact, one can write the polynomials explicitly (they are only defined for $n \geq |m| \geq 0$):

$$R_n^m(r) = \begin{cases} \sum_{l=0}^{(n-|m|)/2} \dfrac{(-1)^l (n-l)!}{l!(\frac{1}{2}(n+|m|)-l)!(\frac{1}{2}(n-|m|)-l)!} r^{n-2l} & \text{for } n-|m| \text{ even,} \\ 0 & \text{for } n-|m| \text{ odd.} \end{cases}$$

$$(10.58)$$

The phase is expanded in the form

$$u(r, \theta) = \sum_{n,m} \alpha_{n,m} Z_n^m(r, \theta).$$

We then substitute this expansion into the minimization problem (10.19) to obtain an infinite-dimensional quadratic minimization problem for the unknown coefficients $\{\alpha_{n,m}\}$. In practice the series is truncated at some finite term, and then, since the functional is quadratic in the unknown coefficients, the minimization problem is reduced to solving a system of linear algebraic equations. Notice that this method has a fundamental practical flaw: since the functional involves *derivatives* of u, and the derivatives of the Zernike functions are *not* orthogonal, we need to evaluate all the inner products of these derivatives. Moreover, this implies that the matrix associated with the linear algebraic system we mentioned above is generically full; in contrast we shall show in Chapter 11 that if we select a clever basis, we can obtain linear algebraic systems that are associated with sparse matrices, whose solution can be computed much faster.

10.2.3 Weak solutions and the Galerkin method

We shall use the following example to illustrate some of the ideas that have been developed in this chapter and also to introduce the concept of weak formulation.

Example 10.16 Consider the minimization problem

$$\min Y(u) = \int_D \left(\frac{1}{2} |\nabla u|^2 + \frac{1}{2} u^2 + fu \right) d\vec{x}, \qquad (10.59)$$

where D is a (bounded) domain in $\mathbb{R}^n$ and f is a given continuous function satisfying without loss of generality $|f| \leq 1$ in D. The first variation is easily found to be

$$\delta Y(u) = \int_D (\nabla u \cdot \nabla \psi + u\psi + f\psi)\, d\vec{x}. \qquad (10.60)$$

We seek a minimizer in the space $H_1(D)$, and take the variation ψ also to belong to this space. Therefore, the condition on the minimizer u is

$$\int_D (\nabla u \cdot \nabla \psi + u\psi + f\psi) \, d\vec{x} = 0 \qquad \forall \psi \in H_1(D). \tag{10.61}$$

If we assume that the minimizer u is a smooth function (i.e. in the class $C^2(\bar{D})$) and that D has a smooth boundary, then we can integrate (10.61) by parts in the usual way and obtain the Euler–Lagrange equation

$$-\Delta u + u = -f \quad \vec{x} \in D, \qquad \partial_n u = 0 \quad \vec{x} \in \partial D. \tag{10.62}$$

Equation (10.61), however, is more general than (10.62) since it also holds under the weaker assumption that u is only once continuously differentiable, or at least is a suitable limit of functions in $C^1(D)$. Therefore, we call (10.61) the *weak formulation* of (10.62). We prove the following statement.

Theorem 10.17 *The weak formulation (10.61) has a unique solution u^*. Moreover, u^* is a minimizer of (10.59).*

Proof Since $|f| \leq 1$, then $\frac{1}{2}u^2 + uf \geq -\frac{1}{2}f^2 \geq -\frac{1}{2}$ for all $\vec{x} \in D$. Therefore, $Y(u) \geq -\frac{1}{2}|D|$ and thus the functional is bounded from below. Let $\{u_n\}$ be a minimizing sequence, i.e.

$$\lim_{n \to \infty} Y(u_n) = I := \inf_{u \in H_1(D)} Y(u).$$

The Cauchy–Schwartz inequality implies that $|\int_D f u d\vec{x}| \leq |D|^{1/2}\|u\|_{L_2(D)}$. Since it suffices to consider u_n such that $Y(u_n) < Y(0) = 0$, it follows that

$$\frac{1}{2}\|u_n\|^2_{L_2(D)} \leq \frac{1}{2}\|u_n\|^2_{H_1(D)} \leq \left| \int_D f u_n \, d\vec{x} \right| \leq |D|^{1/2}\|u_n\|_{L_2(D)}.$$

Therefore, $\|u_n\|_{L_2(D)} < C := 2|D|^{1/2}$, which in turn implies that $\|u_n\|_{H_1(D)} < C$. Thus, Theorem 10.13 implies that $\{u_n\}$ has at least one weakly convergent subsequence $\{u_{n_k}\}$ in $H_1(D)$. We denote its weak limit by u^*.

Using (10.52) and the fact that weak convergence in $H_1(D)$ implies weak convergence in $L^2(D)$ (see Exercise 10.12), it follows that

$$Y(u^*) = \frac{1}{2}\|u^*\|^2_{H_1(D)} + \int_D u^* f \, d\vec{x}$$

$$\leq \frac{1}{2}\liminf_{n \to \infty} \|u_n\|^2 + \lim_{n \to \infty} \int_D u_n f \, d\vec{x} = \lim_{n \to \infty} Y(u_n) = I \leq Y(u^*).$$

Therefore, u^* is a minimizer of the problem.

Now fix $\psi \in H_1$. Then

$$g(\varepsilon) := Y(u^* + \varepsilon\psi) = Y(u^*) + \frac{\varepsilon^2}{2}\|\psi\|^2_{H_1(D)} + \varepsilon\langle u^*, \psi\rangle_{H_1(D)} + \varepsilon\int_D \psi f\, d\vec{x}$$

has a minimum at $\varepsilon = 0$, therefore, $g'(0) = 0$. Hence

$$\langle u^*, \psi\rangle_{H_1(D)} = -\int_D f\psi\, d\vec{x}. \tag{10.63}$$

Since (10.63) holds for all $\psi \in H_1(D)$, we have established the existence of a solution of the weak formulation. To prove the uniqueness of the solution, we assume by contradiction that there exist two solutions u_1^* and u_2^*. We then form their difference $v^* = u_1^* - u_2^*$, and obtain for v^*:

$$\langle v^*, \psi\rangle_{H_1(D)} = 0 \quad \forall\psi \in H_1(D). \tag{10.64}$$

In particular, we can choose $\psi = v^*$, and then (10.64) reduces to $\|v^*\|_{H_1(D)} = 0$, implying $v^* \equiv 0$. $\qquad\square$

If we can prove that v^* is in $C^2(D) \cap C^1(\bar{D})$, then Theorem 10.17 implies the existence of a classical solution to the elliptic boundary value problem (10.62).

Although we have proved that the weak formulation has a unique solution, the proof was not constructive. The limit u^* was identified as a limit of an as yet unknown sequence. We therefore introduce now a practical method for *computing* the solution. The idea is to construct a chain of subspaces $H^{(1)}, H^{(2)}, \dots, H^{(k)}, \dots$ with the property that $H^{(k)} \subset H^{(k+1)}$, and $\dim H^{(k)} = k$, such that their union exhausts the full $H_1(D)$, i.e. there exists a basis $\{\phi_k\}$ of $H_1(D)$ with $\phi_k \in H^{(k)}$. In each subspace $H^{(k)}$, we select a basis $\phi_1^k, \phi_2^k, \dots, \phi_k^k$. We write the weak formulation in $H^{(k)}$ as

$$\langle v^k, \phi_i^k\rangle_{H_1(D)} = -\int_D f\phi_i^k\, d\vec{x} \quad i = 1, 2, \dots, k. \tag{10.65}$$

If we further express the unknown function v^k in terms of the basis ϕ^k, i.e. $v^k = \sum_{j=1}^k \alpha_j^k\phi_j^k$, we obtain for the unknown coefficient vector $\vec{\alpha}^k$ the algebraic equations

$$\sum_{j=1}^k K_{ij}^k\alpha_j^k = d_i \quad i = 1, 2, \dots, k, \tag{10.66}$$

where

$$K_{ij}^k = \langle\phi_i^k, \phi_j^k\rangle_{H_1(D)}, \quad \text{and} \quad d_i = -\int_D f\phi_i^k d\vec{x}. \tag{10.67}$$

It can be shown (although we shall not do it here) that the system (10.66) has a unique solution for all k, and that the sequence v^k converges strongly to u^*.

The practical method we presented for computing u^* is called the *Galerkin method* after the Russian engineer Boris Galerkin (1871–1945). In Exercise 10.10 the reader will show that for the minimization problem at hand the Galerkin method is identical to the Ritz method introduced earlier. This is why these methods are often confused (or maybe just fused...) with each other and go together under the title the Galerkin–Ritz method. We point out, however, that the Galerkin method is more general than the Ritz method in the sense that it is not limited to problems where the weak formulation is derived from a variation of a functional. In fact, given any PDE of the abstract form $L[u] = f$, where L is a linear or nonlinear operator, we can apply the Galerkin method by writing the equation in the form

$$\langle L[u] - f, \psi \rangle = 0 \quad \forall \psi \in H,$$

where H is a suitable Hilbert space. Sometimes, we can then integrate the left hand side by parts and throw some derivatives of u to ψ and thus obtain a formulation that requires less regularity for its solution.

There still remains the important question of how to choose the subspaces H^k that we used in the Galerkin method. A very important class of such subspaces forms a numerical method called *finite elements*, which will be discussed in more detail in Chapter 11.

10.3 Exercises

10.1 Consider the variational problem

$$\min K(y) := \int_0^1 [1 + y'(t)^2] \, dt, \tag{10.68}$$

where $y \in C^1([0, 1])$ satisfies $y(0) = 0$, $y(1) = 1$. Find the Euler–Lagrange equation, the boundary conditions, and the minimizer for this problem. Is the minimizer unique?

10.2 Consider the variational problem

$$\min K(u) := \int_D \left[\frac{1}{2} |\nabla u|^2 + \alpha u_{xx} u_{yy} + (1 - \alpha) u_{xy}^2 \right] \, dx dy, \tag{10.69}$$

where α is a real constant. Find the Euler–Lagrange equation, and the natural boundary conditions for the problem.

10.3 Consider the variational problem

$$\min K(u) := \int_D \left(|\nabla u|^2 + \frac{1}{2} g u^4 \right) \, dx dy, \tag{10.70}$$

where $D \subset \mathbb{R}^2$, and $g(x, y)$ is a given positive function. Find the Euler–Lagrange equation and the natural boundary conditions for this problem.

10.4 Can you guess a third minimal surface spanned by the spatial curve of Figure 10.3?

10.5 A canonical physical model concerns a system whose kinetic energy and potential energy are of the forms

$$E_k = \frac{1}{2} \int_D u_t^2 d\vec{x} \qquad E_p = \int_D \left[\frac{1}{2} |\nabla u|^2 + V(u) \right] d\vec{x}.$$

Here $u(\vec{x}, t)$ is a function that characterizes the system, D is a domain in $\mathbb{R}^3$, and V is a known function.

(a) Write the Lagrangian and the action for the system.

(b) Equating the first variation of the action to zero, find the dynamical PDE obtained by Hamilton's principle.

Comment The PDE that you find in (b) is called the Klein–Gordon equation (see (9.19)).

10.6 Suppose that

$$p, p', q, r \in C([a, b]), \qquad p(x), r(x) > 0, \qquad \forall x \in [a, b].$$

(a) Write the Euler–Lagrange equation for the following *constrained* optimization problem:

$$\min \int_a^b \left[p(x)y'(x)^2 - q(x)y(x) \right] dx, \quad \text{subject to} \int_a^b r(x)y(x)^2 \, dx = 1,$$
$$(10.71)$$

where y satisfies the boundary conditions $y(a) = y(b) = 0$.

Hint Use a Lagrange multiplier method to replace (10.71) with a minimization problem without constraints.

(b) What is the relation between the calculation you performed in (a) and the Rayleigh quotient (6.71)?

10.7 (a) Let D be a domain in $\mathbb{R}^2$. Write the Euler–Lagrange equation for the following *constrained* optimization problem:

$$\min \int_D |\nabla u|^2 \, dxdy, \quad \text{subject to} \int_D u^2 \, dxdy = 1, \quad u = 0 \ x \in \partial D. \quad (10.72)$$

(b) What is the relation between the calculation you performed in (a) and the Rayleigh–Ritz formula (9.53)?

10.8 Use the Hamilton principle and the energy functional for the membrane (see Example 10.3) to compute the equation for the vibration of a membrane with an elasticity constant d and a fixed density ρ in the small slope approximation. What are the eigenfrequencies of the membrane in the rectangle $[0, 1] \times [0, 8]$?

10.9 Consider the vibrations of a rod (equation (10.48)) clamped at $x = 0$ and $x = b$, with $d_2 = 0, d_1 = d$ for some constant d, and $\rho = 1$.

(a) Write separated solutions of the form $u(x, t) = X(x)T(t)$. Denote the eigenvalues of the eigenvalue problem for X by λ_n. Write explicitly the eigenvalue problem.

(b) Show that all eigenvalues are positive.

(c) Show that the eigenvalues are the solutions of the transcendental equation

$$\cosh \alpha b \cos \alpha b = 1, \tag{10.73}$$

where $\alpha = \lambda^{1/4}$. What is the asymptotic behavior of the nth eigenvalue as $n \to \infty$?

10.10 Analyze the minimization problem (10.59) by the Ritz method. Use the same bases $\{\phi^k\}$ as in the Galerkin method, and show that the Ritz method and the Galerkin method give rise to the same algebraic equation.

10.11 Let $\{v_n\}$ be an orthonormal infinite sequence in a given infinite-dimensional Hilbert space H.

(a) Show that $\{v_n\}$ is bounded.

(b) Show that $\{v_n\}$ converges weakly to 0.

(c) Show that $\{v_n\}$ does not admit any subsequence converging strongly to a function in H.

Hint Use the Riemann–Lebesgue lemma (see (6.38)).

10.12 Prove that if $\{u_n\}$ is a weakly converging sequence in $H_1(D)$, then $\{u_n\}$ weakly converges in $L^2(D)$.

11

Numerical methods

11.1 Introduction

In the previous chapters we studied a variety of solution methods for a large number
of PDEs. We point out, though, that the applicability of these methods is limited to
canonical equations in simple domains. Equations with nonconstant coefficients,
equations in complicated domains, and nonlinear equations cannot, in general, be
solved analytically. Even when we can produce an 'exact' analytical solution, it
is often in the form of an infinite series. Worse than that, the computation of each
term in the series, although feasible in principle, might be tedious in practice, and,
in addition, the series might converge very slowly. We shall therefore present in
this chapter an entirely different approach to solving PDEs. The method is based
on replacing the continuous variables by discrete variables. Thus the continuum
problem represented by the PDE is transformed into a discrete problem in finitely
many variables. Naturally we pay a price for this simplification: we can only obtain
an approximation to the exact answer, and even this approximation is only obtained
at the discrete values taken by the variables.

The discipline of numerical solution of PDEs is rather young. The first analysis
(and, in fact, also the first formulation) of a discrete approach to a PDE was presented
in 1929 by the German-American mathematicians Richard Courant (1888–1972),
Kurt Otto Friedrichs (1901–1982), and Hans Lewy (1905–1988) for the special
case of the wave equation. Incidentally, they were not interested in the numerical
solution of the PDE (their work preceded the era of electronic computers by almost
two decades), but rather they formulated the discrete problem as a means for a
theoretical analysis of the wave equation. The Second World War witnessed the
introduction of the first computers that were built to solve problems in continuum
mechanics. Following the war and the rapid progress in the computational power
of computers, it was argued by many scientists that soon people would be able to
solve numerically any PDE. Thus, von Neumann envisioned the ability to obtain

long-term weather prediction by modeling the hydrodynamical behavior of the
atmosphere. These expectations turned out to be too optimistic for several reasons:

(1) Many nonlinear PDEs suffer from inherent instabilities; a small error in estimating the
    equation's coefficients, the initial conditions, or the boundary conditions may lead to
    a large deviation of the solution. Such difficulties are currently investigated under the
    title 'chaos theory'.
(2) Discretizing a PDE turns out to be a nontrivial task. It was discovered that equations
    of different types should be handled numerically differently. This problem led to the
    creation a new branch in mathematics: *numerical analysis*.
(3) Each new generation of computers brings an increase in computational power and has
    been accompanied by an increased demand for accuracy. At the same time scientists
    develop more and more sophisticated physical models. These factors result in a never-
    ending race for improved numerical methods.

We pointed out earlier that a numerical solution provides only an approximation
to the exact solution. In fact, this is not such a severe limitation. In many situations
there is no need to know the solution with infinite accuracy. For example, when
solving a heat conduction problem it is rarely required to obtain an answer with
an accuracy better than a hundredth of a degree. In other words, an exact answer
provides more information than is actually required. Moreover, even if we can write
an exact answer in terms of trigonometric functions or special functions, we can
only evaluate these functions to some finite accuracy.

As we stated above, the main idea of a numerical method is to replace the PDE,
formulated for one unknown real valued function, by a discrete equation in finitely
many unknowns. The discrete problem is called a *numerical scheme*. Thus a PDE
is replaced by an algebraic equation. When the original PDE is linear, we obtain,
in general, a system of linear algebraic equations. We shall demonstrate below
that the accuracy of the solution depends on the number of discrete variables, or,
alternatively, on the number of algebraic equations. Therefore, seeking an accurate
approximation requires us to solve large algebraic systems.

There are several techniques for converting a PDE into a discrete problem. We
have already mentioned in Chapter 10 the Ritz method that is suitable for equations
arising from optimization problems. The main difficulty in the Ritz method is
in finding a good basis for problems in domains that are not simple (where, for
example, the eigenfunctions for the Laplacian are not easy to calculate). The most
popular numerical methods are the *finite difference method (FDM)* and the *finite
elements method (FEM)*. Both methods can be used for most problems, including
equations with constant or nonconstant coefficients, equations in general domains,
and even nonlinear equations. Because of the limited scope of the discussion in this
book we shall only introduce the basic ideas behind these two methods. There is an

on-going debate on whether one of the methods is superior to the other. Our view is that the FDM is simpler to describe and to program (at least for simple equations). The FEM, on the other hand, is somehow 'deeper' from the mathematical point of view, and is more flexible when solving equations in complex geometries.

We end this section by noting that we shall discuss the prototypes of second-order equations (heat, Laplace, wave). In addition we choose simple domains in order to simplify the presentation. Nevertheless, unlike the analytical methods introduced in the preceding chapters, the numerical methods provided here are not limited to symmetric domains and they can be applied in far more complex situations.

11.2 Finite differences

To present the principle of the finite difference approximation, consider a smooth function in two variables $u(x, y)$, defined over the rectangle $D = [0, a] \times [0, b]$. We further define a discrete *grid* (mesh; net) of points in D:

$$(x_i, y_j) = (i\Delta x, j\Delta y) \ 0 \le i \le N - 1, \quad 0 \le j \le M - 1, \tag{11.1}$$

where $\Delta x = a/(N - 1)$, $\Delta y = b/(M - 1)$. Since we are interested in the values taken by u at these points, it is convenient to write

$$U_{i,j} = u(x_i, y_j).$$

We use the Taylor expansion of u around the point (x_i, y_j) to compute $u(x_{i+1}, y_j)$ (see Figure 11.1). We obtain

$$u(x_{i+1}, y_j) = u(x_i, y_j) + \partial_x u(x_i, y_j)\Delta x$$
$$+ \frac{1}{2}\partial_x^2 u(x_i, y_j)(\Delta x)^2 + \frac{1}{6}\partial_x^3 u(x_i, y_j)(\Delta x)^3 + \cdots. \tag{11.2}$$

It follows that

$$\partial_x u(x_i, y_j) = \frac{U_{i+1,j} - U_{i,j}}{\Delta x} + O(\Delta x). \tag{11.3}$$

We obtained the following approximation for the partial derivative of u with respect to x which is called a *forward difference formula*:

$$\partial_x u(x_i, y_j) \sim \frac{U_{i+1,j} - U_{i,j}}{\Delta x}. \tag{11.4}$$

Figure 11.1 A one-dimensional grid.

Similarly, one can derive a *backward difference formula*

$$\partial_x u(x_i, y_j) \sim \frac{U_{i,j} - U_{i-1,j}}{\Delta x}. \tag{11.5}$$

The error induced by the approximation (11.4) is called a *truncation error*. To minimize the truncation error, and thus to obtain a more faithful approximation for the derivative, we need Δx to be very small. Since $\Delta x = O(1/N)$, this requirement implies that N should be very big. We shall see below that working with large values of N (very fine grids) is expensive in terms of computational time as well as in terms of memory requirements. We therefore seek a finite difference approximation for u_x that is more accurate than (11.4). For this purpose write also the Taylor expansion for $u(x_{i-1}, y_j)$, and subtract the two Taylor expansions to obtain

$$\partial_x u(x_i, y_j) = \frac{U_{i+1,j} - U_{i-1,j}}{2\Delta x} + O((\Delta x)^2). \tag{11.6}$$

The approximation

$$\partial_x u(x_i, y_j) \sim \frac{U_{i+1,j} - U_{i-1,j}}{2\Delta x} \tag{11.7}$$

is called a *central finite difference* or, for short, a central difference. For obvious reasons we say that it is a second-order approximation for u_x. Similarly we obtain the central difference for u_y:

$$\partial_y u(x_i, y_j) \sim \frac{U_{i,j+1} - U_{i,j-1}}{2\Delta y}. \tag{11.8}$$

Using a similar method one can also derive second-order central differences for the second derivatives of u:

$$\partial_{xx} u = \frac{U_{i-1,j} - 2U_{i,j} + U_{i+1,j}}{(\Delta x)^2} + O((\Delta x)^2) \tag{11.9}$$

and

$$\partial_{yy} u = \frac{U_{i,j-1} - 2U_{i,j} + U_{i,j+1}}{(\Delta y)^2} + O((\Delta y)^2). \tag{11.10}$$

The computation of a second-order finite difference approximation for the mixed derivative $\partial_{xy} u$ is left for an exercise.

11.3 The heat equation: explicit and implicit schemes, stability, consistency and convergence

The reader might infer from the discussion in the previous section that the construction of a numerical scheme, i.e. converting a PDE to a discrete problem, is a simple task: all one has to do is to replace each derivative by a finite difference approximation, and *voilà* a numerical scheme pops up. It turns out, however, that the matter is not so simple! One should seriously consider several difficulties that frequently arise during the process, and generate an appropriate numerical scheme for each differential problem. In this section we shall present some basic terms and ideas related to the construction of numerical schemes and apply them to derive a variety of schemes for the heat equation.

Consider the problem

$$u_t = k u_{xx} \qquad\qquad 0 < x < \pi, \ \ t > 0, \qquad (11.11)$$

$$u(0, t) = u(\pi, t) = 0 \quad t \geq 0, \quad u(x, 0) = f(x) \quad 0 \leq x \leq \pi, \quad (11.12)$$

where we assume $f(0) = f(\pi) = 0$.

Fix an integer $N > 2$ and a positive number Δt, and set $\Delta x := \pi/(N - 1)$. We define a grid $\{x_i = i \Delta x\}$ on the interval $[0, \pi]$ and another grid $\{t_n = n \Delta t\}$ on the interval $[0, T]$. We further use the notation $U_{i,n} = u(x_i, t_n)$. A simple reasonable difference scheme for the problem (11.11)–(11.12) is based on a first-order difference for the time derivative, and a central difference for the spatial derivative:

$$\frac{U_{i,n+1} - U_{i,n}}{\Delta t} = k \frac{U_{i+1,n} - 2U_{i,n} + U_{i-1,n}}{(\Delta x)^2} \quad 1 \leq i \leq N - 2, \quad n \geq 0. \quad (11.13)$$

Notice that the boundary values are determined by the boundary conditions (11.12), i.e.

$$U_{0,n} = U_{N-1,n} = 0 \ \ n \geq 0.$$

Using simple algebraic manipulations, (11.13) can be written in the form of an explicit expression for the discrete solution at each point at time $n + 1$ in terms of the solution at time n:

$$U_{i,n+1} = U_{i,n} + \alpha(U_{i+1,n} - 2U_{i,n} + U_{i-1,n}), \qquad (11.14)$$

where $\alpha = k \Delta t/(\Delta x)^2$. The initial condition for the PDE becomes an initial condition for the *difference equation* (11.13):

$$U_{i,0} = f(x_i). \qquad (11.15)$$

We have derived a simple algorithm for a numerical solution of the heat equation. Were we to attempt to apply it, however, we would probably obtain meaningless results! The problem is that, unless we are careful in our choice for the differences Δt and Δx, the difference scheme (11.14) is not stable. This means that a small perturbation to the initial condition will grow (very fast) in time. Recalling that the representation of numbers in the computer is always finite, we realize that every numerical solution inevitably includes some *round-off error*. Hence a necessary condition for the validity of a numerical scheme is its stability against small perturbations, including round-off errors.

Let us define more precisely the notion of stability for a numerical scheme (this is analogous to the notion of stability we saw in Chapter 1 in the context of well-posedness). For this purpose, let us denote the vector of unknowns by V. It consists of the approximations to the values of u (the solution of the original PDE) at the grid points where the values of u are not known. Notice that the solution u *is* known at the grid points where the boundary or initial conditions are given. We write the discrete problem in the form $T(V) = F$, where the vector F contains the known parameters of the problem (e.g. initial condition or boundary conditions). When the PDE is linear, so is the numerical scheme. In this case we can write the scheme as $\mathbf{A}V = F$, where we denote the appropriate matrix by $\mathbf{A}$. We shall demonstrate such a matrix notation in the next section.

Definition 11.1 Let $T(V) = F$ be a numerical scheme. Let V^i be two solutions, i.e. $T(V^i) = F^i$, for $i = 1, 2$. We shall say that the scheme is *stable* if for each $\varepsilon > 0$ there exist $\delta(\varepsilon)$ such that $|F^1 - F^2| < \delta$ implies $|V^1 - V^2| < \varepsilon$. In other words, a small change in the problem's data implies a small change in the solution.

We shall demonstrate the stability notion we just defined for the scheme we proposed for the heat equation. The external conditions are given here by the initial conditions $f(x)$ at the grid points. To examine the stability of the scheme for an arbitrary perturbation, we choose an initial condition of the form $f(x) = \sin Lx$ for some positive integer L. Since any solution of the heat equation under consideration is determined by a combination of such solutions, it follows that stability with respect to any such initial condition implies stability for arbitrary perturbations. Conversely, instability even with respect to a single L implies instability with respect to random perturbations. In light of the form of the solution of the heat equation that we found in Chapter 5, we seek a solution for (11.14) in the form $U_{i,n} = A(n) \sin Li\Delta x$. Substituting this function into (11.12) leads to a difference equation for the sequence $A(n)$:

$$A(n + 1) = A(n)\left\{1 + \alpha \frac{\sin[L(i + 1)\Delta x] - 2 \sin Li\Delta x + \sin[L(i - 1)\Delta x]}{\sin Li\Delta x}\right\}.$$

$$(11.16)$$

We use the identity

$$\frac{\sin[L(i + 1)\Delta x] - 2 \sin Li\Delta x + \sin[L(i - 1)\Delta x]}{\sin Li\Delta x} = -4 \sin^2\left(\frac{L\Delta x}{2}\right) \quad (11.17)$$

to simplify the equation for $A(n)$, $n \geq 1$:

$$A(n + 1) = \left[1 - 4\alpha \sin^2 \left(\frac{L\Delta x}{2} \right) \right] A(n).$$

Therefore $\{A(n)\}$ is a geometric sequence and

$$A(n) = \left[1 - 4\alpha \sin^2 \left(\frac{L\Delta x}{2} \right) \right]^n A(0).$$

Consequently,

$$U_{i,n} = \left[1 - 4\alpha \sin^2 \left(\frac{L\Delta x}{2} \right) \right]^n \sin(Li\,\Delta x).$$

Since $1 - 4\alpha \sin^2 L\Delta x < 1$, it follows that a necessary and sufficient condition for the stability of the difference equation (11.14) is

$$1 - 4\alpha \sin^2 \left(\frac{L\Delta x}{2} \right) > -1.$$

Therefore, we cannot choose Δt and Δx arbitrarily; they must satisfy the *stability condition*

$$\Delta t \leq \frac{1}{2k} (\Delta x)^2. \tag{11.18}$$

Recall that we normally select a small value for Δx in order to obtain a finite difference formula that is faithful to the analytic derivative. Thus the stability condition is bad news for us. Although the difference scheme we developed is simple and easy to program, the stability requirement implies that Δt must be *very* small; we have to cover a very large number of time steps to compute the solution at some finite positive time. For this reason many people have tried to upgrade the scheme (11.14) into a faster scheme. However, before considering these alternative schemes, let us examine two further important theoretical aspects: consistency and convergence of a scheme.

Definition 11.2 A numerical scheme is said to be *consistent* if the solution of the PDE satisfies the scheme in the limit where the grid size tends to zero. For example, in the case of the heat equation a scheme will be called consistent if the solution of the heat equation satisfies the scheme in the limit where $\Delta x \to 0$ and $\Delta t \to 0$.

To examine the consistency of the scheme (11.13) we define for each function $v(x, t)$

$$R[v] = \frac{v(x_i, t_{n+1}) - v(x_i, t_n)}{\Delta t} - k \frac{v(x_{i+1}, t_n) - 2v(x_i, t_n) + v(x_{i-1}, t_n)}{(\Delta x)^2}. \tag{11.19}$$

Substituting a solution of the heat equation $u(x, t)$ into the expression R, we obtain (using the finite difference formula from the previous section)

$$R[u] = \frac{1}{2}\Delta t \bar{u}_{tt} - \frac{1}{12}k(\Delta x)^2 \bar{u}_{xxxx}, \qquad (11.20)$$

where $\bar{u}$ denotes the value of u or its derivative at a point near the grid point (x_i, t_n). Hence, if u is sufficiently smooth we have $\lim_{\Delta x, \Delta t \to 0} R[u] = 0$, and the scheme is indeed consistent.

Properties such as stability and consistency are, of course, necessary conditions for an acceptable numerical scheme. The main question in numerical analysis of a PDE is, however, the convergence problem:

Definition 11.3 We say that a numerical scheme *converges* to a given differential problem (a PDE with suitable initial or boundary conditions) if the solution of the discrete numerical problem converges in the limit $\Delta x, \Delta t \to 0$ to the solution of the original differential problem.

It is clear that consistency and stability are necessary conditions for convergence; it turns out that they are also sufficient.

Theorem 11.4 *Any consistent and stable numerical scheme for the problem (11.11)–(11.12) is convergent.*

The usefulness of this theorem stems from the fact that consistency is, in general, easy to verify, while stability is a property of the (discrete) numerical scheme alone (and does not depend on the PDE itself). Therefore, it is easier to check for stability than directly for convergence. We shall skip the proof of Theorem 11.4. We comment that similar theorems hold for other PDEs, such as the Laplace equation and the wave equation.

We have thus derived a convergent numerical scheme for the heat equation with the unfortunate limitation of tiny time steps (if we wish to retain high accuracy). One might suspect that the blame lies with the first-order time difference. To examine this conjecture, and to provide further examples of the notions we have defined, let us consider a scheme that is similar to (11.14), except that the time difference is now of second order:

$$\frac{U_{i,n+1} - U_{i,n-1}}{2\Delta t} = k\frac{U_{i+1,n} - 2U_{i,n} + U_{i-1,n}}{(\Delta x)^2} \qquad 1 \le i \le N - 2, \ n \ge 0.$$
$$(11.21)$$

We face immediately a technical obstacle: the values of $U_{i,n-1}$ are not defined at all for $n = 0$. We can easily overcome this problem, however, by using our previous scheme (11.13) for just the first time step, and then proceeding further in time with (11.21).

Surprisingly enough it turns out that the promising scheme (11.21) would be terrible to work with; it is unstable for *any* choice of the time step! To see that, let us construct, as we did for the scheme (11.14), a product solution for the difference equation of the form $U_{i,n} = A(n) \sin Li \Delta x$. The sequence $A(n)$ satisfies the difference equation

$$A(n+1) = A(n-1) - 8\alpha \sin^2\left(\frac{L\Delta x}{2}\right) A(n).$$

The solution is given by $A(n) = A(0)r^n$, where r is a solution of the quadratic equation

$$r^2 + 8\alpha \sin^2\left(\frac{L\Delta x}{2}\right) r - 1 = 0.$$

Since one of the roots of this quadratic satisfies $r > 1$, the scheme is always unstable.

The problem of finding an efficient stable scheme for (11.13) attracted intense activity in the middle of the twentieth century. One of the popular schemes from this period was proposed by Crank and Nicolson:

$$\frac{U_{i,n+1} - U_{i,n}}{\Delta t} = k \left(\frac{U_{i+1,n} - 2U_{i,n} + U_{i-1,n}}{2(\Delta x)^2} + \frac{U_{i+1,n+1} - 2U_{i,n+1} + U_{i-1,n+1}}{2(\Delta x)^2} \right).$$
$$\tag{11.22}$$

To examine the stability of the *Crank–Nicolson scheme*, let us substitute into it the product solution $U_{i,n} = A(n) \sin Li \Delta x$. We obtain the difference equation

$$A(n+1)\left[1 + 4\alpha \sin^2(L\Delta x/2)\right] = \left[1 - 4\alpha \sin^2(L\Delta x/2)\right] A(n).$$

The solution is of the form $A(n) = r^n$, where

$$r = \frac{1 - 4\alpha \sin^2(L\Delta x/2)}{1 + 4\alpha \sin^2(L\Delta x/2)},$$

implying that the scheme is stable for *any* choice of Δt and Δx. The reader will examine the consistency of the Crank–Nicolson scheme in Exercise 11.3.

It is important to notice a fundamental difference between the scheme (11.22) and the first scheme we presented, (11.13). While (11.13) can be written as an explicit expression for $U_{i,n+1}$ in terms of U_n, this is not the case for (11.22). A scheme of the former character is called an *explicit scheme*, while a scheme of the latter character is called an *implicit scheme*. A rule of thumb in numerical analysis is that implicit schemes are better behaved than explicit schemes (although one should not conclude that implicit schemes are always valid). The better behavior manifests itself, for example, in higher efficiency or higher accuracy.

The advantages of implicit schemes are counterbalanced by one major deficiency: at each time step we need to solve an algebraic system in $N - 2$ unknowns.

Therefore, a major theme in numerical analysis of PDEs is to derive efficient methods for solving large algebraic systems. Even if the PDE is nonlinear (and, therefore, so is the algebraic problem), the solution is often obtained iteratively (e.g. with the Newton method), such that at each iteration one solves a linear system. In Section 11.6, we shall consider solution methods for large algebraic systems and their applications.

We conclude this section by examining yet another numerical scheme for the heat equation that is based on a second-order difference formula for the time variable. Although our first naive approach failed, it was discovered that a slight modification of (11.21) provides a convergent scheme. We thus consider the following scheme proposed by Du-Fort and Frankel:

$$\frac{U_{i,n+1} - U_{i,n-1}}{2\Delta t} = k \frac{U_{i+1,n} - U_{i,n-1} - U_{i,n+1} + U_{i-1,n}}{(\Delta x)^2} \quad 1 \le i \le N-2, \ n \ge 0.$$
(11.23)

The stability and consistency of this scheme (and hence also its convergence) will be proved in Exercise 11.4.

11.4 Laplace equation

We move on to discuss the numerical solution of the Laplace (or, more generally, the Poisson) equation. Let Ω be the rectangle $\Omega = (0, a) \times (0, b)$. We are looking for a function $u(x, y)$ that solves the Dirichlet problem

$$\Delta u(x, y) = f(x, y) \quad (x, y) \in \Omega, \qquad u(x, y) = g(x, y) \quad (x, y) \in \partial\Omega,$$
(11.24)

or the Neumann problem

$$\Delta u(x, y) = f(x, y) \quad (x, y) \in \Omega, \qquad \partial_n u(x, y) = g(x, y) \quad (x, y) \in \partial\Omega.$$
(11.25)

Notice that it is possible to solve (11.24) or (11.25) by the separation of variables method; yet applying this method could be technically difficult since it involves the computation of the Fourier coefficients of the given function f (or g). Moreover, the Fourier series might converge slowly near the boundary.

We define instead a grid over the rectangle Ω (see Figure 11.2):

$$\{(x_i, y_j) = (i\Delta x, j\Delta y) \ i = 0, 1, \ldots, N-1, \ j = 0, 1, \ldots, M-1\}, \quad (11.26)$$

where $\Delta x = a/(N-1)$, $\Delta y = b/(M-1)$. The derivatives of u will be approximated by finite differences of the values of u on the grid's points. We shall use

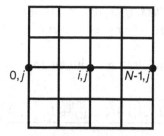

Figure 11.2 The grid for the Laplace equation.

central difference approximations to write a scheme for the Dirichlet problem as

$$\frac{U_{i-1,j} - 2U_{i,j} + U_{i+1,j}}{(\Delta x)^2} + \frac{U_{i,j-1} - 2U_{i,j} + U_{i,j+1}}{(\Delta y)^2} = F_{i,j} := f(x_i, y_j),$$

$$(11.27)$$

$$i = 1, 2, \ldots, N - 2, \quad j = 1, 2, \ldots, M - 2,$$

together with the boundary conditions:

$$
\begin{aligned}
U_{0,j} &:= G_{0,j} \equiv g(0, y_j) & j &= 0, 1, \ldots, M - 1, \\
U_{N-1,j} &:= G_{N-1,j} \equiv g(a, y_j) & j &= 0, 1, \ldots, M - 1, \\
U_{i,0} &:= G_{i,0} \equiv g(x_i, 0) & i &= 0, 1, \ldots, N - 1, \\
U_{i,M-1} &:= G_{i,M-1} \equiv g(x_i, b) & i &= 0, 1, \ldots, N - 1.
\end{aligned}
$$

$$(11.28)$$

Observe that (11.27) determines the value of $U_{i,j}$ in terms of the values of U at the four nearest neighbors of the point (i, j), and the value of f at (i, j). We further point out that the differential problem has been replaced by a linear algebraic system of size $(N - 2) \times (M - 2)$.

Before approaching the question of how the linear system is to be solved, we have to address a number of fundamental theoretical questions. Is the system we obtained solvable? Is the solution unique? Is it stable? To answer these questions we shall prove that the difference scheme we have formulated satisfies a strong maximum principle that is similar to the one we proved in Chapter 7 for the continuous problem:

Theorem 11.5 (The strong maximum principle) *Let $U_{i,j}$ be the solution of the homogeneous system*

$$\frac{U_{i-1,j} - 2U_{i,j} + U_{i+1,j}}{(\Delta x)^2} + \frac{U_{i,j-1} - 2U_{i,j} + U_{i,j+1}}{(\Delta y)^2} = 0,$$

$$(11.29)$$

$$i = 1, 2, \ldots, N - 2, \quad j = 1, 2, \ldots, M - 2,$$

with specified boundary values. If U attains its maximum (minimum) value at an interior point of the rectangle, then U is constant.

Proof We assume for simplicity and without loss of generality that $\Delta x = \Delta y$. Notice that

$$U_{i,j} = \frac{U_{i-1,j} + U_{i+1,j} + U_{i,j-1} + U_{i,j+1}}{4};$$

namely, $U_{i,j}$ is the arithmetic average of its nearest neighbors. Clearly if an average of a set of numbers is greater than or equal to each of the numbers in the set, then all the numbers in the set equal the average. Therefore, if U attains a maximum at some interior point, then the same maximum is also attained by each neighbor of this point. We can continue this process until we cover all the points in the rectangle. Thus U is constant.   □

An important consequence of the maximum principle is the following theorem.

Theorem 11.6 *The difference system (11.27)–(11.28) has a unique solution.*

Proof Let $U^{(1)}$, $U^{(2)}$ be two solutions of the system. Then $U = U^{(1)} - U^{(2)}$ is a solution of the same system with homogeneous boundary conditions and zero right hand side. If $U \neq 0$, then U achieves a maximum or a minimum somewhere inside the rectangle. By the strong maximum principle (Theorem 11.5), U is constant. Since U vanishes on the boundary, it must vanish everywhere. Thus, $U^{(1)} = U^{(2)}$.

The system (11.27) consists of $(N-2) \times (M-2)$ equations in $(N-2) \times (M-2)$ unknowns. A well-known theorem in linear algebra states that if a homogeneous equation possesses only the trivial solution, then nonhomogeneous equation has a unique solution.   □

Another important question is whether the solution of the numerical scheme converges to the solution of the PDE in the limit where Δx and Δy tend to zero. The answer is positive under certain assumptions on the boundary conditions, but a detailed discussion is beyond the scope of the book.

Since (11.27) is a linear system, it can be written in a matrix form. For this purpose we concatenate the two-dimensional arrays $U_{i,j}$ and $F_{i,j}$ into vectors that we denote by V and G, respectively. The components V_k and G_k of these vectors are given by

$$
\begin{aligned}
V_{(j-1)(N-2)+i} &:= U_{i,j} & i, j &= 1, 2, \ldots, N-2, \\
G_{(j-1)(N-2)+i} &:= (\Delta x)^2 F_{i,j} & i, j &= 1, 2, \ldots, N-2.
\end{aligned}
\tag{11.30}
$$

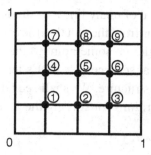

Figure 11.3.

Notice that the vector V consists only of interior points. The system (11.27) can be written now as $AV = b$, where the vector b is determined by G and by the boundary conditions. We demonstrate the matrix notation through the following example.

Example 11.7 Write an explicit matrix equation for the discrete approximation of the Poisson problem

$$\Delta u = 1 \; (x, y) \in \Omega, \qquad u(x, y) = 0 \; (x, y) \in \partial\Omega, \qquad (11.31)$$

for a grid of 3×3 of interior points. The grid's structure, together with the numbering of the interior points is depicted in Figure 11.3. Concatenating the discrete system (11.27) gives rise to the matrix equation

$$
\begin{pmatrix}
-4 & 1 & 0 & 1 & 0 & 0 & 0 & 0 & 0 \\
1 & -4 & 1 & 0 & 1 & 0 & 0 & 0 & 0 \\
0 & 1 & -4 & 0 & 0 & 1 & 0 & 0 & 0 \\
1 & 0 & 0 & -4 & 1 & 0 & 1 & 0 & 0 \\
0 & 1 & 0 & 1 & -4 & 1 & 0 & 1 & 0 \\
0 & 0 & 1 & 0 & 1 & -4 & 0 & 0 & 1 \\
0 & 0 & 0 & 1 & 0 & 0 & -4 & 1 & 0 \\
0 & 0 & 0 & 0 & 1 & 0 & 1 & -4 & 1 \\
0 & 0 & 0 & 0 & 0 & 1 & 0 & 1 & -4
\end{pmatrix}
\begin{pmatrix}
V_1 \\ V_2 \\ V_3 \\ V_4 \\ V_5 \\ V_6 \\ V_7 \\ V_8 \\ V_9
\end{pmatrix}
=
\begin{pmatrix}
1 \\ 1 \\ 1 \\ 1 \\ 1 \\ 1 \\ 1 \\ 1 \\ 1
\end{pmatrix}. \qquad (11.32)
$$

A quick inspection of (11.32) teaches us that the matrix A has some special features:

(1) It is sparse (most of its entries vanish).
(2) The entries that do not vanish concentrate near the diagonal.
(3) The diagonal entry in every row is equal to or greater (in absolute value) than the sum of all the other terms in that row. A matrix with this property is called *diagonally dominated matrix*.

These properties are typical of many linear systems obtained as approximations to PDEs. The sparseness stems from the fact that the differentiation operator is local; it relates the value of a function at some point to the values at near-by points. We conclude that while numerical schemes for PDEs lead to large algebraic systems, these systems have a special structure. We shall see later that this structure enables us to construct efficient algorithms for solving the algebraic systems.

11.5 The wave equation

The numerical solution of hyperbolic equations, such as the wave equation, is more involved than the solution of parabolic and elliptic equations. The reason is that solutions of hyperbolic equations might have singularities. Since the finite difference schemes we presented above are valid only for smooth functions, they may not be adequate for use in hyperbolic equations without some modification. We emphasize that the existence of characteristic surfaces where the solution of hyperbolic equations might be singular is not a mathematical artifact. On the contrary, it is an important aspect of many problems in many scientific and engineering disciplines.

It is impossible to analyze the very important and difficult problem of the numerical solution of PDEs with singularities within our limited framework. Instead we shall briefly consider a finite difference scheme for the wave equation in cases where the solution *is* smooth. Consider, therefore, the wave equation

$$u_{tt} - c^2 u_{xx} = 0 \qquad 0 < x < \pi, \ t > 0, \tag{11.33}$$

with the initial boundary conditions

$$u(0, t) = u(\pi, t) = 0 \quad t > 0, \qquad u(x, 0) = f(x), \ u_t(x, 0) = g(x) \quad 0 \le x \le \pi. \tag{11.34}$$

We construct for (11.33) a second-order explicit finite difference scheme. For this purpose, fix an integer $N > 2$ and a positive number Δt, and set $\Delta x := \pi/(N-1)$. We define a grid $\{x_i = i\Delta x\}$ on the interval $[0, \pi]$, and a grid $\{t_n = n\Delta t\}$ on the interval $[0, T]$. We further introduce the notation $U_{i,n} = u(x_i, t_n)$, and write

$$\frac{U_{i,n+1} - 2U_{i,n} + U_{i,n-1}}{(\Delta t)^2} = c^2 \frac{U_{i+1,n} - 2U_{i,n} + U_{i-1,n}}{(\Delta x)^2} \qquad 1 \le i \le N-2, \ n \ge 0. \tag{11.35}$$

The boundary values are determined by (11.34):

$$U_{0,n} = U_{N-1,n} = 0 \qquad n \ge 0.$$

Let us rewrite the scheme as an explicit expression for the solution at the discrete time $n + 1$ in terms of the solution at times n and $n - 1$:

$$U_{i,n+1} = 2(1 - \alpha)U_{i,n} - U_{i,n-1} + \alpha(U_{i-1,n} + U_{i+1,n}), \tag{11.36}$$

where

$$\alpha = c^2 \left(\frac{\Delta t}{\Delta x} \right)^2.$$

Since the system involves three time steps, we have to compute $U_{i,-1}$ in order to initiate it. For this purpose we use the initial condition u_t and express it by a central second-order difference:

$$U_{i,1} - U_{i,-1} = 2\Delta t g(x_i). \tag{11.37}$$

Solving for $U_{i,-1}$ from (11.37), and using the additional initial condition $U_{i,0} = f(x_i)$, we now have at our disposal all the data required for the difference equation (11.36).

It is straightforward to check that if the solution of the PDE is sufficiently smooth, then the scheme (11.36) is consistent. Is it also stable? We examine the stability of the scheme by the same method we introduced above for the heat equation. For this purpose we analyze the evolution of a fundamental sinusoidal initial wave $\sin Li\Delta x$ in the course of the discrete time argument n. We express the solution of the discrete problem in the form $U_{i,n} = A(n) \sin Li\Delta x$. Substituting this expression into (11.36), we obtain a difference equation for $A(n)$:

$$A(n + 1) = 2(1 - \alpha)A(n) - A(n - 1) + 2\alpha \cos(L\Delta x)A(n). \tag{11.38}$$

We seek solutions to (11.38) of the form $A(n) = r^n$. We find that r must satisfy the quadratic equation

$$r^2 - 2r \left[1 - 2\alpha \sin^2 \left(\frac{L\Delta x}{2} \right) \right] + 1 = 0.$$

Since the product of the roots equals 1, a necessary and sufficient condition that a solution $A(n)$ does not grow as a function of n is that the solution of the quadratic equation would be complex, i.e.

$$\left[1 - 2\alpha \sin^2 \left(\frac{L\Delta x}{2} \right) \right]^2 - 1 \leq 0$$

We thus obtain

$$\frac{\Delta x}{\Delta t} \geq c. \tag{11.39}$$

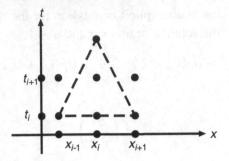

Figure 11.4 The discrete domain of influence.

This is called the *CFL condition* after Courant, Friedrichs, and Lewy. Since the CFL condition enforces values for Δt that are of the order of Δx, it is not as limiting as the corresponding condition for the heat equation. We could have anticipated the condition (11.39) from theoretical (and physical) grounds even without the stability analysis of the discrete scheme. To see that, it is useful to consult Figure 11.4. The triangle bounded between the dashed lines in the drawing is the region of influence of the interval $[(x_{i-1}, t_i), (x_{i+1}, t_i)]$. The CFL condition guarantees that the point (x_i, t_{i+1}) would be within this triangle, as indeed must be the case following our discussion in Chapter 4.

11.6 Numerical solution of large linear algebraic systems

We saw in the previous sections that many numerical schemes give rise to systems of linear algebraic equations (see Example 11.7). Furthermore, the systems we obtained were inherently large and sparse. We shall therefore consider in this section special efficient methods for solving such systems. It is important to realize that numerical linear algebra is a fast growing mathematical discipline; we shall limit ourselves to a brief exposition of some of the basic ideas and classical methods.

Linear systems can be solved, of course, by the Gauss elimination method. The drawback of this method is its high complexity. The complexity of a numerical calculation is defined here as the number of multiplications involved in it. (Truly, this is a somewhat outdated definition that was more relevant to older computers; nevertheless, it is a convenient definition and we shall stick to it.) A direct solution by the Gauss elimination method of a system in K unknowns requires $O(K^3)$ multiplications. Since we normally consider equations with many unknowns (to ensure a good numerical approximation), it is desirable to construct more efficient algorithms. The main idea behind the algorithms we present below is to exploit the special structure of the systems arising in the discrete approximation of PDEs.

Most of the methods that are used for large sparse systems are iterative. (Although we immediately mention that there are also some popular direct methods. We refer the reader to basic linear algebra books for an exposition of matrix decomposition methods.) We shall demonstrate below three iterative methods. The simplest of them all is the *Jacobi method*. A considerable improvement of it is achieved by the *Gauss–Seidel method*, and a far more significant improvement is obtained by the *successive over relaxation* (SOR) method. Although we shall verify that the SOR method is far superior to the Jacobi or to the Gauss–Seidel method, we prefer to start by presenting the simpler methods that are easier to understand. Moreover, it turns out that sometimes the complexity is not the main consideration; there are some applications for which the Gauss–Seidel method is preferred over the SOR method for deep reasons that we cannot discuss here. We emphasize again that the methods we are about to present do not necessarily work for any matrix! We consider them here for a special class of matrices that are sparse and have a certain relation between the diagonal terms and the off-diagonal terms.

We start with the Jacobi method. To fix ideas, we shall use as a prototype the Crank–Nicolson scheme for the heat equation. Rewrite (11.22) in the form

$$U_{i,n+1} = \frac{\alpha}{2}(U_{i+1,n+1} - 2U_{i,n+1} + U_{i-1,n+1}) + r_{i,n}, \tag{11.40}$$

where

$$r_{i,n} = \frac{\alpha}{2}(U_{i+1,n} - 2U_{i,n} + U_{i-1,n}) + U_{i,n}.$$

The values of $r_{i,n}$ are known at the nth step for all i, and the unknowns are the values of $U_{i,n+1}$ for all relevant indices i (i.e. for $1 \leq i \leq N - 2$). We fix n, and solve (11.40) iteratively. The solution at the pth iteration will be denoted by $V_{i,n+1}^p$. The process starts at some guess $V_{i,n+1}^0$. For example, we can choose $V_{i,n+1}^0 = U_{i,n}$. In the Jacobi method, we update at each step $V_{i,n+1}^{p+1}$ by solving (11.40), using the values of $V_{i-1,n+1}^{(p)}$ and $V_{i+1,n+1}^{(p)}$ (which are known from the previous iteration). We therefore obtain the following recursive equation:

$$V_{i,n+1}^{(p+1)} = \frac{\alpha}{2\alpha + 2}(V_{i-1,n+1}^{(p)} + V_{i+1,n+1}^{(p)}) + \frac{1}{\alpha + 1}r_{i,n}. \tag{11.41}$$

A close inspection of (11.41) reveals that while scanning the vector $V_{n+1}^{(p+1)}$, we update $V_{i,n+1}^{(p+1)}$ using $V_{i-1,n+1}^{(p)}$, although at this stage we already know the updated value $V_{i-1,n+1}^{(p+1)}$. We therefore intuitively expect an improvement in the convergence if we incorporate in the iterative process a more updated value. We thus write

$$V_{i,n+1}^{(p+1)} = \frac{\alpha}{2\alpha + 2}(V_{i-1,n+1}^{(p+1)} + V_{i+1,n+1}^{(p)}) + \frac{1}{\alpha + 1}r_{i,n}. \tag{11.42}$$

This is known as the Gauss–Seidel formula. We shall verify below that the Gauss–Seidel method is twice as fast as the Jacobi method. Furthermore, in the Gauss–Seidel algorithm there is no need to use two vectors to describe two successive steps in the iteration: it is possible now to update $V_{n+1}^{(p)}$ in a single vector. Hence the Gauss–Seidel method is superior to the Jacobi method from all perspectives.

As we noted above, one can improve upon the Gauss–Seidel method by a method called SOR . To formulate this algorithm, it is convenient to rewrite the Gauss–Seidel formula as

$$V_{i,n+1}^{(p+1)} = V_{i,n+1}^{(p)} + \left[\frac{\alpha}{2\alpha + 2}(V_{i-1,n+1}^{(p+1)} + V_{i+1,n+1}^{(p)}) + \frac{1}{\alpha + 1}r_{i,n} - V_{i,n+1}^{(p)} \right].$$

$$(11.43)$$

The meaning of this notation is that the term in the square brackets is the change obtained in passing from $V_{i,n+1}^{(p)}$ to $V_{i,n+1}^{(p+1)}$. In the SOR method we multiply this term by a *relaxation* parameter ω:

$$V_{i,n+1}^{(p+1)} = V_{i,n+1}^{(p)} + \omega \left[\frac{\alpha}{2\alpha + 2}(V_{i-1,n+1}^{(p+1)} + V_{i+1,n+1}^{(p)}) + \frac{1}{\alpha + 1}r_{i,n} - V_{i,n+1}^{(p)} \right].$$

$$(11.44)$$

In the special case where $\omega = 1$ we recover the Gauss–Seidel method. Surprisingly, it turns out that for a clever choice of the parameter ω in the interval $(1, 2)$ the scheme (11.44) converges much faster than the Gauss–Seidel scheme.

To analyze the iterative methods we have presented, we shall verify that they indeed converge (under suitable conditions), and examine their rate of convergence (so that we can select an efficient method). For this purpose it is convenient to write the equations in a matrix form

$$\mathbf{A}V = b.$$

$$(11.45)$$

The Crank–Nicolson method, for example, can be written (for the special choice $N = 7$) as a system of the type (11.45), where

$$V_i = U_{i,n+1}, \quad b_i = r_{i,n} \qquad i = 1, 2, \ldots, 5$$

$$(11.46)$$

and

$$\mathbf{A} = \begin{pmatrix} 1+\alpha & -\alpha/2 & 0 & 0 & 0 \\ -\alpha/2 & 1+\alpha & -\alpha/2 & 0 & 0 \\ 0 & -\alpha/2 & 1+\alpha & -\alpha/2 & 0 \\ 0 & 0 & -\alpha/2 & 1+\alpha & -\alpha/2 \\ 0 & 0 & 0 & -\alpha/2 & 1+\alpha \end{pmatrix}.$$

$$(11.47)$$

To write the iterative methods presented above, we express $\mathbf{A}$ as

$$\mathbf{A} = \mathbf{L} + \mathbf{D} + \mathbf{S},$$

where $\mathbf{L}, \mathbf{D}$, and $\mathbf{S}$ are matrices whose nonzero entries are below the diagonal, on the diagonal, and above the diagonal, respectively. Observing that the Jacobi method can be written as

$$\mathbf{A}_{ii}V_i^{(p+1)} = -\sum_{j\neq i}\mathbf{A}_{ij}V_j^{(p)} + b_i,$$

we obtain

$$V^{(p+1)} = -\mathbf{D}^{-1}(\mathbf{L} + \mathbf{S})V^{(p)} + \mathbf{D}^{-1}b. \tag{11.48}$$

Similarly, the Gauss–Seidel method is equivalent to

$$\sum_{j\leq i}\mathbf{A}_{ij}V_j^{(p+1)} = -\sum_{j>i}\mathbf{A}_{ij}V_j^{(p)} + b_i,$$

or in a matrix formulation

$$V^{(p+1)} = -(\mathbf{D} + \mathbf{L})^{-1}\mathbf{S}V^{(p)} + (\mathbf{D} + \mathbf{L})^{-1}b. \tag{11.49}$$

The reader will be asked to show in Exercise 11.8 that the SOR method is equivalent to

$$V^{(p+1)} = -(\mathbf{D} + \omega\mathbf{L})^{-1}\left\{[(1 - \omega)\mathbf{D} - \omega\mathbf{S}]V^{(p)} + \omega b\right\}. \tag{11.50}$$

The iterative process for each of the methods we have presented is of the general form

$$V^{(p+1)} = \mathbf{M}V^{(p)} + \mathbf{Q}b. \tag{11.51}$$

Obviously the solution V of (11.45) satisfies (11.51), namely

$$V = \mathbf{M}V + \mathbf{Q}b.$$

To investigate the convergence of each method we define $\Xi^{(p+1)}$ to be the difference between the $(p + 1)$th iteration and the exact solution V; we further construct for Ξ a difference equation:

$$\Xi^{(p+1)} = V^{(p+1)} - V = \mathbf{M}V^{(p)} + \mathbf{Q}b - \mathbf{M}V - \mathbf{Q}b = \mathbf{M}(V^{(p)} - V) = \mathbf{M}\Xi^{(p)}. \tag{11.52}$$

It remains to examine whether the sequence $\{\Xi^{(p)}\}$, defined through the difference equation

$$\Xi^{(p+1)} = \mathbf{M}\Xi^{(p)},$$

indeed converges to zero, and, if so, to find the rate of convergence. For simplicity we assume here that the matrix $\mathbf{M}$ is diagonalizable. We denote its eigenvalues by λ_i and its eigenvectors by w_i. We expand $\{\Xi^{(p)}\}$ by these eigenvectors, and obtain

$$\Xi^{(0)} = \sum_i \beta_i w_i$$

for the initial condition, and (using (11.52))

$$\Xi^{(p)} = \sum_i \beta_i \lambda_i^p w_i \tag{11.53}$$

for the subsequent terms.

Define the *spectral radius* of the matrix $\mathbf{M}$:

$$\lambda(\mathbf{M}) = \max_i |\lambda_i|.$$

It is readily seen from (11.53) that the iterative scheme converges if $\lambda(\mathbf{M}) < 1$. Moreover, the rate of convergence itself is also determined by $\lambda(\mathbf{M})$. We provide now an example for the computation of the spectral radius for a particular equation.

Example 11.8 Compute the spectral radius $\lambda(\mathbf{M})$, where $\mathbf{M}$ is the matrix associated with the Jacobi method for the Crank–Nicolson scheme. The matrix is given by

$$\mathbf{M} = -\mathbf{D}^{-1}(\mathbf{L} + \mathbf{S}) = \frac{-1}{1+\alpha} \begin{pmatrix} 0 & -\alpha/2 & \cdot & \cdot & \cdot \\ -\alpha/2 & 0 & -\alpha/2 & \cdot & \cdot \\ \cdot & \cdot & \cdot & \cdot & \cdot \\ \cdot & \cdot & -\alpha/2 & 0 & -\alpha/2 \\ \cdot & \cdot & \cdot & -\alpha/2 & 0 \end{pmatrix}. \tag{11.54}$$

Let w be an eigenvector of $-(L + S)$. The entries of w satisfy the equation

$$\frac{\alpha}{2} w_{j-1} + \frac{\alpha}{2} w_{j+1} = \lambda w_j \qquad j = 1, 2, \ldots, K, \tag{11.55}$$

where we extend the natural K components by setting $w_0 = w_{K+1} = 0$. Equation (11.55) has solutions w^k of the form

$$w_j^k = \sin kj \Delta x \qquad j, k = 1, 2, \ldots, K,$$

corresponding to the eigenvalue $\lambda_k = \alpha \cos k \Delta x$. Since D is a diagonal matrix, we obtain that the spectral radius is

$$\lambda_{\text{Jacobi}} = \frac{\alpha}{1+\alpha} \cos \Delta x \sim \frac{\alpha}{1+\alpha} \left[1 - \frac{(\Delta x)^2}{2} \right]. \tag{11.56}$$

Similarly, one can show that the Gauss–Seidel spectral radius for the same scheme is

$$\lambda_{\text{Gauss–Seidel}} \sim \frac{\alpha}{1+\alpha}[1 - (\Delta x)^2].$$

The spectral radius for the SOR method depends on the parameter ω. It can be shown that for the special case of the Crank–Nicolson scheme for the heat equation the optimal choice is $\omega \approx 1.8$. For this value one obtains

$$\lambda_{\text{SOR}} \sim \frac{\alpha}{1+\alpha}(1 - 2\Delta x).$$

One can derive similar results for elliptic PDEs. The main difference is that the term $\alpha/(1+\alpha)$ is absent in the elliptic case. Thus, one can show that the spectral radii for second-order difference schemes for the Laplace equation are

$$\lambda_{\text{Jacobi}} \sim 1 - \frac{\gamma_1}{K},$$

$$\lambda_{\text{Gauss–Seidel}} \sim 1 - \frac{2\gamma_1}{K},$$

and, for an appropriate choice of ω,

$$\lambda_{\text{SOR}} \sim 1 - \frac{\gamma_2}{\sqrt{K}},$$

where the constants γ_1, γ_2 depend on the domain Ω.

We finally remark that there exist several sophisticated methods (for example the multi-grid method) that accelerate the solution process well beyond the methods we presented here.

11.7 The finite elements method

The finite elements method (FEM) is a special case of the Galerkin method that was presented in Chapter 10. To recall this method and to introduce the essentials of the FEM, we shall demonstrate the theory for a canonical elliptic problem:

$$-\Delta u = f \quad \vec{x} \in D, \qquad u = 0 \quad \vec{x} \in \partial D, \tag{11.57}$$

where f is a given function and D is a domain in $\mathbb{R}^2$. Multiplying both sides by a test function ψ and integrating by parts we obtain

$$\int_D \nabla u \cdot \nabla \psi \, d\vec{x} = \int_D f\psi \, d\vec{x}. \tag{11.58}$$

When we use a weak formulation we should specify our function spaces. The natural space for u is the Hilbert space that is obtained from the completion of the

C^1 functions in D that have a compact support there (i.e. that vanish on ∂D). The condition on the support of u comes from the homogeneous Dirichlet boundary conditions. This space is denoted by $\dot{H}_1$. The test function ψ is selected in some suitable Hilbert space. For example, we can select ψ also to lie in $\dot{H}_1$. To proceed with the Galerkin method we should define a sequence of Hilbert spaces $H^{(k)}$, select a basis $\phi_1, \phi_2, \ldots, \phi_k$ in each one of them and write $u = \sum_{i=1}^{k} \alpha_i \phi_i$. As we saw in Chapter 10 this leads to the linear algebraic equation (10.66) for the unknowns α_i, where the entries of the matrix K and the vector d are given (similarly to (10.67)) by

$$K_{ij} = \int_D \nabla \phi_i \cdot \nabla \phi_j \, d\vec{x}, \qquad d_i = \int_D f \phi_i \, d\vec{x}. \qquad (11.59)$$

Remark 11.9 The FEM was invented by Courant in 1943. It was later extensively developed by mechanical engineers to solve problems in structural design. This is why the matrix K is called the *stiffness matrix* and the vector d is called the *force vector*. The mathematical justification in terms of the general Galerkin method came later.

The special feature of the FEM lies in the choice of the family ϕ_i. The idea is to *localize* the test functions ϕ_i to facilitate the computation of the stiffness matrix. There are many variants of the FEM, and we only describe one of them (the most popular one) here. Similar to the discretization of the domain we used in the FDM, we divide D into many smaller regions. We use triangles for this division. This provides a great deal of geometric flexibility that makes the FEM a powerful tool for solving PDEs in complex geometries. In Figure 11.5 we have drawn two examples of triangulations. The initial step in the triangulation involves the numbering of the triangles T_j and the vertices V_i. The numbering is arbitrary in principle, but, as will become clear shortly, a clever numbering can be important in practice.

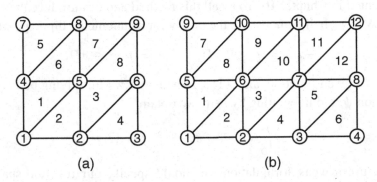

(a)   (b)

Figure 11.5 Two examples of triangulation.

Each test function is constructed to be linear in each triangle and continuous at the vertices. The shape taken by the test functions in the triangles is called an *element*. In principle we can choose other more complex elements than linear functions. This will make the problem harder to solve, but may yield higher accuracy (somewhat similar to writing high-order finite difference schemes). To determine the actual linear shape of ϕ_i in each triangle we impose the conditions

$$\phi_i(V_j) = \begin{cases} 1 & i = j, \\ 0 & i \neq j. \end{cases} \tag{11.60}$$

Since the general linear function in the plane has three coefficients, and since these coefficients are uniquely determined in each triangle in terms of the value of the function at the vertices, the set of conditions (11.60) determines each ϕ_i uniquely. Obviously, if T is a triangle that does not have V_i as a vertex, then ϕ_i is identically zero there. This implies that when the number of vertices is large the stiffness matrix K is quite sparse. We also see that if we number the vertices in a reasonable way, then the nonzero entries of K will not be far from the diagonal. This will considerably simplify the complexity and stability of the algebraic system $K\alpha = d$.

Another important consequence of our choice of test functions is that if we use $U(V_i)$ to denote the numerical approximation of the exact solution u at V_i, then we obtain at once the identification $\alpha_i = U(V_i) := U_i$; namely the unknowns α_i in the expansion of u are exactly the values of the approximant U at the vertices.

It remains to compute the matrix K. While we could use the definition (11.59), this would require us to compute ϕ_i in all the triangles where it does not vanish. Instead we shall employ a popular quicker way that uses the variational characterization of the problem (11.57). We have already seen in Chapter 10 examples where the Galerkin method and the Ritz method yield the same algebraic equation. It is easy to cast (11.57) as a variational problem; it is exactly the Euler–Lagrange equation for minimizing the functional

$$F(u) = \int_D \left(\frac{1}{2}|\nabla u|^2 - fu\right) d\vec{x} \tag{11.61}$$

over all functions u that vanish on ∂D. Expressing the approximate minimizer as $u = \sum_{i=1}^k U_i\phi_i$, the functional F is converted to

$$F_k = \frac{1}{2}U^t K U - U^t \cdot d, \tag{11.62}$$

where K and d are given in (11.59), and $a \cdot b$ is the standard inner product of the vectors a and b in $\mathbb{R}^k$. The minimization problem is now k-dimensional.

Now, since each of the ϕ_i is a linear function over each triangle T_j, it follows that the approximant $u = \sum_{i=1}^k U_i\phi_i$ is also a linear function. Therefore, we can easily

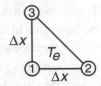

Figure 11.6 A representative triangle.

compute the integrals in (11.61) directly in terms of the unknowns U_i. Notice that the choice of linear elements leads to particularly simple computations since the gradient of u is constant in each element.

The computation of K and d is straightforward to program; to demonstrate it explicitly, we shall restrict ourselves further to cases where D is a square or a rectangle partitioned into identical isosceles right triangles. The length of each of the equal sides of the triangle is Δx. We start the computation with a canonical representative triangle T_e (Figure 11.6). Denoting the value of the linear approximation of u at the vertices $U_{e_1}, U_{e_2}, U_{e_3}$ we obtain

$$\int_{T_e} |\nabla U|^2 \, d\vec{x} = \frac{1}{2} \left[(U_{e_2} - U_{e_1})^2 + (U_{e_3} - U_{e_1})^2 \right]. \tag{11.63}$$

Notice that in the formula above we actually had two factors of $(\Delta x)^2$. One of them, due to the numerical integration, is in the numerator, while the other one, due to the gradient, is in the denominator. Therefore, these factors cancel each other. Formula (11.63) can also be written as a quadratic form

$$\frac{1}{2} \int_{T_e} |\nabla u|^2 \, d\vec{x} = \frac{1}{2} U_e^t K_e U_e, \quad \text{where} \quad K_e = \begin{pmatrix} 1 & -\frac{1}{2} & -\frac{1}{2} \\ -\frac{1}{2} & \frac{1}{2} & 0 \\ -\frac{1}{2} & 0 & \frac{1}{2} \end{pmatrix}. \tag{11.64}$$

Similarly we integrate the second term in the integrand of (11.61). In general even in a simple domain such as the unit square, and even in the case of identical triangles we need to perform the integration $\int_{T_e} f u \, d\vec{x}$ numerically. There are many numerical integration schemes for this purpose. One simple integration formula is

$$\int_{T_e} f u \, d\vec{x} \approx (\Delta x)^2 \frac{f(C_e)}{6} (U_{e_1} + U_{e_2} + U_{e_3}), \tag{11.65}$$

where C_e is the center of gravity of the triangle T_e. It remains to perform integrations such as (11.63) and (11.65) for every triangle and to assemble the results into one big matrix K and one big vector d.

We first demonstrate the assembly for the triangulation (a) in Figure 11.5. Since there is only one internal vertex (vertex 5 in the drawing), there is only one unknown – U_5. Therefore, there is only one test function ϕ_5. We use the canonical formula (11.64) for all the triangles that have V_5 as a vertex. This means that triangles 5 and 4 do not participate in the computation. We write

$$\int_D |\nabla u|^2 d\vec{x} \approx \begin{pmatrix} U_4 \\ U_1 \\ U_5 \end{pmatrix}^t K_e \begin{pmatrix} U_4 \\ U_1 \\ U_5 \end{pmatrix} + \begin{pmatrix} U_2 \\ U_1 \\ U_5 \end{pmatrix}^t K_e \begin{pmatrix} U_2 \\ U_1 \\ U_5 \end{pmatrix} + \begin{pmatrix} U_5 \\ U_2 \\ U_6 \end{pmatrix}^t K_e \begin{pmatrix} U_5 \\ U_2 \\ U_6 \end{pmatrix}$$
$$+ \begin{pmatrix} U_5 \\ U_4 \\ U_8 \end{pmatrix}^t K_e \begin{pmatrix} U_5 \\ U_4 \\ U_8 \end{pmatrix} + \begin{pmatrix} U_8 \\ U_5 \\ U_9 \end{pmatrix}^t K_e \begin{pmatrix} U_8 \\ U_5 \\ U_9 \end{pmatrix} + \begin{pmatrix} U_6 \\ U_5 \\ U_9 \end{pmatrix}^t K_e \begin{pmatrix} U_6 \\ U_5 \\ U_9 \end{pmatrix}.$$

(11.66)

Since the boundary conditions imply that $U_i = 0$ for $i \neq 5$, we obtain $K = (4)$. The computation of the force vector d is based on (11.65). Suppose that f is constant (say 1), then we obtain at once that each of the relevant six triangles contributes exactly $(\Delta x)^2/6$ to the entry d_5 of d which is the only nonzero entry. Therefore, $d_5 = (\Delta x)^2 U_5$. Thus, after optimization, we obtain $4U_5 = (\Delta x)^2$, and since $\Delta x = \frac{1}{2}$, the numerical solution in this triangulation is $U_5 = 1/16$.

We proceed to the more "interesting" triangulation depicted in Figure 11.5(b). Now there are two internal vertices (6 and 7); thus there are two unknowns U_6 and U_7, and K is a 2×2 matrix. A little algebra gives

$$K = \begin{pmatrix} 4 & -1 \\ -1 & 4 \end{pmatrix}, \qquad d = (\Delta x^2) \begin{pmatrix} 1 \\ 1 \end{pmatrix}.$$

(11.67)

Finally we look at the general case of a rectangle divided into identical isosceles right triangles. Instead of writing the full matrix K, we derive the equation for $U_{i,j}$ – the numerical solution at the vertex (i, j) (see Figure 11.7). We use the computation in (11.63). The vertex (i, j) appears in six of the triangles in the drawing. Summing all the contributions to the term $\int_D |\nabla u|^2 d\vec{x}$ in energy that involve $U_{i,j}$ gives

$$4U_{i,j}^2 - U_{i,j}(U_{i,j-1} + U_{i-1,j} + U_{i+1,j} + U_{i,j+1}).$$

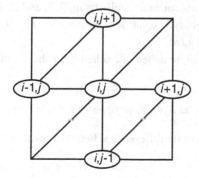

Figure 11.7 Two relevant triangles for a given vertex.

Similarly, the term $\int_D fu \, d\vec{x}$ in the energy contributes $(\Delta x)^2 U_{i,j}$. Therefore, the equation for the minimizer $U_{i,j}$ is

$$4U_{i,j} - U_{i,j-1} - U_{i-1,j} - U_{i+1,j} - U_{i,j+1} = (\Delta x)^2. \tag{11.68}$$

The observant reader will realize that this is exactly the equation we obtained in (11.27) by the FDM. Does it mean that the two methods are the same? They certainly give rise to the same algebraic system for the PDE we are considering in the current example and for the present triangulation. This fact should boost our confidence in these algebraic equations! While there are indeed equations and domains for which both methods yield the same discrete equations, this is certainly not always the case. Even for the present domain and triangulation, we would have obtained different discrete equations had we solved the equation $-\Delta u = f(x, y)$, where f is not constant (see Exercise 11.15).

11.8 Exercises

11.1 Consider the rectangular grid (11.1) and assume $\Delta x = \Delta y$. Find a second-order difference scheme for u_{xy}.

11.2 Prove (11.9) and (11.10).

11.3 Prove that the Crank–Nicolson scheme is consistent.

11.4 Prove that, under suitable conditions, the Du-Fort–Frankel scheme is stable and consistent.

11.5 Consider the heat equation

$$u_t = u_{xx} \quad 0 < x < \pi, \quad t > 0, \tag{11.69}$$

$$u(0, t) = u(\pi, t) = 0, \quad u(x, 0) = x(\pi - x). \tag{11.70}$$

(a) Solve (11.69)–(11.70) numerically (in spatial grids of 25, 61, and 101 points) using the Crank–Nicolson scheme. Compute for each one of the grids the solution at the point $(x, t) = (\pi/4, 2)$.

(b) Solve the same problem analytically using 2, 7, and 20 Fourier terms. Construct a table to compare the analytic solution at the point $(x, t) = (\pi/4, 2)$ with the numerical solutions found in part (a).

11.6 (a) Write an explicit finite difference scheme for the problem

$$u_t = u_{xx} \quad 0 < x < 1, \quad t > 0, \tag{11.71}$$

$$u(0, t) = u_x(1, t) = 0, \quad u(x, 0) = f(x). \tag{11.72}$$

(b) Write an implicit finite difference scheme for problem

$$u_t = u_{xx} \quad 0 < x < 1, \quad t > 0, \tag{11.73}$$

$$u(0, t) = u_x(1, t) = 0, \quad u(x, 0) = f(x). \tag{11.74}$$

11.7 Solve the problem

$$u_t = u_{xx} + 5t^4 \qquad 0 \le x \le 1, \ \ t > 0, \tag{11.75}$$
$$u(0, t) = u(1, t) = t^5, \quad u(x, 0) = 0. \tag{11.76}$$

using scheme (11.13). Use $\Delta x = \Delta t = 0.1$. Compute $u(\frac{1}{2}, 3)$. Compare your answer with the analytical solution of the same equation and explain what you observe.

11.8 Derive (11.50).

11.9 Show that if $F_{i,j}$ is positive at all the grid points, then the solution $U_{i,j}$ of (11.27) cannot attain a positive maximal value at an interior point.

11.10 (a) Let D be the unit rectangle $D = \{(x, y) | \ 0 < x < 1, \ \ 0 < y < 1\}$. Solve

$$\Delta u(x, y) = 1 \quad (x, y) \in D, \qquad u(x, y) = 0 \quad (x, y) \in \partial D \tag{11.77}$$

for $N = 3, 4$ manually, and for $N = 11, 41, 91$ using a computer (write the code for this problem). For each choice of grid find an approximation to $u(\frac{1}{2}, \frac{1}{2})$ and to $u(\frac{1}{10}, \frac{1}{10})$.

(b) Solve the problem of part (a) by the method of separation of variables. Evaluate the solution at the points $u(\frac{1}{2}, \frac{1}{2})$, and $u(\frac{1}{10}, \frac{1}{10})$, using a Fourier series with 2, 5, and 20 coefficients. Compare the numerical solution you found in part (a) with the analytical solution of part (b).

11.11 Let D be the unit square. Solve

$$\Delta u(x, y) = 0 \quad (x, y) \in D, \qquad u(x, y) = 1 + \frac{1}{5}\sin x \quad (x, y) \in \partial D, \tag{11.78}$$

for $N = 3, 11, 41, 80$. In each case find an approximation for $u(\frac{1}{2}, \frac{1}{2})$

11.12 Write a finite difference scheme for the equation $\Delta u = 1$ in the rectangle $T = \{(x, y) | \ 0 < x < 3, \ \ 0 < y < 2\}$ under the Dirichlet condition $u = 0$ on the boundary of T. Use a discrete grid with step size $\Delta x = \Delta y = 1$. Solve the algebraic equations without using a computer, and find an approximation for $u(1, 1)$ and $u(2, 1)$.

11.13 Consider the discrete Dirichlet problem for the Laplace equation on a rectangular uniform $N \times N$ grid; there are $(N - 2)^2$ unknowns, and $4(N - 2)$ boundary points. Prove that the space of all solutions to the discrete Dirichlet problem is of dimension $4(N - 2)$.

11.14 Let $w(x, y)$ be a smooth given vector field in the unit square D. Generalize the scheme (11.27) to write a second-order finite difference scheme for the equation

$$\Delta u + w \cdot \nabla u = 0$$

under the Dirichlet conditions $u = 1$ on the boundary of D.

11.15 Consider problem (11.57), where $f(x, y) = x^2 + x^2 y$ and D is the unit square.

(a) Write explicitly an FEM scheme in which the triangulation consists of identical isosceles right triangles with 16 vertices.

(b) Now write for the same set of vertices an FDM scheme. Is it the same as the scheme you obtained in (a)?

(c) Solve the equations you derived in (a).

11.16 Consider the ODE

$$u''(x) = f(x) \ x < 0 < L, \qquad x(0) = 1 \quad x(L) = 0. \tag{11.79}$$

Divide the interval $(0, L)$ into N identical subintervals with vertices $x_0 = 0, x_1, \ldots, x_{N+1} = L$. Consider the basis functions ϕ_i where ϕ_i is linear at each subinterval (x_{i-1}, x_i), with

$$\phi_i(x_j) = \begin{cases} 1 & i = j, \\ 0 & i \neq j. \end{cases} \tag{11.80}$$

(a) Use one of the methods we introduced above to construct an FEM scheme for the Poisson equation in a rectangle to construct an FEM scheme for the ODE (11.79). Compute explicitly the stiffness matrix K.

(b) Solve the ODE analytically and numerically for $f(x) = \sin 2x$ and for $N = 4, 10, 20, 40$. Discuss the error in each of the numerical solutions compared with the exact solution.

12

Solutions of odd-numbered problems

Here we give numerical solutions and hints for most of the odd-numbered problems. Extended solutions to the problems are available for course instructors using the book from solutions@cambridge.org.

Chapter 1

1.1 (a) Write $u_x = af'$, $u_y = bf'$. Therefore a and b can be any constants such that $a + 3b = 0$.

1.3 (a) Integrate the first equation with respect to x to get $u(x, y) = x^3 y + xy + F(y)$, where $F(y)$ is still undetermined. Differentiate this solution with respect to y and compare with the equation for u_y to conclude that F is a constant function. Finally, using the initial condition $u(0, 0) = 0$, obtain $F(y) = 0$.

(b) The compatibility condition $u_{xy} = u_{yx}$ does not hold. Therefore there does not exist a function u satisfying both equations.

1.5 (a) $u(x, t) = f(x + kt)$ for any differentiable function f.

(b), (c). Equations (b) and (c) do not have such explicit solutions. Nevertheless, if selecting $f(s) = s$, then (b) is solved by $u = x + ut$ that can be written explicitly as $u = x/(1 - t)$, which is well defined if $t \neq 1$.

1.7 (a) Substitute $v(s, t) = u(x, y)$, and use the chain rule to get

$$u_x = v_s + v_t \qquad u_y = -v_t,$$

and

$$u_{xx} = v_{ss} + v_{tt} + 2v_{st} \qquad u_{xy} = -v_{ll} - v_{st} \qquad u_{yy} - v_{tt}.$$

Therefore, $u_{xx} + 2u_{xy} + u_{yy} = v_{ss}$, and the equation becomes $v_{ss} = 0$.

(b) The general solution is $u(x, y) = f(x - y) + xg(x - y)$.

(c) Proceeding similarly, obtain for $v(s, t) = u(x, y)$ the equation $v_{ss} + v_{tt} = 0$.

Chapter 2

2.1 (a) The characteristics are $y = x + c$.

(b) The solution is $u(x, y) = f(x - y) + y$.

2.3 (a) The parametric solution is

$$x(t) = x_0 e^t \qquad y(t) = y_0 e^t \qquad u(t) = u_0 e^{pt},$$

and the characteristics are the curves $x/y = $ constant.

(b) $u(x, y) = (x^2 + y^2)^2$ is the unique solution.

(c) The initial curve $(s, 0, s^2)$ is a characteristic curve (see the characteristic equations). Thus, there exist infinitely many solutions: $u(x, y) = x^2 + ky^2 \quad \forall k \in \mathbb{R}$.

2.5 (a) The projection on the (x, y) plane of each characteristic curve has a positive direction and it propagates with a strictly positive speed in the square.

(b) On each characteristic line u equals $u(t) = f(s)e^{-t}$, therefore u preserves its sign along characteristics.

(c) Since $\nabla u(x_0, y_0) = 0$ at critical points, it follows from the PDE that $u(x_0, y_0) = 0$.

(d) Follows from part (c) and (b).

2.7 The parametric solution is

$$(x(t, s), y(t, s), u(t, s)) = \left(t + s, t, \frac{1}{1 - t} \right),$$

implying $u = 1/(1 - y)$.

2.9 (a) The transversality condition holds, implying a unique solution near the initial curve.

(b) The solution of the characteristic equations is

$$x(t, s) = s - 2 \sin s \left(e^{-t/2} - 1 \right) \qquad y(t, s) = t \quad u(t, s) = \sin s e^{-t/2}.$$

(c) The solution passing through Γ_1 is

$$x(t, s) = s \quad y(t, s) = s + t \quad u(t, s) = 0,$$

namely, $u(x, y) = 0$.

(d) Such a curve must be a characteristic curve. It follows that it can be represented as $\{(n\pi, t, 0) \mid t \in \mathbb{R}\}$, where $n \in \mathbb{Z}$.

2.11 The Jacobian satisfies $J \equiv 0$. Since $u \equiv 0$ is a solution of the problem, there exist infinitely many solutions. To compute other solutions, define a new Cauchy problem such as

$$(y^2 + u)u_x + yu_y = 0, \quad u(x, 1) = x - \frac{1}{2}.$$

Now the Jacobian satisfies $J \equiv 1$. The parametric form of the solution is

$$x(t, s) = (s - \tfrac{1}{2})t + \tfrac{1}{2}e^{2t} + s - \tfrac{1}{2},$$

$$y(t, s) = e^t,$$

$$u(t, s) = s - \tfrac{1}{2}.$$

It is convenient in this case to express the solution as a graph of the form

$$x(y, u) = \frac{y^2}{2} + u \ln y + u.$$

2.13 The transversality condition is violated for all s. "Guess" a solution of the form $u = u(x)$, to find $u = \sqrt{2(x - 1)}$. This means that there are infinitely many solutions. To find them, define a new Cauchy problem; for instance, select the problem

$$u u_x + x u_y = 1, \quad u \left(x + \tfrac{3}{2}, \tfrac{7}{6}\right) = 1.$$

The parametric representation of the solution to the new problem is

$$x(t, d) = \tfrac{1}{2}t^2 + t + d + \tfrac{3}{2},$$

$$y(t, d) = \tfrac{1}{6}t^3 + \tfrac{1}{2}t^2 + (d + \tfrac{3}{2})t + \tfrac{7}{6},$$

$$u(t, d) = t + 1.$$

Finally,

$$y(x, u) = \frac{(u - 1)^3}{6} + \frac{(u - 1)^2}{2} + (u - 1)\left[x - \frac{(u - 1)^2}{2} - (u - 1)\right] + \frac{7}{6}.$$

2.15 (a)

$$u(x, y) = y \frac{1 - y^{x/y - y}}{x - y^2}.$$

(b),(d) The transversality condition holds everywhere. The explicit solution shows that u is not defined at the origin. This does not contradict the local existence theorem, since this theorem only guarantees a solution in a neighborhood of the original curve $(y = 1)$.

2.17 (a) The parametric surface representation is

$$x = x_0 e^t \qquad y = y_0 + t \qquad u = u_0 + t,$$

and the characteristic curve passing through the point $(1, 1, 1)$ is $(e^t, 1 + t, 1 + t)$.

(b) The direction of the projection of the initial curve on the (x, y) plane is $(1, 0)$. The direction of the projection of the characteristic curve is $(s, 1)$. Since the directions are not parallel, there exists a unique solution.

(c) $u(x, y) = \sin(x/e^y) + y$. It is defined for all x and y.

2.19 (a) $u(x, y) = x^2 y^2 / [4(y - x)^2 - xy(y - 2x)]$.

(b) The projection of the initial curve on the (x, y) plane is in the direction $(1, 2)$. The direction of the projection of the characteristic curve (for points on the initial curve) is $s^2(1, 4)$. The directions are not parallel, except at the origin where the characteristic direction is degenerate.

(c) The characteristic that starts at the points $(0, 0, 0)$ is degenerate.

(d) The solution is not defined on the curve $4(y - x)^2 = xy(y - 2x)$ that passes through the origin.

2.21 (a) $u(x, y) = 2x^{3/2} y^{1/2} - xy$.

(b) The transversality condition holds. The solution is defined only for $y > 0$.

2.23 (a) $u = (x - ct)/[1 + t(x - ct)]$.

(b), (c) The observer that starts at a point $x_0 > 0$ sees the solution $u(x_0 + ct, t) = x_0/(1 + x_0 t)$. Therefore, if $x_0 > 0$, the observed solution decays, while if $x_0 < 0$ the solution explodes in a finite time. If $x_0 = 0$ the solution is 0.

2.25 The transversality condition is violated identically. However, the characteristic direction is $(1, 1, 1)$, and so is the direction of the initial curve. Therefore the initial curve is itself a characteristic curve, and there exist infinitely many solutions. To find solutions, consider the problem $u_x + u_y = 1$, $u(x, 0) = f(x)$, for an arbitrary f satisfying $f(0) = 0$. The solution is $u(x, y) = y + f(x - y)$. It remains to fix five choices for f.

2.27 (a) $u(x, y) = (6y - y^2 - 2x)/[2(3 - y)]$.

(b) A straightforward calculation verifies $u(3x, 2) = 4 - 3x$.

(c) The transversality condition holds in this case. Therefore the problem has a unique solution, and from (b) the solution is the same as in (a).

Chapter 3

3.1 (a) The equation is parabolic. The required transformation is

$$y = t, \quad x = \frac{s - t}{3}.$$

(b)

$$u(x, y) = \frac{(3x + y)y^4}{324} - \frac{y^5}{540} + y\phi(3x + y) + \psi(3x + y).$$

(c)

$$u(x, y) = \frac{(3x + y)y^4}{324} - \frac{y^5}{540} + y\left[\cos(x + \frac{y}{3}) - \frac{1}{3}\cos(x + \frac{y}{3})\right] + \sin(x + y/3).$$

3.3 (a) Writing $w(s, t) = u(x, y)$, the canonical form is $w_{st} + \frac{1}{4}w_t = 0$.
(b) Using $W := w_t$, the general solution is $u(x, y) = f(y - 4x)e^{-y/4} + g(y)$, for arbitrary functions $f, g \in C^2(\mathbb{R})$.
(c) $u(x, y) = (-y/2 + 4x)e^{-y/4}$.
3.5 (a) The equation is hyperbolic when $xy > 0$, elliptic when $xy < 0$, and parabolic when $xy = 0$ (but this is not a domain!).
(b) The characteristic equation is $y'^2 = y/x$.

(1) When $xy > 0$ there are two real roots $y' = \pm\sqrt{y/x}$. Suppose for instance that $x, y > 0$. Then the solution is $\sqrt{y} \pm \sqrt{x} = $ constant. Define the new variables $s(x, y) = \sqrt{y} + \sqrt{x}$ and $t(x, y) = \sqrt{y} - \sqrt{x}$.
(2) When $xy < 0$ there are two complex roots $y' = \pm i\sqrt{|y/x|}$. Choose $y' = i\sqrt{|y/x|}$. The solution of the ODE is $2\text{sign}(y)\sqrt{|y|} = i2\text{sign}(x)\sqrt{|x|} + $ constant. Divide by $2\text{sign}(y) = -2\text{sign}(x)$ to obtain $\sqrt{|y|} + i\sqrt{|x|} = $ constant. Define the new variables $s(x, y) = \sqrt{|x|}$ and $t(x, y) = \sqrt{|y|}$.

3.7 (a) The equation is hyperbolic for $q > 0$, i.e. for $y > 1$. The equation is elliptic for $q < 0$, i.e. for $y < -1$. The equation is parabolic for $q = 0$, i.e. for $|y| \leq 1$.
(b) The characteristics equation is $(y')^2 - 2y' + (1 - q) = 0$; its roots are $y'_{1,2} = -1 \pm \sqrt{q}$.

(1) The hyperbolic regime $y > 1$. There are two real roots $y'_{1,2} = 1 \pm 1$. The solutions of the ODEs are $y_1 = $ constant, $y_2 = 2x + $ constant. Hence the new variables are $s(x, y) = y$ and $t(x, y) = y - 2x$.
(2) The elliptic regime $y < -1$. The two roots are imaginary: $y'_{1,2} = 1 \pm i$. Choose one of them, $y' = 1 + i$, to obtain $y = (1 + i)x + $ constant. The new variables are $s(x, y) = y - x$, $t(x, y) = x$.
(3) The parabolic regime $|y| \leq 1$. There is a single real root $y' = 1$; The solution of the resulting ODE is $y = x + $ constant. The new variables are $s(x, y) = x$, $t(x, y) = x - y$

3.11 (a)

$$u(x, y) = \frac{1}{2}[f(1 - \cos x - x + y) + f(1 - \cos x + x + y)] + \frac{1}{2}\left(\int_{1-\cos x - x + y}^{1-\cos x + x + y} g(s)ds\right).$$

(b) The solution is classic if it is twice differentiable. Thus, one should require that f would be twice differentiable, and that g would be differentiable.

Chapter 4

4.3 (a)

$$u(x, 1) = \begin{cases} 0 & x < -3, \\ \dfrac{1 - (x + 2)^2}{2} & -3 \le x \le -1, \\ x + 1 & -1 \le x \le 0, \\ 1 & 0 \le x \le 1, \\ \dfrac{1 - (x - 2)^2}{2} + 1 & 1 \le x \le 3, \\ 4 - x & 3 \le x \le 4, \\ 0 & x > 4. \end{cases}$$

(b) $\lim_{t \to \infty} u(5, t) = 1$.

(c) The solution is singular at the lines: $x \pm 2t = \pm 1, 2$.

(d) The solution is continuous at all points.

4.5 (a) The backward wave is

$$u_r(x, t) = \begin{cases} 12(x + t) - (x + t)^2 & 0 \le x + t \le 4, \\ 0 & x + t < 0, \\ 32 & x + t > 4. \end{cases}$$

and the forward wave is

$$u_p(x, t) = \begin{cases} -4(x - t) - (x - t)^2 & 0 \le x - t \le 4, \\ 0 & x - t < 0, \\ -32 & x - t > 4. \end{cases}$$

(d) The explicit representation formulas for the backward and forward waves of (a) imply that the limit is 32, since for t large enough $5 + t > 4$ *and* $5 - t < 0$.

4.7 (a) Consider a forward wave $u = u_p(x, t) = \psi(x - t)$. Then

$$u_p(x_0 - a, t_0 - b) + u_p(x_0 + a, t_0 + b) = \psi(x_0 - t_0 - a + b) + \psi(x_0 - t_0 + a - b)$$
$$= u_p(x_0 - b, t_0 - a) + u_p(x_0 + b, t_0 + a).$$

A similar equality is obtained for a backward wave $u = u_r(x, t) = \phi(x + t)$. Since every solution of the wave equation is a linear combination of forward and backward waves, the statement follows.

(b) $u(x_0 - ca, t_0 - b) + u(x_0 + ca, t_0 + b) = u(x_0 - cb, t_0 - a) + u(x_0 + cb, t_0 + a)$.

(c)

$$u(x, t) = \begin{cases} \dfrac{f(x+t)+f(x-t)}{2} + \dfrac{1}{2}\displaystyle\int_{x-t}^{x+t} g(s)\,ds & t \leq x, \\[3mm] \dfrac{f(x+t)-f(t-x)}{2} + \dfrac{1}{2}\displaystyle\int_{t-x}^{x+t} g(s)\,ds + h(t-x) & t \geq x. \end{cases}$$

(d) $h(0) = f(0)$, $h'(0) = g(0)$, $h''(0) = f''(0)$. If these conditions are not satisfied the solution is singular along the line $x - t = 0$.

(e)

$$u(x, t) = \begin{cases} \dfrac{f(x+ct)+f(x-ct)}{2} + \dfrac{1}{2c}\displaystyle\int_{x-ct}^{x+ct} g(s)\,ds & ct \leq x, \\[3mm] \dfrac{f(x+ct)-f(ct-x)}{2} + \dfrac{1}{2c}\displaystyle\int_{ct-x}^{x+ct} g(s)\,ds + h\left(t - \dfrac{x}{c}\right) & ct \geq x. \end{cases}$$

The corresponding compatibility conditions are $h(0) = f(0)$, $h'(0) = g(0)$, $h''(0) = c^2 f''(0)$. If these conditions are not satisfied the solution is singular along the line $x - ct = 0$.

4.9 $u(x, t) = x^2 + t + 3t^2/2$.

4.11 D'Alembert's formula implies

$$P(x, t) = \frac{1}{2}[f(x+4t) + f(x-4t)] + \frac{1}{8}[H(x+4t) - H(x-4t)],$$

where $H(x) = \int_0^x g(s)ds$. Hence

$$H(x) = \begin{cases} x & |x| \leq 1, \\ 1 & x > 1, \\ -1 & x < -1. \end{cases}$$

Notice that at $x_0 = 10$:

$$f(10 + 4t) = 0, \quad f(10 - 4t) \leq 10, \quad |H(t)| \leq 1, \quad t > 0.$$

Therefore, $P(10, t) \leq 5 + \frac{1}{4} = \frac{21}{4} < 6$, and the structure will not collapse.

4.13 (a) The solution is not classical when $x \pm 2t = -1, 0, 1, 2, 3$.

(b) $u(1, 1) = 1/3 + e - e^3/2 - e^{-1}/2$.

4.15 $u(x, t) = \int v(x, t)dx + f(t) = \frac{1}{2}[\sin(x - t) - \sin(x + t)] + f(t)$, where $f(t)$ is an arbitrary function.

4.17 (a) $u(x, t) = x + \frac{1}{2}t\sin(x + t) + \frac{1}{4}\cos(x - t) - \frac{1}{4}\cos(x + t)$.

(b) $v(x, t) = \frac{1}{2}t\sin(x + t) + \frac{1}{4}\cos(x + t) - \frac{1}{4}\cos(x - t)$.

(c) The function $w = \frac{1}{2}\cos(x + t) - \frac{1}{2}\cos(x - t) - x$ solves the homogeneous wave equation $w_{tt} - w_{xx} = 0$, and satisfies the initial conditions $w(x, 0) = x$, $w_t(x, 0) = \sin x$.

(d) w is an odd function of x.

4.19 The unique solution is $u(x, t) = 1 + 2t$.

Chapter 5

5.1 $u(x, t) = (4/\pi) \sum_{n=1}^{\infty} (1/n) \left[\cos \left(\frac{1}{2} n\pi \right) - (-1)^n \right] e^{-17n^2 t} \sin nx$.

5.3 (a)

$$u(x, t) = \sum_{n=1}^{\infty} \left(A_n \cos \frac{c\pi n t}{L} + B_n \sin \frac{c\pi n t}{L} \right) \sin \frac{n\pi x}{L},$$

$$A_n = \frac{2}{L} \int_0^L f(x) \sin \left(\frac{n\pi x}{L} \right) dx \qquad n \geq 1,$$

$$B_n = \frac{2}{c n \pi} \int_0^L g(x) \sin \left(\frac{n\pi x}{L} \right) dx \qquad n \geq 1.$$

5.5 (a)

$$u(x, t) = \frac{A_0}{2} + \sum_{n=1}^{\infty} A_n e^{-k\pi^2 n^2 t/L^2} \cos \left(\frac{n\pi x}{L} \right),$$

where

$$A_n = \frac{2}{L} \int_0^L f(x) \cos \left(\frac{n\pi x}{L} \right) dx \qquad n \geq 0.$$

(c) The obtained function is a classical solution of the equation for all $t > 0$, since if f is continuous the exponential decay implies that for every $\varepsilon > 0$ the series and all its derivatives converge uniformly for all $t > \varepsilon > 0$. For the same reason, the series (without $A_0/2$) converges uniformly to zero (as a function of x) in the limit $t \to \infty$. Thus,

$$\lim_{t \to \infty} u(x, t) = \frac{A_0}{2}.$$

It is instructive to compute A_0 by an alternative method. Notice that

$$\frac{d}{dt} \int_0^L u(x, t) dx = \int_0^L u_t(x, t) dx = k \int_0^L u_{xx}(x, t) dx$$
$$= k \left[u_x(L, t) - u_x(0, t) \right] = 0,$$

where the last equality follows from the Neumann boundary condition. Hence,

$$\int_0^L u(x, t)dx = \int_0^L u(x, 0)dx = \int_0^L f(x)dx$$

holds for all $t > 0$. Since the uniform convergence of the series implies the convergence of the integral series, you can infer

$$\frac{A_0}{2} = \frac{\int_0^L f(x)dx}{L}.$$

A physical interpretation The quantity $\int_0^L u(x, t)dx$ was shown to be conserved in a one-dimensional insulated rod. The quantity $ku_x(x, t)$ measures the heat flux at a point x and time t. The homogeneous Neumann condition amounts to stating that there is zero flux at the rod's ends. Since there are no heat sources either (the equation is homogeneous), the temperature tends to equalize its gradient, and therefore it converges to a constant temperature, such that the total stored energy is the same as the initial energy.

5.7 To obtain a homogeneous equation write $u = v + w$, where $w = w(t)$ satisfies

$$w_t - kw_{xx} = A\cos\alpha t, \qquad w(x, 0) \equiv 0.$$

Therefore,

$$w(t) = \frac{A}{\alpha}\sin\alpha t .$$

Solving for v, the complete solution is

$$u(x, t) = 3/2 + 1/2\cos 2\pi x\, e^{-4k\pi^2 t} + \frac{A}{\alpha}\sin\alpha t.$$

5.9 (a)

$$u(x, t) = \sum_{n=1}^{\infty} B_n e^{(-n^2+h)t}\sin nx,$$

where

$$B_n = \frac{2}{\pi}\int_0^\pi x(\pi - x)\sin nx dx = -\frac{4[(-1)^n - 1]}{\pi n^3}.$$

(b) The limit $\lim_{t\to\infty} u(x, t)$ exists if and only if $h \le 1$. When $h < 1$ the series converges uniformly to 0. If $h = 1$, the series converges to $B_1\sin x$.

5.11 (a) The domain of dependence is the interval $[\frac{1}{3} - \frac{1}{10}, \frac{1}{3} + \frac{1}{10}]$ along the x axis.
(b) Part (a) implies that the domain of dependence does not include the boundary. Therefore, D'Alembert's formula can be used to compute $u(\frac{1}{3}, \frac{1}{10}) = -\frac{1}{2}\frac{65}{15^3} = -\frac{13}{1350}.$

(c)

$$u(x, t) = 1 - \cos 4\pi x \cos 4\pi t.$$

5.13

$$u(x, t) = e^{-t} \sum_{n=0}^{\infty} B_n e^{-(2n+1)^2 t\pi^2/4} \sin\left(\frac{2n+1}{2}\pi x\right),$$

where

$$B_n = 2 \int_0^1 x(2 - x) \sin\left(\frac{2n+1}{2}\pi x\right) dx = \frac{32}{(2n+1)^3\pi^3}.$$

5.17 Let u_1 and u_2 be a pair of solutions. Set $v = u_1 - u_2$. We need to show that $v \equiv 0$. Thanks to the superposition principle v solves the homogeneous system

$$v_{tt} - c^2 v_{xx} + hv = 0 \qquad -\infty < x < \infty, \ t > 0,$$
$$\lim_{x \to \pm\infty} v(x, t) = \lim_{x \to \pm\infty} v_x(x, t) = \lim_{x \to \pm\infty} v_t(x, t) = 0 \qquad t \geq 0,$$
$$v(x, 0) = v_t(x, 0) = 0 \qquad -\infty < x < \infty.$$

Let $E(t)$ be as suggested in the problem. The initial conditions imply $E(0) = 0$. Formally differentiating $E(t)$ by t we write

$$\frac{dE}{dt} = \int_{-\infty}^{\infty} \left(v_t v_{tt} + c^2 v_x v_{xt} + hvv_t\right) dx,$$

assuming that all the integrals converge (we ought to be careful since the integration is over the entire real line).

We compute

$$\int_{-\infty}^{\infty} v_x v_{xt} dx = -\int_{-\infty}^{\infty} v_t v_{xx} dx + \int_{-\infty}^{\infty} \frac{\partial(v_x v_t)}{\partial x} dx.$$

Using the homogeneous boundary conditions

$$\int_{-\infty}^{\infty} \frac{\partial(v_x v_t)}{\partial x} dx = \lim_{x \to \infty} v_x(x, t)v_t(x, t) - \lim_{x \to -\infty} v_x(x, t)v_t(x, t) = 0,$$

hence, $\int_{-\infty}^{\infty} v_x v_{xt} dx = -\int_{-\infty}^{\infty} v_{xx} v_t dx$. Conclusion:

$$\frac{dE}{dt} = \int_{-\infty}^{\infty} v_t \left(v_{tt} - c^2 v_{xx} + hv\right) dx = 0.$$

We have verified that $E(t) = E(0) = 0$ for all t. The positivity of h implies that $v \equiv 0$.

5.19 (b) We consider the homogeneous equation

$$(y^2 v_x)_x + (x^2 v_y)_y = 0 \quad (x, y) \in D,$$
$$v(x, y) = 0 \quad (x, y) \in \Gamma.$$

Multiply the equation by v and integrate over D:

$$\int\int_D v \left[(y^2 v_x)_x + (x^2 v_y)_y\right] dx dy = 0.$$

Using the identity of part (a):

$$\int\int_D v \left[(y^2 v_x)_x + (x^2 v_y)_y\right] dx dy = -\int\int_D \left[(y v_x)^2 + (x v_y)^2\right]$$
$$+ \int\int_D \operatorname{div}\left(y^2 v v_x, x^2 v v_y\right) dx dy.$$

Using further the divergence theorem (see Formula (2) in Section A.2):

$$\int\int_D \operatorname{div}\left(v y^2 v_x, x^2 v v_y\right) dx dy = \int_\Gamma v y^2 v_x dy - v x^2 v_y dx = 0,$$

where in the last equality we used the homogeneous boundary condition $v \equiv 0$ on Γ. We infer that the energy integral satisfies

$$E := \int\int_D \left[(y v_x)^2 + (x v_y)^2\right] dx dy = 0,$$

hence $v_x = v_y = 0$ in D. We conclude that $v(x, y)$ is constant in D, and then the homogeneous boundary condition implies that this constant must vanish.

Chapter 6

6.1 (b) Use part (a) to set $\lambda = \mu^2$ and write

$$u(x) = A \sin \mu x + B \cos \mu x.$$

The boundary conditions lead to the transcendental equation

$$\frac{2\mu}{\mu^2 - 1} = \tan \mu.$$

(c) In the limit $\lambda \to \infty$ (or $\mu \to \infty$), μ_n satisfies the asymptotic relation $\mu_n \sim n\pi$ (where $n\pi$ is the root of the nth branch of $\tan \mu$). Therefore, $\lambda_n \approx n^2 \pi^2$ when $n \to \infty$.

6.3 (a) The eigenvalues are

$$\lambda_n = \left(\frac{n\pi}{\ln b}\right)^2 + \frac{1}{4} > \frac{1}{4} \quad n = 1, 2, 3. \ldots,$$

and the eigenfunctions are

$$v_n(x) = x^{-1/2} \sin\left(\frac{n\pi}{\ln b} \ln x\right) \qquad n = 1, 2, 3, \ldots$$

(b)

$$u(x, t) = \sum_{n=1}^{\infty} C_n e^{-\lambda_n t} x^{-1/2} \sin\left(\frac{n\pi}{\ln b} \ln x\right).$$

The constants C_n are determined by the initial data:

$$u(x, 0) = f(x) = \sum_{n=1}^{\infty} C_n x^{-1/2} \sin\left(\frac{n\pi}{\ln b} \ln x\right).$$

This is a generalized Fourier series expansion for $f(x)$, and

$$C_n = \frac{\langle f, v_n \rangle}{\langle v_n, v_n \rangle},$$

where $\langle \cdot, \cdot \rangle$ denotes the appropriate inner product.

6.5 (a)

$$u_n(x) = (x + 1)^{-1/2} \sin\left[\frac{n\pi \ln(x + 1)}{\ln 2}\right], \qquad \lambda_n = \frac{n^2 \pi^2}{\ln^2 2} + 1/4 \quad n = 1, 2, \ldots.$$

6.7 (a) Verify first that all eigenvalues are greater than $1/4$. Then find

$$u_n(x) = x^{-1/2} \sin(n\pi \ln x), \qquad \lambda_n = n^2 \pi^2 + 1/4 \qquad n = 1, 2, 3, \ldots.$$

6.9 (a) Perform two integration by parts for the expression $\int_{-1}^{1} u''v \, dx$, and use the boundary conditions to handle the boundary terms.

(b) Let u be an eigenfunction associated with the eigenvalue λ. Write the equation that is conjugate to the one satisfied by u:

$$\bar{u}'' + \bar{\lambda}\bar{u} = 0.$$

Obviously $\bar{u}$ satisfies the same boundary conditions as u. Multiply respectively by $\bar{u}$ and by u, and integrate over the interval $[-1, 1]$. Use part (a) to get

$$\lambda \int_{-1}^{1} |u(x)|^2 \, dx = \bar{\lambda} \int_{-1}^{1} |u(x)|^2 \, dx.$$

Hence λ is real.

(c) Verify first that all the eigenvalues are positive. Then, the eigenvalues are $\lambda_n = \left[(n + \frac{1}{2})\pi\right]^2$ and the eigenfunctions are

$$u_n(x) = a_n \cos\left(n + \frac{1}{2}\right)\pi x + b_n \sin\left(n + \frac{1}{2}\right)\pi x.$$

(d) It follows from part (c) that the multiplicity is 2, and a basis for the eigenspace is

$$\left\{\cos\left(n+\tfrac{1}{2}\right)\pi x, \ \sin\left(n+\tfrac{1}{2}\right)\pi x\right\}.$$

(e) Indeed, the multiplicity is not 1, but this is not a regular Sturm–Liouville problem!

6.11

$$u(x,t) = e^{-t} + \sum_{n=2}^{10} 3n e^{(-4n^2-1)t}\cos 2nx + \left(2t - 2 + 2e^{-t} + 3\cos 2x\right).$$

This is a finite sum of elementary smooth functions, and therefore it is a classical solution.

6.13 To obtain a homogeneous problem, write

$$u = v + \frac{xt}{\pi} + 2\left(1 - \frac{x^2}{\pi^2}\right).$$

v is a solution for the system

$$\begin{cases} v_t - v_{xx} = xt - \dfrac{4}{\pi^2} & 0 < x < \pi, \ t > 0, \\ v(0,t) = v(\pi,t) = 0 & t \geq 0, \\ v(x,0) = 0 & 0 \leq x \leq \pi. \end{cases}$$

Solving for v obtain u:

$$u(x,t) = \sum_{n=1}^{\infty}\left\{-\frac{(2\pi^3+8)(-1)^{n+1}+8}{n^5\pi^3}\left(1 - e^{-n^2 t}\right)\right.$$

$$\left. + \left[\frac{2(-1)^{n+1}}{n^3}\right]t\right\}\sin(nx) + \frac{xt}{\pi} + 2\left(1 - \frac{x^2}{\pi^2}\right).$$

6.15 (a) To generate a homogeneous boundary condition write $u(x,t) = v(x,t) + x + t^2$. The initial-boundary value problem for v is

$$\begin{aligned} v_t - v_{xx} &= (9t+31)\sin(3x/2) & 0 < x < \pi, \\ v(0,t) &= v_x(\pi,t) = 0 & t \geq 0, \\ v(x,0) &= 3\pi & 0 \leq x \leq \pi/2. \end{aligned}$$

Its solution is

$$v(x, t) = \sum_{n=0}^{\infty} \frac{12}{2n + 1} e^{-(n+1/2)^2 t} \sin[(n + 1/2)x]$$

$$+ \left\{ 9e^{-9t/4} \left[\frac{4}{9} \left(t - \frac{4}{9} \right) e^{9t/4} + \left(\frac{4}{9} \right)^2 \right] + \frac{31 \times 4}{9} (1 - e^{-9t/4}) \right\} \sin \left(\frac{3x}{2} \right).$$

Finally,

$$u(x, t) = x + t^2 + v(x, t).$$

(b) The solution is classical in the domain $(0, \pi) \times (0, \infty)$. On the other hand, the initial condition does not hold at $x = 0, t = 0$ since it conflicts there with the boundary condition.

6.17 The solution u is given by

$$u(x, t) = x \sin(t) + 1 + t + e^{-4\pi^2 t} \cos 2\pi x.$$

It is clearly classical.

6.19 To obtain a homogeneous boundary condition write $u = w + x/\pi$, and obtain for w:

$$\begin{aligned}
w_t - w_{xx} + hw &= -\frac{hx}{\pi} & 0 < x < \pi, \ t > 0, \\
w(0, t) = w(\pi, t) &= 0 & t \geq 0, \\
w(x, 0) = u(x, 0) - v(x) &= -\frac{x}{\pi} & 0 \leq x \leq \pi.
\end{aligned}$$

The solution for w is

$$w(x, t) = \sum_{n=1}^{\infty} \frac{2(-1)^n}{n\pi} \left[\left(1 - \frac{h}{n^2 + h} \right) e^{-(n^2+h)t} + \frac{h}{n^2 + h} \right] \sin nx.$$

This solution is not classical at $t = 0$, since the sine series does not converge to $-x/\pi$ in the closed interval $[0, 1]$.

6.21

$$u(x, t) = \pi e^{-4t} \cos 2x + \frac{1}{2001^4} e^{-2001^2 t} \cos 2001x + \left(\frac{t}{2001^2} - \frac{1}{2001^4} \right) \cos 2001x.$$

6.23 (a)

$$u(x, t) = -\frac{e^{-17^2 \pi^2 t} \cos 17\pi x}{3 + 17^2 \pi^2} + 3e^{-42^2 \pi^2 t} \cos 42\pi x + \frac{e^{3t} \cos 17\pi x}{3 + 17^2 \pi^2}.$$

(b) The general solution has the form

$$u(x, t) = A_0 + \sum_{n=1}^{\infty} A_n e^{-n^2 \pi^2 t} \cos n\pi x.$$

The function $f(x) = 1/(1 + x^2)$ is continuous in $[0, 1]$, implying that A_n are all bounded. Therefore, the series converges uniformly for all $t > t_0 > 0$, and

$$\lim_{t \to \infty} u(x, t) = A_0 = \int_0^\pi \frac{dx}{1 + x^2} = \frac{\pi}{4}.$$

6.25

$$u = \frac{L}{2} - \frac{4L}{\pi^2} \sum_{m=1}^\infty \frac{e^{-k \frac{(2m-1)^2 \pi^2}{L^2} t}}{(2m - 1)^2} \cos \frac{(2m - 1)\pi x}{L} + \frac{\alpha}{\omega} \sin \omega t.$$

6.27 The PDE is equivalent to

$$r u_t = r u_{rr} + 2 u_r.$$

Set

$$w(r, t) := u(r, t) - a,$$

and obtain for w:

$$\begin{cases} r w_t = r w_{rr} + 2 w_r & 0 < r < a, \ t > 0, \\ w(a, t) = 0 & t \geq 0, \\ w(r, 0) = r - a & 0 \leq r \leq a. \end{cases}$$

Solve for w by the method of separation of variables and obtain

$$w(r, t) = \sum_{n=1}^\infty A_n e^{-\frac{n^2 \pi^2 t}{a^2}} \frac{1}{r} \sin \frac{n \pi r}{a}.$$

The initial conditions then imply

$$w(r, 0) = \sum_{n=1}^\infty A_n \sin \frac{n \pi r}{a} = r (r - a).$$

Therefore, A_n are the (generalized) Fourier coefficients of $r(r - a)$, i.e.

$$A_n = \frac{2}{a} \int_0^a r (r - a) \sin \frac{n \pi r}{a} dr = -\frac{4 a^2}{n^3 \pi^3} [1 - (-1)^n].$$

Chapter 7

7.1 Select $\vec{\psi} = v \vec{\nabla} u$ in Gauss' theorem

$$\int_D \vec{\nabla} \cdot \vec{\psi}(x, y) \, dx dy = \int_{\partial D} \vec{\psi}(x(s), y(s)) \cdot \hat{n} ds.$$

7.3

$$u(x, y) = \frac{4}{\pi} \sum_{l=1}^\infty \sin[(2l - 1)y] \frac{\sinh \left[\sqrt{k + (2l - 1)^2} (\pi - x) \right]}{(2l - 1) \sinh \left[\sqrt{k + (2l - 1)^2} \pi \right]}.$$

7.5 It needs to be shown that

$$M(r_1) < M(r_2) \qquad \forall\ 0 < r_1 < r_2 < R.$$

Let $B_r = \{(x, y) \mid x^2 + y^2 \le r^2\}$ be a disk of radius r. Choose arbitrary $0 < r_1 < r_2 < R$. Since $u(x, y)$ is a nonconstant harmonic function in B_R, it must be a nonconstant harmonic function in each subdisk. The strong maximum principle implies that the maximal value of u in the disk B_{r_2} is obtained only on the disk's boundary. since all the points in B_{r_1} are internal to B_{r_2},

$$u(x, y) < \max_{(x,y) \in \partial B_{r_2}} u(x, y) = M(r_2), \qquad \forall\ (x, y) \in B_{r_1}.$$

In particular,

$$M(r_1) = \max_{(x,y) \in \partial B_{r_1}} u(x, y) < M(r_2).$$

7.7 (b)

$$u(r, \theta) = 3 - r^2 \cos^2 \theta + \frac{r^2}{2} + r\ \sin \theta,$$

or, in Cartesian coordinates,

$$u(x, y) = 3 + y + \frac{1}{2}(y^2 - x^2).$$

7.9 $n = 0$: A homogeneous harmonic polynomial is of the form $P_0(x, y) = c$ and the dimension of V_0 is 1.

$n \ge 1$: A homogeneous harmonic polynomial has the following form in polar coordinates:

$$P_n(r, \theta) = r^n (A_n \cos n\theta + B_n \sin n\theta).$$

Therefore the homogeneous harmonic polynomials of order $n \ge 1$ are spanned by two basis functions:

$$v_1(r, \theta) = r^n \cos n\theta ; \qquad v_2(r, \theta) = r^n \sin n\theta,$$

and the dimension of V_n (for $n \ge 1$) is 2.

7.11

$$u(r, \theta) = \frac{4}{r} \sin \theta,$$

or, in Cartesian coordinates,

$$u(x, y) = \frac{4y}{x^2 + y^2}.$$

7.13 Outline of the proof:

(i) Motivated by the Poisson formula, define the function

$$g(\varphi) = \frac{a^2 - r^2}{a^2 - 2ar\,\cos(\theta - \varphi) + r^2}$$

in the interval $[-\pi, \pi]$.

(ii) Prove

$$\frac{a-r}{a+r} \le g(\varphi) \le \frac{a+r}{a-r}. \tag{12.1}$$

(iii) Use (12.1) to show

$$\left(\frac{a-r}{a+r}\right) \frac{1}{2\pi} \int_{-\pi}^{\pi} f(\varphi)\,d\varphi \le u(r, \theta) \le \left(\frac{a+r}{a-r}\right) \frac{1}{2\pi} \int_{-\pi}^{\pi} f(\varphi)\,d\varphi.$$

(iv) The result now follows from the mean value theorem.

7.17 (a)

$$u(x, t) = \sum_{n=1}^{\infty} B_n e^{-2n^2 t} \sin nx,$$

where

$$B_n = \frac{2}{\pi} \int_0^{\pi} f(x) \sin nx\, dx = \frac{12(-1)^n}{n^3}.$$

(b) Use Corollary 7.18.

7.19 (a) The mean value theorem for harmonic functions implies

$$u(0, 0) = \frac{1}{2\pi} \int_{-\pi}^{\pi} u(R, \theta)\,d\theta = \frac{1}{2\pi} \int_{-\pi/2}^{\pi/2} \sin^2 2\theta\,d\theta = \frac{1}{4}.$$

(b) This is an immediate consequence of the strong maximum principle. The principle implies

$$u(r, \theta) \le \max_{\psi \in [-\pi/2, \pi/2]} u(R, \psi) = 1$$

for all $r < R$, and the equality holds if and only if u is constant. Clearly the solution is not a constant function, and therefore $u < 1$ in D. The inequality $u > 0$ is obtained from the strong maximum principle applied to $-u$.

7.21 The function $w(x, t) = e^{-t} \sin x$ is a solution of the problem

$$
\begin{aligned}
w_t - w_{xx} &= 0 & (x, t) \in Q_T, \\
w(0, t) = w(\pi, t) &= 0 & 0 \le t \le T, \\
w(x, 0) &= \sin x & 0 \le x \le \pi.
\end{aligned}
$$

On the parabolic boundary $0 \leq u(x, t) \leq w(x, t)$, and therefore, from the maximum principle $0 \leq u(x, t) \leq w(x, t)$ in the entire rectangle Q_T.

Chapter 8

8.1 (a) Use polar coordinates (r, θ) for (x, y), and (R, ϕ) for (ξ, η), to obtain

$$\frac{\partial G_R(x, y; \xi, \eta)}{\partial \xi} = \frac{\xi(1 - r^2/R^2)}{2\pi[R^2 - 2Rr\cos(\theta - \phi) + r^2]},$$

and similarly for $\partial/\partial\eta$. The exterior unit normal at a point (ξ, η) on the sphere is $(\xi, \eta)/R$, therefore,

$$\frac{\partial G_R(x, y; \xi, \eta)}{\partial r} = \frac{R^2 - r^2}{2\pi R[R^2 - 2Rr\cos(\theta - \phi) + r^2]}.$$

(b) $\lim_{R \to \infty} G_R(x, y; \xi, \eta) = \infty$.

8.3 (a) The solution for the Poisson equation with zero Dirichlet boundary condition is known from Chapter 7 to be

$$w(r, \theta) = \frac{\tilde{f}_0(r)}{2} + \sum_{n=1}^{\infty} [\tilde{f}_n(r)\cos n\theta + \tilde{g}_n(r)\sin n\theta]. \tag{12.2}$$

Substituting the coefficients $\tilde{f}_n(r)$, $\tilde{g}_n(r)$ into (12.2), we obtain

$$w(r, \theta) = \frac{1}{2} \int_0^r K_1^{(0)}(r, a, \rho)\delta_0(\rho)\rho \, d\rho + \frac{1}{2} \int_r^a K_2^{(0)}(r, a, \rho)\delta_0(r)\rho \, d\rho$$

$$+ \sum_{n=1}^{\infty} \left(\int_0^r K_1^{(n)}(r, a, \rho)[\delta_n(\rho)\cos n\theta + \varepsilon_n(r)\sin n\theta]\rho \, d\rho \right)$$

$$+ \sum_{n=1}^{\infty} \left(\int_r^a K_2^{(n)}(r, a, \rho)[\delta_n(r)\cos n\theta + \varepsilon_n(r)\sin n\theta]\rho \, d\rho \right).$$

Recall that the coefficients $\delta_n(\rho)$, $\varepsilon_n(r)$ are the Fourier coefficients of the Function F, hence

$$\delta_n(\rho) = \frac{1}{\pi} \int_0^{2\pi} F(\rho, \varphi)\cos n\varphi \, d\varphi, \quad \varepsilon_n(r) = \frac{1}{\pi} \int_0^{2\pi} F(\rho, \varphi)\sin n\varphi \, d\varphi.$$

Substitute these coefficients, and interchange the order of summation and integration to obtain

$$w(r, \theta) = \int_0^a \int_0^{2\pi} G(r, \theta; \rho, \varphi)F(\rho, \varphi)\,d\varphi\rho \, d\rho,$$

where G is given by

$$
G(r, \theta; \rho, \varphi) = \frac{1}{2\pi}
\begin{cases}
\log \dfrac{r}{a} + \displaystyle\sum_{n=1}^{\infty} \frac{1}{n} \left[\left(\frac{r}{a}\right)^n - \left(\frac{a}{r}\right)^n \right] \left(\frac{\rho}{a}\right)^n \cos n(\theta - \varphi) & \text{if } \rho < r, \\[4mm]
\log \dfrac{\rho}{a} + \displaystyle\sum_{n=1}^{\infty} \frac{1}{n} \left[\left(\frac{\rho}{a}\right)^n - \left(\frac{a}{\rho}\right)^n \right] \left(\frac{r}{a}\right)^n \cos n(\theta - \varphi) & \text{if } \rho > r.
\end{cases}
$$

(b) To calculate the sum of the above series use the identities

$$
\sum_{n=1}^{\infty} \frac{1}{n} z^n \cos n\alpha = \int_0^z \sum_{n=1}^{\infty} \zeta^{n-1} \cos n\alpha \, d\zeta
$$

$$
= \int_0^z \frac{\cos \alpha - \zeta}{1 + \zeta^2 - 2\zeta \cos \alpha} \, d\zeta = -\frac{1}{2} \log(1 + z^2 - 2z \cos \alpha).
$$

8.5 (a) On the boundary of $\mathbb{R}_+^2$ the exterior normal derivative is $\partial/\partial y$. Therefore,

$$
\left. \frac{\partial G(x, y; \xi, \eta)}{\partial y} \right|_{y=0} = \frac{\eta}{\pi[(x - \xi)^2 + \eta^2]} \qquad x \in \mathbb{R}, \ (\xi, \eta) \in \mathbb{R}_+^2 .
$$

(b) The function

$$
G(x, y; \xi, \eta) = -\frac{1}{4\pi} \ln \left\{ \frac{[(x - \xi)^2 + (y - \eta)^2][(x + \xi)^2 + (y + \eta)^2]}{[(x - \xi)^2 + (y + \eta)^2][(x + \xi)^2 + (y - \eta)^2]} \right\}
$$

satisfies all the required properties.

8.7 (b) Since

$$
2\pi \int_0^1 \exp \left(\frac{1}{|r|^2 - 1} \right) r \, dr \approx 0.4665,
$$

the normalization constant c is approximately 2.1436.

8.9 By Exercise 5.20, the kernel K (as a function of (x, t)) is a solution of the heat equation for $t > 0$.

Set $\rho(x) := (1/\sqrt{\pi})e^{-x^2}$, and consider

$$
\rho_\varepsilon(x) := \varepsilon^{-1} \rho \left(\frac{x - y}{\varepsilon} \right).
$$

By Exercise 8.7, ρ_ε approximates the delta function as $\varepsilon \to 0_+$.

Take $\varepsilon - \sqrt{4kt}$, then $\rho_\varepsilon(x) = K(x, y, t)$. Thus, $K(x, y, 0) = \delta(x - y)$.

8.11 *Hint* For $(x, y) \in D_R$, let

$$
(\tilde{x}, \tilde{y}) := \frac{R^2}{x^2 + y^2}(x, y)
$$

be the inverse point of (x, y) with respect to the circle ∂B_R, and set

$$r = \sqrt{(x - \xi)^2 + (y - \eta)^2}, \; r^* = \sqrt{\left(x - \frac{R^2}{\rho^2}\xi\right)^2 + \left(y - \frac{R^2}{\rho^2}\eta\right)^2}, \; \rho = \sqrt{\xi^2 + \eta^2}.$$

Finally, verify (as was done in Exercise 8.1) that the function

$$G_R(x, y; \xi, \eta) = -\frac{1}{2\pi} \ln \frac{Rr}{\rho r^*} \qquad (\xi, \eta) \neq (x, y)$$

is the Green function in D_R.

8.13 Fix $(\xi, \eta) \in B_R$, and define for $(x, y) \in B_R \setminus (\xi, \eta)$

$$N_R(x, y; \xi, \eta) = \begin{cases} -\dfrac{1}{2\pi} \ln \dfrac{rr^*\rho}{R^3} & (\xi, \eta) \neq (0, 0), \\ -\dfrac{1}{2\pi} \ln \dfrac{r}{R} & (\xi, \eta) = (0, 0), \end{cases}$$

where

$$r = \sqrt{(x - \xi)^2 + (y - \eta)^2}, \; r^* = \sqrt{\left(x - \frac{R^2}{\rho^2}\xi\right)^2 + \left(y - \frac{R^2}{\rho^2}\eta\right)^2}, \; \rho = \sqrt{\xi^2 + \eta^2}.$$

Verify that

$$\Delta N_R(x, y; \xi, \eta) = -\delta(x - \xi, y - \eta),$$

and that N_R satisfies the boundary condition

$$\frac{\partial N_R(x, y; \xi, \eta)}{\partial r} = \frac{1}{2\pi R}.$$

Finally, check that N_R satisfies the normalization (8.34).

Chapter 9

9.1 (b) From the eikonal equation itself $u_z(0, 0, 0) = \pm\sqrt{1 - u_x^2(0, 0, 0) - u_y^2(0, 0, 0)} = \pm 1$, where the sign ambiguity means that there are two possible waves, one propagating into $z > 0$, and one into $z < 0$.

The characteristic curves (light rays) for the equations are straight lines perpendicular to the wavefront. Therefore the ray that passes through $(0, 0, 0)$ is in the direction $(0, 0, 1)$. This implies $u_x(0, 0, z) = u_y(0, 0, z) = 0$ for all z, and hence $u_{xz}(0, 0, z) = u_{yz}(0, 0, z) = 0$. Differentiating the eikonal equation by z and using the last identity implies $u_{zz}(0, 0, 0) = 0$. The result for the higher derivatives is obtained similarly by further differentiation.

9.3 *Hint* Verify that the proposed solution (9.26) indeed satisfies (9.23) and (9.25), and that $u_r(0, t) = 0$.

9.5 $u(r, t) = 2 + (1 + r^2 + c^2 t^2)t$.

9.7 The representation (9.35) for the spherical mean makes it easier to interchange the order of integration. For instance,

$$\frac{\partial}{\partial a} M_h(a, \vec{x}) = \frac{1}{4\pi} \int_{|\vec{\eta}|=1} \nabla h(\vec{x} + a\vec{\eta}) \cdot \vec{\eta} \, ds_\eta.$$

Use Gauss' theorem (recall that the radius vector is orthogonal to the sphere) to express the last term as

$$\frac{a}{4\pi} \int_{|\eta|<1} \Delta_x h(\vec{x} + a\vec{\eta}) \, d\vec{\eta}.$$

To return to a surface integral notation rewrite the last expression as

$$\frac{a^{-2}}{4\pi} \Delta_x \int_{|\vec{x}-\vec{\xi}|<a} h(\vec{\xi}) d\vec{\xi} = \frac{a^{-2}}{4\pi} \Delta_x \int_0^a d\alpha \int_{|\vec{x}-\vec{\xi}|=\alpha} h(\vec{\xi}) ds_\xi$$

$$= a^{-2} \Delta_x \int_0^a \alpha^2 M_h(\alpha, \vec{x}) d\alpha.$$

Multiply the two sides by a^2 and differentiate again with respect to the variable a to obtain the Darboux equation.

9.9 Use the same method as in Subsection 9.5.2 to find

$$\lambda_{l,n,m} = \pi^2 \left(\frac{l^2}{a^2} + \frac{n^2}{b^2} + \frac{m^2}{c^2} \right), \quad u_{l,n,m}(x, y, z) = \sin \frac{l\pi x}{a} \sin \frac{n\pi y}{b} \sin \frac{m\pi z}{c},$$

for $l, n, m = 1, 2, \ldots$.

9.11 *Hint* Differentiate (9.76) with respect to r to obtain one recursion formula, and differentiate with respect to θ to obtain another recursion formula. Combining the two recursion formulas leads to (9.77).

9.13 (a) Multiply Legendre equations for v_i by v_j, $i \neq j$, subtract and integrate over $[-1, 1]$. It follows that

$$\int_{-1}^1 \left\{ v_2 \left[(1 - t^2)v_1' \right]' - v_1 \left[(1 - t^2)v_2' \right]' \right\} ds = (\mu_2 - \mu_1) \int_{-1}^1 v_1(s)v_2(s) \, ds.$$

Integrating the left hand side by parts implies $(\mu_2 - \mu_1) \int_{-1}^1 v_1(s)v_2(s) \, ds = 0$.
Since $\mu_1 \neq \mu_2$, it follows that $\int_{-1}^1 v_1(s)v_2(s) \, ds = 0$.

(b) Suppose that the Legendre equation admits a smooth solution v on $[-1, 1]$ with $\mu \neq k(k + 1)$. By part (a), v is orthogonal to the space of all polynomials. Weierstrass' approximation theorem implies that v is orthogonal to the space $E(-1, 1)$. This implies $v = 0$.

9.15 Write the general homogeneous harmonic polynomial as in Corollary 9.24, and express it in the form $Q = r^n F(\phi, \theta)$. Substitute Q into the spherical form of the Laplace equation (see (9.85)) to get that F satisfies

$$\frac{1}{\sin\phi}\frac{\partial}{\partial\phi}\left(\sin\phi\frac{\partial F}{\partial\phi}\right) + \frac{1}{\sin^2\phi}\frac{\partial^2 F}{\partial\theta^2} = -n(n+1)F.$$

Therefore F is a spherical harmonic (or a combination of spherical harmonics).

9.17 (a) By Exercise 9.13, each Legendre polynomial P_n is an n-degree polynomial which is orthogonal to P_m for $n \neq m$. These together with the normalization $P_n(1) = 1$ determine the Legendre polynomials uniquely.

Set

$$Q_n(t) := \frac{1}{2^n n!}\frac{d^n}{dt^n}(t^2 - 1)^n.$$

Clearly, Q_n is an n-degree polynomial. Repeatedly integrating by parts implies the orthogonality of Q_n. Since $Q_n(1) = 1$ it follows that $P_n = Q_n$.

(b) Compute

$$\int_{-1}^{1} P_n(t)^2 dt = \frac{1}{2^{2n} n!^2}\int_{-1}^{1}\left[\frac{d^n}{dt^n}(t^2 - 1)^n\right]^2 dt$$

$$= \frac{(2n!)}{2^{2n} n!^2}\int_{-1}^{1}(t^2 - 1)^n \, dt = \frac{2}{2n+1}. \tag{12.3}$$

The general case of associated Legendre functions can be proved similarly using (9.100) and (12.3).

9.19 *Hint* Note that for $y \in B_R$ the function $\Gamma(\sqrt{(|\vec{x}||\vec{y}|/R)^2 + R^2 - 2\vec{x}\cdot\vec{y}})$ is harmonic in B_R, and that on ∂B_R the identity $\partial/\partial n = \partial/\partial r$ holds.

9.21 *Hints* (a) Substitute $\vec{y} = \vec{0}$ into the Poisson integral formula.

(b) Prove that $U(r) := (1/N\omega_N r^{N-1})\int_{\partial B_r} u(\vec{x})\, d\sigma_{\vec{x}}$ is the constant function.

(c) The proof of the strong maximum principle for domains in $\mathbb{R}^N$ is exactly the same as for planar domains, and the weak maximum principle is a direct consequence of it.

9.23 *Hints* (i) Write $\vec{x} = (x', x_N)$, and let $\tilde{x} := (x', -x_N)$. Then $\Gamma(\tilde{x}; \vec{y})$ is harmonic as a function of $\vec{x}$ in $\mathbb{R}^2_+$, while $\Delta_{\vec{x}}\Gamma(\tilde{x}; \vec{y}) = -\delta(\tilde{x} - \vec{y})$.

(ii) Notice that for $\vec{y} \in \partial\mathbb{R}^N_+$, the identity $\partial/\partial n = \partial/\partial y_N$ holds.

9.25 *Hint* The general formula for the eigenfunction expansion is (9.178). The specific cases of the rectangle and the disk are solved in (9.61) and (9.80) respectively.

Chapter 10

10.1 The first variation is $\delta K = 2 \int_0^1 y' \psi' dt$, where ψ is the variation function. Therefore the Euler–Lagrange equation is $y'' = 0$, and the solution is $y_M(t) = t$. Expand the functional with respect to the variation ψ about $y = y_M$, to get $K(u_M + \psi) = K(u_M) + \int_0^1 (\psi')^2 dt$. This shows that y_M is a minimizer, and it is indeed unique.

10.3 The Euler–Lagrange equation is $\Delta u - g u^3 = 0$, $x \in D$, while u satisfies the natural boundary conditions $\partial_n u = 0$ on ∂D

10.5 *Hints* The action is

$$J = \int_{t_1}^{t_2} \int_D \left[\frac{1}{2} u_t^2 - \frac{1}{2} |\nabla u|^2 - V(u) \right] dx.$$

The Euler–Lagrange equation is $u_{tt} - \Delta u + V'(u) = 0$.

10.7 (a) Introduce a Lagrange multiplier λ, and solve the minimization problem

$$\min \int_D |\nabla u|^2 dx dy + \lambda \left(1 - \int_d u^2 dx dy \right),$$

for all u that vanish on ∂D. Equate the first variation to zero to find the Euler–Lagrange equation

$$\Delta u = -\lambda u \quad x \in D, \qquad u = 0 \quad x \in \partial D.$$

10.9 *Hints* The eigenvalue problem is

$$X^{(iv)}(x) - \lambda X(x) = 0, \qquad X(0) = X'(0) = X(b) = X'(b) = 0.$$

Multiply both sides by X and integrate over $(0, b)$. Perform two integrations by parts and use the boundary conditions to derive

$$\int_0^b (X'')^2 dx = \lambda \int_0^b X^2 dx.$$

Therefore $\lambda > 0$.

The solution satisfying the boundary conditions at $x = 0$ is

$$X(x) = A(\cosh \alpha x - \cos \alpha x) + B(\sinh \alpha x - \sin \alpha x).$$

Enforcing the boundary condition at $x = b$ implies that a necessary and sufficient condition for a nontrivial solution is indeed given by condition (10.73).

10.11 *Hints* (i) $\|v_n\| = 1$.

(ii) By the Riemann–Lebesgue lemma, $\lim_{n \to \infty} \langle v_n, v \rangle = 0$.

(iii) Note that strong convergence implies $\|v_n\| \to \|v\|$.

Chapter 11

11.1

$$\partial_{xy}u(x_i, y_j) = \frac{U_{i+1,j+1} - U_{i-1,j+1} - U_{i+1,j-1} + U_{i-1,j-1}}{4\Delta x \Delta y}.$$

11.5 *Hints* The analytic solution is

$$u(x, t) = \frac{8}{\pi} \sum_{m=1}^{\infty} \frac{e^{-(2m-1)^2 t}}{(2m-1)^3} \sin(2m-1)x.$$

In comparing the numerical solution and analytic solution at $(x, t) = (\pi/4, 2)$, observe that very few Fourier terms are needed to capture the right answer.

11.7 The analytic solution is $u(x, t) = t^5$. The numerical procedure blows up because of the violation of the stability condition.

11.9 Let (i, j) be the index of an internal maximum point. Both terms on the left hand side of (11.27) are dominated by $U_{i,j}$. Therefore, if $U_{i,j}$ is positive, the left hand side is negative which is a contradiction.

11.13 Let p_i, $i = 1, \ldots, 4(N-2)$ be the set of boundary point. For each i define the harmonic function T_i, such that $T_i(p_i) = 1$, while $T_i(p_j) = 0$ if $j \neq i$. Clearly the set $\{T_i\}$ spans all solutions to the Laplace equation in the grid. It also follows directly from the construction that the set $\{T_i\}$ is linearly independent.

Appendix: Useful formulas

A.1 Trigonometric formulas

(1) $\int x^n \sin ax \, dx = -\dfrac{x^n}{a} \cos ax + \dfrac{n}{a} \int x^{n-1} \cos ax \, dx.$

(2) $\int x^n \cos ax \, dx = \dfrac{x^n}{a} \sin ax - \dfrac{n}{a} \int x^{n-1} \sin ax \, dx.$

(3) $\int e^{ax} \sin bx \, dx = \dfrac{e^{ax}}{a^2 + b^2}(a \sin bx - b \cos bx).$

(4) $\int e^{ax} \cos bx \, dx = \dfrac{e^{ax}}{a^2 + b^2}(a \cos bx + b \sin bx).$

(5) $\sin(\alpha \pm \beta) = \sin\alpha \, \cos\beta \pm \sin\beta \, \cos\alpha.$

(6) $\cos(\alpha \pm \beta) = \cos\alpha \, \cos\beta \mp \sin\alpha \, \sin\beta.$

(7) $\cos\alpha \, \cos\beta = \dfrac{1}{2}[\cos(\alpha + \beta) + \cos(\alpha - \beta)].$

(8) $\sin\alpha \, \sin\beta = \dfrac{1}{2}[\cos(\alpha - \beta) - \cos(\alpha + \beta)].$

(9) $\sin\alpha \, \cos\beta = \dfrac{1}{2}[\sin(\alpha + \beta) + \sin(\alpha - \beta)].$

(10) $\sin\alpha + \sin\beta = 2 \sin\dfrac{\alpha + \beta}{2} \cos\dfrac{\alpha - \beta}{2}.$

(11) $\cos\alpha + \cos\beta = 2 \cos\dfrac{\alpha + \beta}{2} \cos\dfrac{\alpha - \beta}{2}.$

(12) $\tan\alpha + \tan\beta = \dfrac{\sin(\alpha + \beta)}{\cos\alpha \, \cos\beta}.$

(13) $\sin^3 \alpha = \dfrac{1}{4}(3 \sin\alpha - \sin 3\alpha).$

(14) $\cos^3 \alpha = \dfrac{1}{4}(3 \cos\alpha + \cos 3\alpha).$

A.2 Integration formulas

(1) $\displaystyle\iint_D \nabla \cdot w \, dx dy = \int_{\partial D} w \cdot n \, d\sigma.$

(2) $\displaystyle\iint_D (Q_x - P_y) \, dx dy = \oint (P \, dx + Q \, dy).$

(3) $\displaystyle\iint_D (v \Delta u + \nabla u \cdot \nabla v) \, dx dy = \int_{\partial D} v \frac{\partial u}{\partial n} \, d\sigma.$

(4) $\displaystyle\iint_D (v \Delta u - u \Delta v) \, dx dy = \int_{\partial D} \left(v \frac{\partial u}{\partial n} - u \frac{\partial v}{\partial n} \right) d\sigma.$

(5) $\displaystyle\frac{\partial}{\partial t} \int_{a(t)}^{b(t)} G(\xi, t) \, d\xi = G(b(t), t) b'(t) - G(a(t), t) a'(t) + \int_{a(t)}^{b(t)} \frac{\partial}{\partial t} G(\xi, t) \, d\xi.$

A.3 Elementary ODEs

(1) The general solution of the linear ODE $y' + P(x)y = Q(x)$ is given by

$$y(x) = e^{-\int P(x) \, dx} \left(\int Q(x) e^{\int P(x) \, dx} \, dx + c \right).$$

(2) The general solution of the ODE

$$y'' + \lambda y = 0 \qquad \lambda \in \mathbb{R},$$

is given by

$$y(x) = \begin{cases} \alpha e^{\sqrt{-\lambda}x} + \beta e^{-\sqrt{-\lambda}x} = \tilde{\alpha} \cosh(\sqrt{-\lambda}x) + \tilde{\beta} \sinh(\sqrt{-\lambda}x) & \lambda < 0, \\ \alpha + \beta x & \lambda = 0, \\ \alpha \cos(\sqrt{\lambda}x) + \beta \sin(\sqrt{\lambda}x) & \lambda > 0, \end{cases}$$

where $\alpha, \beta, \tilde{\alpha}, \tilde{\beta}$ are arbitrary real numbers.

(3) Let $A, B, C \in \mathbb{R}$, and let r_1, r_2 be the roots of the (quadratic) indicial equation $Ar(r - 1) + Br + C = 0$. Then the general solution of the Euler (equidimensional) equation:

$$Ax^2 y'' + Bx y' + Cy = 0,$$

is given by

$$y(x) = \begin{cases} \alpha x^{r_1} + \beta x^{r_2} & r_1, r_2 \in \mathbb{R}, \ r_1 \neq r_2, \\ \alpha x^{r_1} + \beta x^{r_1} \log x & r_1, r_2 \in \mathbb{R}, \ r_1 = r_2, \\ \alpha x^{\lambda} \cos(\mu \log x) + \beta x^{\lambda} \sin(\mu \log x) & r_1 = \lambda + i\mu \in \mathbb{C}, \end{cases}$$

where α, β are arbitrary real numbers.

A.4 Differential operators in polar coordinates

We use the notation e_r and e_θ to denote unit vectors in the radial and angular direction, respectively, and e_z to denote a unit vector in the z direction. A vector $\vec{u}$ is expressed as $\vec{u} = u_1 e_r + u_2 e_\theta$. We also use $V(r, \theta)$ to denote a scalar function.

$$\nabla V = \frac{\partial V}{\partial r} e_r + \frac{1}{r} \frac{\partial V}{\partial \theta} e_\theta.$$

$$\nabla \cdot \vec{u} = \frac{1}{r} \frac{\partial (r u_1)}{\partial r} + \frac{1}{r} \frac{\partial u_2}{\partial \theta}.$$

$$\nabla \times \vec{u} = \left[\frac{1}{r} \frac{\partial (r u_2)}{\partial r} - \frac{1}{r} \frac{\partial u_1}{\partial \theta} \right] e_z.$$

$$\Delta V = \vec{\nabla} \cdot \vec{\nabla} V = V_{rr} + \frac{1}{r} V_r + \frac{1}{r^2} V_{\theta\theta}.$$

A.5 Differential operators in spherical coordinates

We use the notation e_r, e_θ, and e_ϕ to denote unit vectors in the radial, vertical angular direction, and horizontal angular direction, respectively. A vector $\vec{u}$ is expressed as $\vec{u} = u_1 e_r + u_2 e_\theta + u_3 e_\phi$. We also use $V(r, \theta, \phi)$ to denote a scalar function.

$$\nabla V = \frac{\partial V}{\partial r} e_r + \frac{1}{r} \frac{\partial V}{\partial \theta} e_\theta + \frac{1}{r \sin \theta} \frac{\partial V}{\partial \phi} e_\phi.$$

$$\nabla \cdot \vec{u} = \frac{1}{r^2} \frac{\partial (r^2 u_1)}{\partial r} + \frac{1}{r} \frac{\partial (\sin \theta \, u_2)}{\partial \theta} + \frac{1}{r \sin \theta} \frac{\partial u_3}{\partial \phi}.$$

$$\nabla \times \vec{u} = \frac{1}{r \sin \theta} \left[\frac{\partial (\sin \theta \, u_3)}{\partial \theta} - \frac{\partial u_2}{\partial \phi} \right] e_r + \frac{1}{r} \left[\frac{1}{\sin \theta} \frac{\partial u_1}{\partial \phi} - \frac{\partial (r u_3)}{\partial r} \right] e_\theta$$

$$+ \frac{1}{r} \left[\frac{\partial (r u_2)}{\partial r} - \frac{\partial u_1}{\partial \theta} \right] e_\phi.$$

$$\Delta V = \vec{\nabla} \cdot \vec{\nabla} V = \frac{1}{r^2} \frac{\partial}{\partial r} \left(r^2 \frac{\partial V}{\partial r} \right) + \frac{1}{r^2} \left[\frac{1}{\sin \phi} \frac{\partial}{\partial \phi} \left(\sin \phi \frac{\partial V}{\partial \phi} \right) + \frac{1}{\sin^2 \phi} \frac{\partial^2 V}{\partial \theta^2} \right].$$

References

[1] D.M. Cannell, *George Green; Mathematician and Physicist 1793–1841,* second edition. Philadelphia, PA: Society for Industrial and Applied Mathematics (SIAM), 2001.

[2] G.F. Carrier and C.E. Pearson, *Partial Differential Equations, Theory and Technique,* second edition. Boston, MA: Academic Press, 1988.

[3] W. Cheney and D. Kincaid, *Numerical Mathematics and Computing.* Pacific Grove, CA: Brooks Cole, Monterey, 1985.

[4] R. Courant and D. Hilbert, *Methods of Mathematical Physics,* Vols. I,II, New York, NY: John Wiley & Sons, 1996.

[5] N.H. Fletcher and T.D. Rossing, *The Physics of Musical Instruments.* New York, NY: Springer-Verlag, 1998.

[6] I. Gohberg and S. Goldberg, *Basic Operator Theory.* Boston, MA: Birkhäuser, 2001.

[7] P.R. Halmos, *Introduction to Hilbert Space and the Theory of Spectral Multiplicity.* Providence, RI: American Mathematical Society – Chelsea Publications, 1998.

[8] A.L. Hodgkin and A.F. Huxley, "A quantitative description of membrane current and its application to conduction and excitation in nerve", *Journal of Physiology* **117**, 500–544, 1952.

[9] E.L. Ince, *Ordinary Differential Equations.* Mineda, NY: Dover, 1944.

[10] D. Jackson, *Classical Electrodynamics,* second edition. New York, NY: Wiley, 1975.

[11] F. John, *Partial Differential Equations,* reprint of the fourth edition, Applied Mathematical Sciences Vol. 1. Berlin: Springer-Verlag, 1991.

[12] J.D. Murray, *Mathematical Biology,* second edition. Berlin: Springer-Verlag, 1993.

[13] A. Pinkus and S. Zafrani, *Fourier Series and Integral Transforms.* Cambridge: Cambridge University Press, 1997.

[14] M.H. Protter and H.F. Weinberger, *Maximum Principles in Differential Equations,* corrected reprint of the 1967 original. New York, NY: Springer-Verlag, 1984.

[15] R.D. Richtmyer and K.W. Morton, *Difference Methods for Initial Value Problems,* reprint of the second edition. Malabar, FL: Robert E. Krieger, 1994.

[16] M. Schatzman, *Numerical Analysis – A Mathematical Introduction.* Oxford: Oxford University Press, 2002.

[17] L. Schiff, *Quantum Mechanics.* Tokyo, Mcgraw-Hill, 1968.

[18] G.D. Smith, *Numerical Solutions of Partial Differential Equations, Finite Difference Methods,* third edition, Oxford Applied Mathematics and Computing Science Series. New York, NY: Oxford University Press, 1985.

[19] I.N. Sneddon, *Elements of Partial Differential Equations*. New York, NY: McGraw-Hill, 1957.

[20] J.L. Troutman, *Variational Calculus and Optimal Control,* second edition. Undergraduate Texts in Mathematics. New York, NY: Springer-Verlag, 1996.

[21] G.N. Watson, *A Treatise on the Theory of Bessel Functions*. Cambridge: Cambridge University Press, 1966.

[22] G.B. Whitham, *Linear and Nonlinear Waves*. New York, NY: John Wiley, 1974.

[23] E. Zauderer, *Partial Differential Equations of Applied Mathematics,* second edition. New York, NY: John Wiley & Sons, 1989.

Index

acoustics, 7–11
action, 292, 293
adjoint operator, 213
admissible surface, 283
asymptotic behavior
  Bessel function, 249
  eigenvalue, 154, 245
  solution, 155

backward
  difference, 312
  heat operator, 213
  wave, 77
Balmer, Johann Jacob, 263, 266
basis, 297, 299, 301, 302, 305, 310, 330, 336
  Zernike, 302
Bernoulli, Daniel, 98
Bessel equation, 133, 248, 257, 262, 269
Bessel function, 248
  asymptotic, 249
  properties, 249
Bessel inequality, 138
Bessel, Friedrich Wilhelm, 248
biharmonic
  equation, 16
  operator, 213, 290
Born, Max, 263
boundary conditions, 18–20
  Dirichlet, 18, 108, 174
  first kind, 108
  mixed, 19, 109
  natural, 288, 307
  Neumann, 19, 108, 109, 175
  nonhomogeneous, 164
  nonlocal, 19
  oblique, 19
  periodic, 109
  Robin, 19, 109, 175
  second kind, 108
  separated, 108, 130, 133
  third kind, 19, 109
Brown, Robert, 13
Brownian motion, 13–14

cable equation, 119–123
  semi-infinite, 121
calculus of variations, 282–308
canonical form, 66, 231
  elliptic, 66, 70–73
  hyperbolic, 66–69
  parabolic, 66, 69–70
  wave equation, 76
Cauchy problem, 24, 27, 55, 76, 78, 176, 224, 229, 236, 241
Cauchy sequence, 298
Cauchy, Augustin Louis, 24
Cauchy–Schwartz inequality, 137
central difference, 312
CFL condition, 324
change of coordinates, 65
characteristic curve, 27, 229
characteristic equations, 27, 67, 226
characteristic function, 137
characteristic projection, 68
characteristic strip, 54, 228
characteristic surface, 229
characteristic triangle, 82
characteristics, 31, 67, 68, 77
  method, 25–63, 226
clarinet, 267–269
classical Fourier system, 135
classical solution, 3, 43, 79, 175
classification of PDE, 3, 64–75, 228–234
compact support, 211, 234, 330
compatibility condition, 55, 99, 109, 167, 188, 252, 287
complementary error function, 129
complete orthonormal sequence, 138, 299
compression wave, 46
conservation laws, 8, 9, 41–50
consistent numerical scheme, 315
constitutive law, 6, 11, 12, 295
convection equation, 11, 17
convergence
  in distribution sense, 212
  in norm, 137
  in the mean, 137

convergence (*cont.*)
    numerical scheme, 316
    strong, 297
    weak, 300
convex functional, 291
Courant, Richard, 309
Crank–Nicolson method, 317
curvature, 295

δ function, 211, 224, 272
d'Alembert formula, 79–97, 208, 234
d'Alembert, Jean, 12, 98
Darboux equation, 238, 279
Darboux problem, 95
Darboux, Gaston, 237, 238
degenerate states, 247, 257, 266
delta function, 211, 224, 272
diagonally dominated matrix, 321
difference equation, 13, 313, 314
difference scheme, 13, 313, 319, 322, 329, 331
differential operator, 4
diffusion coefficient, 7, 124
diophantic equations, 247
dirac distribution, 211, 224, 272
Dirichlet condition, 18, 100
Dirichlet functional, 284, 285
Dirichlet integral, 284, 285, 287, 300
Dirichlet problem, 174, 209–218, 285
    ball, 257, 262
    cylinder, 261
    disk, 195
    eigenfunction expansion, 273
    eigenvalue, 243
    exterior domain, 198
    numerical solution, 318–322
    rectangle, 188
    sector, 198
    spectrum, 242
    stability, 182
    uniqueness, 181, 183
Dirichlet, Johann Lejeune, 18
dispersion relation, 122
distribution, 211, 223
    convergence, 212
    Dirac, 211, 224, 272
divergence theorem, 7, 8, 182, 362
domain of dependence, 83, 89
drum, 260, 269
Du-Fort–Frankel method, 318
Duhamel principle, 127, 222, 276

eigenfunction, 101, 131
    expansion, 114, 130–172
    orthogonality, 143, 243
    principal, 152
    properties, 243–245
    real, 146, 243
    zeros, 154, 245
eigenvalue, 131
    asymptotic behavior, 154, 245
    existence, 147, 244

    multiplicity, 102, 132, 135, 146, 243, 244, 247, 257
    principal, 152, 244
    problem, 101, 131, 242–258
    properties, 243–245
    real, 145, 243
    simple, 102, 132, 146, 245
eigenvector, 132
eikonal equation, 15, 26, 50–52, 57, 233, 292
element (for FEM), 331
elliptic equation, 65, 173–183, 209, 231, 232, 305, 329
elliptic operator, 65
energy integral, 116–119
energy level, 263
energy method, 116–119, 182
entropy condition, 47
equation
    elliptic, 65, 209, 231, 232, 305, 329
    homogeneous, 4
    hyperbolic, 65, 231, 232
    Klein–Gordon, 233
    nonhomogeneous, 4
    parabolic, 65, 209, 231, 232
error function, 129
    complementary, 129
Euler equation, 41
Euler fluid equations, 9
Euler equidimensional equation, 362
Euler, Leonhard, 9, 41
Euler–Lagrange equation, 285, 331
even extension, 94, 235, 238
expansion wave, 45
explicit numerical scheme, 317

FDM, 310–324
FEM, 306, 310, 329–334
    element, 331
Fermat principle, 292
Fermat, Pierre, 26
finite difference method (FDM), 310–324
finite differences, 311–312
finite elements method (FEM), 306, 310, 329–334
    element, 331
first variation, 284
first-order equations, 23–63
    existence, 36–38
    high dimension, 226–228
    Lagrange method, 39–41, 62
    linear, 24
    nonlinear, 52–58
    uniqeness, 36–38
flare, 268
flexural rigidity, 289
flute, 267, 268
formally selfadjoint operator, 213
formula
    Poisson, 202
    Rayleigh-Ritz, 152, 244, 308
forward difference, 311
forward wave, 77
Fourier classical system, 135
Fourier coefficients, 103, 138

Fourier expansion, 98, 139
    convergence, 148, 244
    convergence on average, 139
    convergence in norm, 139
    convergence in the mean, 139
    generalized, 139, 244
Fourier law, 6
Fourier series, 103, 258
Fourier, Jean Baptiste Joseph, 5, 98, 103, 139, 148
Fourier–Bessel coefficients, 252
Fourier–Bessel series, 251
Fresnel, Augustin, 26
Friedrichs, Kurt Otto, 309
Frobenius–Fuchs method, 248, 254, 265
function
    characteristic, 137
    error, 129
    harmonic
        mean value, 179–180, 274, 280
    piecewise continuous, 136
    piecewise differentiable, 136
    real analytic, 71
functional, 283
    bounded below, 304
    convex, 291
    Dirichlet, 284, 285
    first variation, 284
    linear, 284
    second variation, 291
fundamental equations of mathematical physics, 66
fundamental solution, 213, 224, 270
    Laplace, 178, 209, 271
    uniqueness, 213

Galerkin method, 303–306
Galerkin, Boris, 306
Gauss theorem, 7, 8, 176, 182, 362
Gauss–Seidel method, 325, 326
Gaussian kernel, 224
general solution, 40, 76–78, 230, 239, 362
generalized Fourier coefficients, 103, 138
generalized Fourier expansion, 139
generalized Fourier series, 103, 258
generalized solution, 77, 104, 112
geometrical optics, 14–15, 287
Gibbs phenomenon, 150, 192, 195
Gibbs, Josiah Willard, 150
Green's formula, 88, 143, 144, 160, 182, 210, 243, 271, 284
Green's identity, 182, 210, 271, 284
Green's representation formula, 211, 219, 271
Green's function, 209–221, 272
    ball, 274
    definition, 214
    Dirichlet problem, 209–218
    disk, 217, 224
    exterior of disk, 225
    half-plane, 218, 224
    half-space, 275
    higher dimensions, 269–275
    monotonicity, 217, 273

Neumann problem, 219–221
    positivity, 216, 273
    properties, 273
    rectangle, 281
    symmetry, 215, 273
    uniqueness, 273
Green, George, 208
grid, 311
ground state, 152
    energy, 152
guitar, 267

Hadamard example, 176
Hadamard method of descent, 241
Hadamard, Jacques, 2, 176, 239
Hamilton characteristic function, 26
Hamilton principle, 292
Hamilton, William Rowan, 1, 15, 25, 26, 39, 292, 293
Hamiltonian, 292–296
harmonic function, 173
harmonic polynomial
    homogeneous, 177, 205
Harnack inequality, 205
heat equation
    Dirichlet problem, 185
        uniqueness, 118
    maximum principle, 184
    numerical solution, 312–318
    separation of variables, 99–109, 259
    stability, 185
heat flow, 6, 109, 130
heat flux, 6, 19, 175
heat kernel, 129, 221–224, 275–278, 281
    properties, 276–278
Heisenberg, Werner, 263
Helmholtz equation, 204, 281
Hilbert space, 298
Hilbert, David, 298
homogeneous equation, 4
Huygens' principle, 239
Huygens, Christian, 26
hydrodynamics, 7–11
hydrogen atom, 263–266
hyperbolic equation, 65, 231, 232
hyperbolic operator, 65

ill-posed problem, 2, 82, 176
implicit numerical scheme, 317
induced norm, 136
inequality
    Bessel, 138
    Cauchy–Schwartz, 137
    Harnack, 205
    triangle, 136
initial condition, 1, 24, 79, 99, 226, 229, 313
initial curve, 26
initial value problem, 17, 24, 89
inner product, 136
    induced norm, 136
    space, 136

insulate, 99
insulated boundary condition, 109, 125
integral surface, 28
inverse point
  circle, 217, 356
  line, 218
  sphere, 274
iteration, 318, 325–327

Jacobi method, 325
jump discontinuity, 136

Kelvin, Lord, 122

Lagrange identity, 142, 146
Lagrange method, 39–41, 62
Lagrange multiplier, 307, 359
Lagrange, Joseph-Louis, 16, 39, 283, 294
Lagrangian, 292–296
Laguerre equation, 265
Laplace equation, 15, 173–206
  ball, 262
  cylinder, 261
  eigenvalue problem, 242–258
  fundamental solution, 178
  Green's function, 209–218
    higher dimension, 269–275
  maximum principle, 178–181
  numerical solution, 318–322
  polar coordinates, 177
  separation of variables, 187–201, 245–258,
    261–263
Laplace, Pierre-Simon, 16, 283
Laplacian, 16
  cylindrical coordinates, 261
  polar coordinates, 363
  spectrum, 245
    ball, 257
    disk, 251
    rectangle, 245
  spherical coordinates, 363
least squares approximation, 287
Legendre associated equation, 255
Legendre equation, 254
Legendre polynomial, 254
Legendre, Adrien-Marie, 254
Lewy, Hans, 309
linear equation, 3
  first-order, 24
linear functional, 284
linear operator, 4
linear PDE, 3
Liouville, Joseph, 131, 147

Maupertuis, Pierre, 294
maximum principle
  heat equation, 184
  numerical scheme, 319
  strong, 180, 274
  weak, 178
mean value principle, 179–180, 204, 274, 280

membrane, 11, 16, 122, 260–261, 266, 269, 288, 289,
  308
mesh, 311
minimal surface, 16, 282–287
  equation, 16, 285, 286
minimizer, 283, 284, 304, 331
  existence, 299
  uniqueness, 291
minimizing sequence, 300
modes of vibration, 267, 269
Monge, Gaspard, 53
multiplicity, 102, 132, 135, 243, 244, 247, 257
musical instruments, 266–269

natural boundary conditions, 288, 307
Navier, Claude, 9
net, 311
Neumann boundary conditions, 19, 108, 109, 131,
  174, 193, 242
Neumann function, 219–221, 224
Neumann problem, 110, 125, 175, 183, 195, 203,
  219–221, 318
Neumann, Carl, 19
Newtonian potential, 211, 271
nodal lines, 245
nodal surfaces, 245
nonhomogeneous boundary conditions, 164
nonhomogeneous equation, 4, 114–116,
  159–164
norm, 136
normal modes, 267, 269
numerical methods, 309–336
  linear systems, 324–329
numerical scheme, 310
  consistent, 315
  convergence, 316
  explicit, 317
  implicit, 317
  stability, 314
  stability condition, 315

odd extension, 93
operator, 4
  elliptic, 231
  formally self-adjoint, 213
  hyperbolic, 231
  parabolic, 231
  symmetric, 143
order of PDE, 3
organ, 268
orthogonal projection, 138
orthogonal sequence, 137
orthogonal vectors, 137
orthonormal sequence, 137
  complete, 138, 148, 299
orthonormal system
  complete, 244
outward normal vector, 6

parabolic boundary, 184
parabolic equation, 65, 209, 231, 232

parabolic operator, 65
parallelogram identity, 94
Parseval identity, 138
PDE, 1
    classification, 3, 64–75, 228–234
    linear, 3
    order, 3
    quasilinear, 3, 9, 24–50
    semilinear, 3
    system, 3
periodic eigenvalue problem, 134, 196, 253
periodic problem, 171
periodic solution, 91
periodic Sturm–Liouville problem, 134, 196, 253
piecewise continuous function, 136
piecewise differentiable function, 136
pipes
    closed, 268
    open, 268
Planck constant, 17, 263
Planck quantization rule, 265
plate equation, 289
Plateau, Joseph Antoine, 283
Poisson equation, 13, 174, 219, 289
    separation of variables, 199
Poisson formula, 201–204
Poisson kernel, 202, 215, 223, 224, 272–274,
    280
    Neumann, 203
Poisson ratio, 289
Poisson, Simeon, 174
principal eigenfunction, 152
principal eigenvalue, 152, 244
principal part, 64, 228
product solutions, 99, 100

quasilinear equation, 3, 9, 24–50

random motion, 13–14
Rankine–Hugoniot condition, 47, 49
Rayleigh quotient, 151–154, 244, 302
Rayleigh, Lord, 152
Rayleigh–Ritz formula, 152, 244, 308
real analytic function, 71
reflection principle, 217, 218, 224, 225, 275, 280
refraction index, 15, 50
region of influence, 83, 324
regular Sturm–Liouville problem, 133
resonance, 164, 261
Riemann–Lebesgue lemma, 138
Ritz method, 301–303
Rodriguez formula, 280
round-off error, 313
Runge, Carl, 26
Rydberg constant, 263

scalar equation, 3
Schrödinger equation, 16
    hydrogen atom, 263–266
Schrödinger operator, 151, 152
Schrödinger, Erwin, 16, 263

second variation, 291
second-order equation
    Cauchy problem, 229–234
    classification, 64–75, 228–234
semi-infinite cable, 121
semi-infinite string, 93, 94
semilinear equation, 3
separated solutions, 99, 100
separation of variables, 98–172, 245–263
shock wave, 41–50
similarity solution, 129
simple eigenvalue, 102, 132, 146, 245
singular Sturm–Liouville problem, 133, 254
soap film, 283
Sobolev space, 299
Sobolev, Sergei, 299
solution
    classical, 3, 79
    even, 91
    general, 40, 76–78, 230, 239, 362
    generalized, 77, 104, 112
    odd, 91
    periodic, 91
    strong, 3
    weak, 3, 41–50, 296–301
Sommerfeld, Arnold, 26
SOR method, 325, 326
spectral radius, 328
spectrum, 151, 152, 242
    hydrogen atom, 263–266
    Laplacian
        ball, 257
        disk, 251
        rectangle, 245
spherical harmonics, 256, 262, 266
spherical mean, 237
square wave, 150
stability
    CFL condition, 324
    Dirichlet problem, 182
    heat equation, 185
    numerical scheme, 314
    wave equation, 90
stiffness matrix, 330
Stokes, George Gabriel, 9
string, 11–12, 19, 79, 87, 98, 109, 117, 130, 164,
    266–267, 294
    semi-infinite, 93, 94
strip equations, 54, 228
strong convergence, 297
strong maximum principle, 180, 274
strong solution, 3
Sturm, Jacques Charles, 131, 147
Sturm–Liouville asymptotic behavior
    eigenvalue, 154
    solution, 155
Sturm–Liouville eigenfunctions, 141–158
Sturm–Liouville eigenvalues, 141–158
Sturm–Liouville operator, 132
Sturm–Liouville problem, 131, 133–135, 141–158
    periodic, 134, 196, 253

Sturm–Liouville problem (*cont.*)
   regular, 133
   singular, 133, 254
superposition principle, 4, 89, 92, 99, 103
   generalized, 104
support, 85
symmetric operator, 143
system, 3

telegraph equation, 128, 170
temperature, 6, 8, 18, 99, 123–124
   equilibrium, 16, 174
tension, 11, 295
test function, 154, 329–331
Thomson, William, 119, 122
trace formula, 276
transcendental equation, 155
transport equation, 8, 17, 121
transversality condition, 30, 227, 228
   generalized, 55
traveling waves, 77
triangle inequality, 136
Tricomi equation, 68
truncation error, 312
Turing, Alan Mathison, 245

uniqueness, 36–38, 82, 87, 182, 291
   Dirichlet problem, 181
   energy method, 116–119
   Fourier expansion, 115

variational methods, 282–308
viscosity, 9
von Neumann, John, 298

wave compression, 46
wave equation, 10–12, 14, 26, 76–97, 266, 295, 309
   Cauchy problem, 78–82
   domain of dependence, 83, 89
   general solution, 76–78
   graphical method, 84
   nonhomogeneous, 87–92
   numerical solution, 322–324
   parallelogram identity, 94
   radial solution, 234–236
   region of influence, 83, 324
   separation of variables, 109–114, 260–261
   stability, 90
   three-dimensional, 234–241
   two-dimensional, 241–242
wave expansion, 45
wave number, 15, 233
wave speed, 12, 76, 77
weak convergence, 300
weak solution, 3, 41–50, 296–301
Webster's horn equation, 268
weight function, 132
well-posedness, 2, 9, 30, 81, 82, 90, 176
Weyl formula, 155, 245
Weyl, Herman, 155
wine cellars, 123–124

Young, Thomas, 26

Zeeman effect, 266
Zernike basis, 302
Zernike, Frits, 302

Printed in the United States
by Bookmasters

Printed in the United States
By Bookmasters